Wastewater Based Microbial Biorefinery for Bioenergy Production

Wastewater Based Microbial Biorefinery for Bioenergy Production

Editor

Shashi Kant Bhatia

MDPI • Basel • Beijing • Wuhan • Barcelona • Belgrade • Manchester • Tokyo • Cluj • Tianjin

Editor
Shashi Kant Bhatia
Konkuk University
Korea

Editorial Office
MDPI
St. Alban-Anlage 66
4052 Basel, Switzerland

This is a reprint of articles from the Special Issue published online in the open access journal *Sustainability* (ISSN 2071-1050) (available at: https://www.mdpi.com/journal/sustainability/special_issues/Wastewater_Based).

For citation purposes, cite each article independently as indicated on the article page online and as indicated below:

LastName, A.A.; LastName, B.B.; LastName, C.C. Article Title. *Journal Name* **Year**, *Volume Number*, Page Range.

ISBN 978-3-0365-1950-0 (Hbk)
ISBN 978-3-0365-1951-7 (PDF)

Contents

About the Editor

Shashi Kant Bhatia working as an Associate Professor at Dept. of Biological Engineering Konkuk University, Seoul, South Korea, and have more than ten year's experience in the area of biowastes valorization into bioenergy, biochemicals, and biomaterials. Dr. Bhatia has contributed extensively to the industrial press and serving as an Editorial Board member of *Sustainability* (MDPI) journal, Associate Editor in *Frontier of Microbiology*, *Microbial Cell Factories*, and *PLoS ONE* journals. He has published more than 100 research/review articles on industrial biotechnology, bioenergy production, biomaterial, biotransformation, microbial fermentation, and enzyme technology in international scientific peer-reviewed journals and holds twelve international patents. He is also selected in the Top 2% Scientist list of Stanford University. He did his M.Sc. and Ph.D. in Biotechnology from Himachal Pradesh University (India). He has qualified JNU combined Biotechnology entrance, DBT-JRF, GATE, CSIR exams. Dr. Bhatia has worked as a Brain Pool Post Doc Fellow at Konkuk University (2014–2016), South Korea. He can be reached by email at shashibiotechhpu@gnmail.com.

Preface to "Wastewater Based Microbial Biorefinery for Bioenergy Production"

This book is on the topic of wastewater to bioenergy. Book provides information on recent technological advancements in wastewater-based biorefinery for bioenergy production. The intent is to provide overall information to readers on possible approaches in wastewater treatment with simultaneous energy production. There are ten chapters in this book that discuss about the characterization of wastewaters, compositions, challenges in treatment, resource recovery, and bioenergy production. The first chapter describes types of wastewaters, their source, and their compositions. An overview of various possible approaches for resource recovery from wastewater and production of biofuels such as biogas, bioethanol, biodiesel, microbial fuel cell, and direct conversion of sludge wastes using gasification, pyrolysis process is discussed. Chapter 2 discusses that cost associated with wastewater has become a key issue in wastewater management technology. This chapter provides a bibliometric analysis of publications in the Web of Science database on wastewater treatment costs in the period 1950–2020. The findings of this study reveal the leading countries in this field of research (China, USA, India, Spain, and the UK). Chapter 3; utilization of microalgae provides a sustainable approach for wastewater treatment and bioenergy production simultaneously. Microalgae can remove nutrients from the wastewater whilst capturing carbon dioxide from the atmosphere. The resulting biomass is employed to generate biofuels, which can run fuel cell vehicles of zero-emission, power combustion engines, and power plants. By cultivating microalgae in wastewater, eutrophication can be prevented, thereby enhancing the quality of the effluent. Chapter 4 discuss about microbial fuel cell experimental setup for bioelectricity production using seafood waste as raw material. Chapter five represents a recent update on biohydrogen production using wastewater-based microbial electrolysis cells (MEC). The present limiting issues for effective scaling up of the manufacturing process include the high manufacturing costs of microbial electrolysis cells, their high internal resistance and methanogenesis, and membrane/cathode biofouling. The challenges in the evolution of microbial electrolysis cell technology in terms of hydrogen yield, operational aspects that impact total hydrogen output in optimization studies, and important information on the efficiency of the processes are discussed. Chapter 6 discuss about a combined approach of anaerobic digestion (AD) and MEC on waste treatment and energy production. The use of MEC supported anaerobic digestion process could increase overall energy production and organic removal. Chapter 7 reports discuss about a method to reduce hydrogen sulfide production during anaerobic digestion and conclude that the addition of waterworks sludge containing iron to a digester for the removal of dissolved hydrogen sulfide is a technically and economically good alternative when producing biogas. Chapter eight is about biofilm a challenge in wastewater treatment that cause many serious problems, such as chronic infections, food contamination, and equipment corrosion. Biofilm formation and growth are complex due to interactions among physicochemical and biological processes under operational and environmental conditions. In this chapter advanced numerical modeling techniques using the lattice Boltzmann method (LBM) are enabling the prediction of biofilm formation and growth and microbial community structures is discussed in detail. Chapter 9 discuss problems related to environmental pollution caused by enhanced energy consumption and emphasize that by improving the efficiency of energy utilization this problem can be overcome. The last chapter defines the water and wastewater sector as an important lifeline upon which other economic sectors depend and securing this sector's critical infrastructure is therefore important for any country's economy. This chapter provides a case report on South Africa strategy to contextualize the water and wastewater sector's cybersecurity responsibilities within the national cybersecurity legislative and policy environment.

Shashi Kant Bhatia
Editor

MDPI

Editorial

Wastewater Based Microbial Biorefinery for Bioenergy Production

Shashi Kant Bhatia [1,2]

[1] Institute for Ubiquitous Information Technology and Application, Konkuk University, Seoul 05029, Korea; shashibiotechhpu@gmail.com

[2] Department of Biological Engineering, Konkuk University, Seoul 05029, Korea

A continuous increase in global population is demanding more development and industrialization, which leads to the production of various waste such as municipal wastewater, agricultural waste, industrial waste, medical waste, electronic wastes, etc. [1–3]. Waste management is important to maintain human health and the sustainability of the environment [4,5]. Various technologies have been reported to manage various kinds of wastes and produce valuable products [6–9]. Wastewater is produced from households and the textile, municipal, dairy, and pharmaceutical industries [10]. Management of wastewater is a challenging task as it is rich in nutrients and its release in the open may cause eutrophication and pose a threat to the environment. Generally, wastewater is treated using various physical and chemical methods, but these methods are costly and often result in sludge production and secondary water pollution. The use of microbe-based biowaste to bioenergy conversion technology can provide an economic way to manage wastewater and produce bioenergy simultaneously. Various microbe-mediated methods like anaerobic digestion (AD), microbial fuel cells (MFC), dark fermentation, etc. can be used for resource recovery from wastewater and energy production. Wastewater to bioenergy is a hot topic and many articles have been published in this area and the number is continuously increasing with each passing year.

According to the Scopus database, 1354 articles were found published during 2016–2021 using the search keywords wastewater and bioenergy. In terms of number of publications, China (433), India (213), the United States (173), South Korea (83), and Australia (73) make up the top five list. Figure 1 illustrates the visualization network of countries involved in this research area. Distinct clusters can be identified with different colors assigned to the individual country. The size of the circle corresponds to the frequency of publications while the thickness of the lines represents the co-authorship link and relative strength.

The editorial aim of this editorial is to highlight the key aspects of all contributions and provide an update to a broader scientific audience. In this Special Issue, we are publishing ten studies, which include six research articles and four review articles (Figure 2 provides a word cloud of published articles).

The review article contributed to this Special Issue by Bhatia et al. provides an overall overview of renewable energy products produced through wastewater valorization technologies [11]. Wastewaters were categorized into fifteen types based on their source. Wastewater from industrial effluent contains a significant quantity of harmful chemicals and heavy metals and lower biological content as compared to household wastewater. Different strategies reported for resource recovery from wastewater such as anaerobic digestion (AD), fermentation, microbial fuel cell (MFC), combustion, gasification, and pyrolysis of sludge for bioenergy production were reviewed in brief. The need for integrated wastewater treatment with benefits and challenges was also discussed.

Citation: Bhatia, S.K. Wastewater Based Microbial Biorefinery for Bioenergy Production. *Sustainability* **2021**, *13*, 9214. https://doi.org/10.3390/su13169214

Received: 15 August 2021
Accepted: 16 August 2021
Published: 17 August 2021

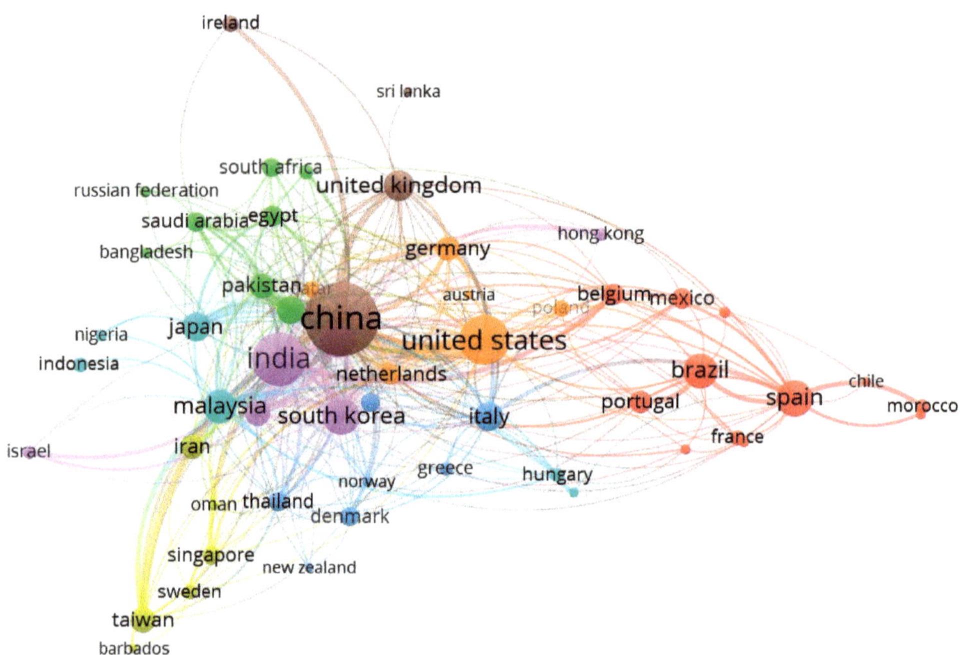

Figure 1. Co-occurrence mapping based on publication numbers (min. number of occurrences: three). The map was generated using Vosviewer.

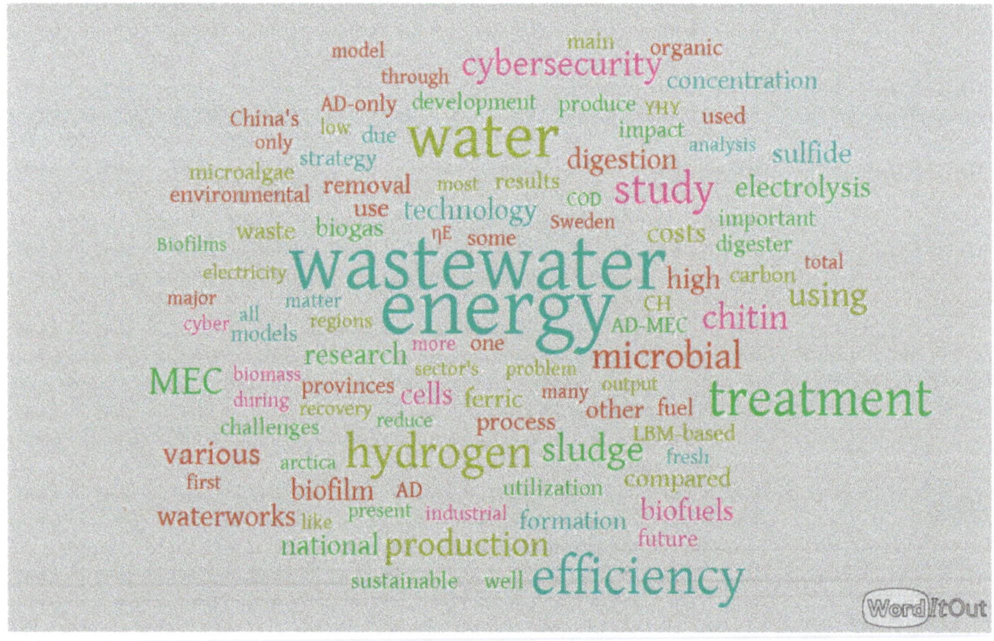

Figure 2. Word cloud of articles published in the Special Issue. The most frequent words are displayed with a larger font. The image was generated using worditout.com.

Wastewater treatment cost is also an issue for proper wastewater management. Gallego-valero et al. prepared a bibliometric analysis of publications on wastewater treatment cost from Web of Science published in the period 1950–2020 [12]. This study concluded that leading countries involved in wastewater research include China, the USA, India, Spain, and the UK based on the number of publications. The leading research institutes in terms of the number of related articles are the Chinese Academy of Sciences, China (561), followed by the Indian Institute of Technology (445). Different technologies are in fashion for resource recovery from wastewater and bioenergy production. Contribution by Jeyaseelan et al. discussed algal system utilization in wastewater treatment and bioenergy production [13]. Algae have high photosynthetic activity, are more effective in carbon dioxide capture, and produce 5–10 times more biomass compared to terrestrial plants. Algae are able to accumulate fatty acids up to 40–50% of their biomass and are used for biodiesel production. Microalgae like *Chlorella* are highly efficient in the removal of nutrients and are able to remove nitrogen and phosphorus up to 98% and 89%, respectively. Still, the use of this technology for biodiesel production is a challenge as algae cultivation and harvesting is costly. Residual algal biomass after oil extraction can also be used for bioenergy production using anaerobic digestion and fermentation processes. Microbial fuel cells (MFC) are also getting attention for wastewater treatment and bioelectricity production. Microorganisms biodegrade and oxidize the organic materials in the anodic chamber and produce electrons and protons that travel through external circuits and proton exchange membranes, respectively, to the cathode, producing current. Gurav et al. explored seafood processing chitin waste as a raw material for *Oceanisphaera aractica* YHY01 to produce electricity in MFC [14]. Microbial fuel cells demonstrated stable electricity generation until 216 h (0.228 mA/cm^2). N-acetyl-D-glucosamine (GlcNAc) was the major by-product and other organic acids such as lactate, acetate, propionate, and butyrate were the main metabolites produced in MFC. Cyclic voltammetry (CV) showed the role of outer membrane-bound cytochrome in the electron transfer mechanism. Biohydrogen is a clean and renewable source of energy. Wastewater can be used as a feedstock for hydrogen production using microbial electrolysis cells (MEC). In the MEC process, microbes are used in anodic chambers, which utilize organic material and convert chemical energy into electrical energy and produce energy that is subsequently used in cathodic parts to produce hydrogen. Dange et al. discuss the thermodynamics and electrochemistry of hydrogen-producing MEC technology in their contribution [15]. The MEC reactor architecture, materials used in electrode preparation, and various factors that affect hydrogen production are also discussed. The authors conclude that high MEC manufacturing cost, high internal resistance, methanogenesis, and membrane/cathode biofouling are the main limiting factors in the scale-up of this technology. To increase hydrogen production Hassanein et al., combined MEC technology with anaerobic digestion [16]. A comparative study between MEC and AD-MEC was conducted, and it was concluded that cumulative H_2 and CH_4 production and COD removal were higher in AD-MEC as compared to AD-only. During the AD process, biogas is produced by the digestion of organic material by methanogenic bacteria. Most of the substrates are rich in sulfur content and methanogenic microbes reduce it into hydrogen sulfide, which negatively affect methanogens activity and poison the digester. Ferric salt is used in AD to remove hydrogen sulfide production. Persson et al. used ferric oxide containing waterworks sludge in the AD process and reported a reduction in operational cost up to 50% [17]. This technology is one step closer to the circular economy as it helps to replace the use of virgin chemicals with the by-product waterworks sludge and reduces the carbon footprint of the waterworks. Biofilm is important for MFC wastewater treatment but is also related to problems such as infections, food contamination, and equipment corrosion. However, biofilm production is a complex process due to the interaction among physicochemical and biological processes. In the contribution by Delavar et al., (2021) an overview of advanced numerical modeling techniques using the lattice Boltzmann method (LBM) for prediction of biofilm formation and growth of microbial community structure is provided [18]. The main features and drawbacks of LBM-based biofilm models from the ecological and biotechnological perspective with challenges and future directions are also discussed. With industrial development and economic growth, environmental problems caused by excessive energy consumption become prominent. Improvement of efficiency of energy utilization can help to solve this problem. Huang et al. used a data envelopment analysis (DEA) model considering various input factors like labor force, capital stock, energy consumption and carbon emission to calculate the energy utilization efficiency of China's provinces [19]. This study concluded that urbanization openness and industrial structure have a negative effect on energy efficiency, while marketization has a significant positive impact. Water and wastewater treatment is an important lifeline for the economic development of any country, and it is very important to secure these sectors. Malatji et al. has provided a report on South Africa's cybersecurity strategy aimed at addressing cyber terrorism and cybercriminal activities [20]. In this study, they found that along with the National Cybersecurity Policy Framework, the Electronic Communications and Transactions Act, Critical Infrastructure Protection Act, and other supporting legislation, a wastewater sector's computer security incident response team can be established without the need to propose any new laws or amend existing ones.

Considering the content of various articles contributed to this Special Issue, it is clear that wastewater-based microbial biorefinery for bioenergy production is an emerging field and has the potential to meet the challenge of wastewater treatment and increased energy demands. Keeping in view the present status of the coronavirus pandemic, it is very important to treat wastewater. There is an essential need to increase research activity in this area to make wastewater to bioenergy technologies more efficient and economic. The main aim of various articles included in this Special Issue is to reach a broad readership and develop understanding in researchers and stakeholders related to wastewater to bioenergy technology.

Author Contributions: Conceptualization, writing—original draft preparation, visualization, writing—review and editing, S.K.B. The author has read and agreed to the published version of the manuscript.

Funding: This study was supported by the Research Program to solve social issues of the National Research Foundation of Korea (NRF)s funded by the Ministry of Science and ICT (grant number 2021R1F1A1050325).

Acknowledgments: The authors would like to acknowledge the KU Research Professor Program of Konkuk University, Seoul, South Korea.

Conflicts of Interest: The authors declare no conflict of interest.

References

1. Bhatia, S.K.; Joo, H.-S.; Yang, Y.-H. Biowaste-to-bioenergy using biological methods—A mini-review. *Energy Convers. Manag.* **2018**, *177*, 640–660. [CrossRef]
2. Duan, Y.; Pandey, A.; Zhang, Z.; Awasthi, M.K.; Bhatia, S.K.; Taherzadeh, M.J. Organic solid waste biorefinery: Sustainable strategy for emerging circular bioeconomy in China. *Ind. Crop. Prod.* **2020**, *153*, 112568. [CrossRef]
3. Saratale, G.D.; Bhosale, R.; Shobana, S.; Banu, J.R.; Pugazhendhi, A.; Mahmoud, E.; Sirohi, R.; Bhatia, S.K.; Atabani, A.; Mulone, V.; et al. A review on valorization of spent coffee grounds (SCG) towards biopolymers and biocatalysts production. *Bioresour. Technol.* **2020**, *314*, 123800. [CrossRef] [PubMed]
4. Arora, K.; Kaur, P.; Kumar, P.; Singh, A.; Patel, S.K.S.; Li, X.; Yang, Y.-H.; Bhatia, S.K.; Kulshrestha, S. Valorization of Wastewater Resources Into Biofuel and Value-Added Products Using Microalgal System. *Front. Energy Res.* **2021**, *9*, 119. [CrossRef]
5. Bhatia, S.K.; Gurav, R.; Choi, T.-R.; Kim, H.J.; Yang, S.-Y.; Song, H.-S.; Park, J.Y.; Park, Y.-L.; Han, Y.-H.; Choi, Y.-K.; et al. Conversion of waste cooking oil into biodiesel using heterogenous catalyst derived from cork biochar. *Bioresour. Technol.* **2020**, *302*, 122872. [CrossRef] [PubMed]
6. Ginni, G.; Kavitha, S.; Kannah, Y.; Bhatia, S.K.; Kumar, A.; Rajkumar, M.; Kumar, G.; Pugazhendhi, A.; Chi, N.T.L. Valorization of agricultural residues: Different biorefinery routes. *J. Environ. Chem. Eng.* **2021**, *9*, 105435. [CrossRef]
7. Bhatia, S.K.; Palai, A.K.; Kumar, A.; Bhatia, R.K.; Patel, A.K.; Thakur, V.K.; Yang, Y.-H. Trends in renewable energy production employing biomass-based biochar. *Bioresour. Technol.* **2021**, *340*, 125644. [CrossRef] [PubMed]

8. Bhatia, S.K.; Otari, S.V.; Jeon, J.-M.; Gurav, R.; Choi, Y.-K.; Bhatia, R.K.; Pugazhendhi, A.; Kumar, V.; Banu, J.R.; Yoon, J.-J.; et al. Biowaste-to-bioplastic (polyhydroxyalkanoates): Conversion technologies, strategies, challenges, and perspective. *Bioresour. Technol.* **2021**, *326*, 124733. [CrossRef] [PubMed]
9. Kumar, G.; Ponnusamy, V.K.; Bhosale, R.R.; Shobana, S.; Yoon, J.-J.; Bhatia, S.K.; Banu, R.; Kim, S.-H. A review on the conversion of volatile fatty acids to polyhydroxyalkanoates using dark fermentative effluents from hydrogen production. *Bioresour. Technol.* **2019**, *287*, 121427. [CrossRef] [PubMed]
10. Bhatia, S.K.; Mehariya, S.; Bhatia, R.K.; Kumar, M.; Pugazhendhi, A.; Awasthi, M.K.; Atabani, A.E.; Kumar, G.; Kim, W.; Seo, S.-O.; et al. Wastewater based microalgal biorefinery for bioenergy production: Progress and challenges. *Sci. Total Environ.* **2021**, *751*, 141599. [CrossRef] [PubMed]
11. Bhatia, R.K.; Sakhuja, D.; Mundhe, S.; Walia, A. Renewable Energy Products through Bioremediation of Wastewater. *Sustainability* **2020**, *12*, 7501. [CrossRef]
12. Gallego-Valero, L.; Moral-Parajes, E.; Román-Sánchez, I.M. Wastewater Treatment Costs: A Research Overview through Bibliometric Analysis. *Sustainability* **2021**, *13*, 5066. [CrossRef]
13. Jayaseelan, M.; Usman, M.; Somanathan, A.; Palani, S.; Muniappan, G.; Jeyakumar, R.B. Microalgal Production of Biofuels Integrated with Wastewater Treatment. *Sustainability* **2021**, *13*, 8797. [CrossRef]
14. Gurav, R.; Bhatia, S.K.; Choi, T.-R.; Kim, H.-J.; Lee, H.-J.; Cho, J.-Y.; Ham, S.; Suh, M.-J.; Kim, S.-H.; Kim, S.-K.; et al. Seafood Processing Chitin Waste for Electricity Generation in a Microbial Fuel Cell Using Halotolerant Catalyst *Oceanisphaera arctica* YHY1. *Sustainability* **2021**, *13*, 8508. [CrossRef]
15. Dange, P.; Pandit, S.; Jadhav, D.; Shanmugam, P.K.; Gupta, P.; Kumar, S.; Kumar, M.; Yang, Y.-H.; Bhatia, S. Recent Developments in Microbial Electrolysis Cell-Based Biohydrogen Production Utilizing Wastewater as a Feedstock. *Sustainability* **2021**, *13*, 8796. [CrossRef]
16. Hassanein, A.; Witarsa, F.; Lansing, S.; Qiu, L.; Liang, Y. Bio-electrochemical Enhancement of Hydrogen and Methane Production in a Combined Anaerobic Digester (AD) and Microbial Electrolysis Cell (MEC) from Dairy Manure. *Sustainability* **2020**, *12*, 8491. [CrossRef]
17. Persson, T.; Persson, K.M.; Åström, J. Ferric Oxide-Containing Waterworks Sludge Reduces Emissions of Hydrogen Sulfide in Biogas Plants and the Needs for Virgin Chemicals. *Sustainability* **2021**, *13*, 7416. [CrossRef]
18. Delavar, M.A.; Wang, J. Lattice Boltzmann Method in Modeling Biofilm Formation, Growth and Detachment. *Sustainability* **2021**, *13*, 7968. [CrossRef]
19. Huang, G.; Pan, W.; Hu, C.; Pan, W.-L.; Dai, W.-Q. Energy Utilization Efficiency of China Considering Carbon Emissions—Based on Provincial Panel Data. *Sustainability* **2021**, *13*, 877. [CrossRef]
20. Malatji, M.; Marnewick, A.L.; Solms, S.V. Cybersecurity Policy and the Legislative Context of the Water and Wastewater Sector in South Africa. *Sustainability* **2021**, *13*, 291. [CrossRef]

Review

Renewable Energy Products through Bioremediation of Wastewater

Ravi Kant Bhatia [1], Deepak Sakhuja [1], Shyam Mundhe [1] and Abhishek Walia [2,*]

1 Department of Biotechnology, Himachal Pradesh University, Summer Hill, Shimla 171005 (H.P.), India; ravibiotech07@gmail.com (R.K.B.); deepaksakhuja6@gmail.com (D.S.); shyammundhe143@gmail.com (S.M.)
2 Department of Microbiology, College of Basic Sciences, CSKHPKV, Palampur 176062 (H.P.), India
* Correspondence: sunny_0999walia@yahoo.co.in; Tel.: +91-9815358847

Received: 24 July 2020; Accepted: 8 September 2020; Published: 11 September 2020

Abstract: Due to rapid urbanization and industrialization, the population density of the world is intense in developing countries. This overgrowing population has resulted in the production of huge amounts of waste/refused water due to various anthropogenic activities. Household, municipal corporations (MC), urban local bodies (ULBs), and industries produce a huge amount of waste water, which is discharged into nearby water bodies and streams/rivers without proper treatment, resulting in water pollution. This mismanaged treatment of wastewater leads to various challenges like loss of energy to treat the wastewater and scarcity of fresh water, beside various water born infections. However, all these major issues can provide solutions to each other. Most of the wastewater generated by ULBs and industries is rich in various biopolymers like starch, lactose, glucose lignocellulose, protein, lipids, fats, and minerals, etc. These biopolymers can be converted into sustainable biofuels, i.e., ethanol, butanol, biodiesel, biogas, hydrogen, methane, biohythane, etc., through its bioremediation followed by dark fermentation (DF) and anaerobic digestion (AD). The key challenge is to plan strategies in such a way that they not only help in the treatment of wastewater, but also produce some valuable energy driven products from it. This review will deal with various strategies being used in the treatment of wastewater as well as for production of some valuable energy products from it to tackle the upcoming future demands and challenges of fresh water and energy crisis, along with sustainable development.

Keywords: effluent; anaerobic digestion; incineration; Co-pyrolysis; syngas; biodiesel; biofuel

1. Introduction

All living beings could not live without water. Humans require water for not only to sustain their life, but also to accomplish their day to day activities. But now, a day's pure water gets out of reach of humans because of the addition of various harmful and toxic pollutants in water sources. Beside this basic necessity of the present world, the management of resulting effluent or wastewater is another challenge [1,2]. While technologies for the recovery of wastewater resources have been discussed extensively by the scientific community in recent decades, their large-scale implementation in municipal wastewater treatment facilities (WWTPs) still requires serious consideration. This can be demonstrated mainly by technical and non-technical reasons for doing so. Wastewater management plays a significant part in sustainable urban planning [3]. It is a well-known worldwide reality that the energy demand is increasingly growing due to rapid population growth that has also increased the rate of generation of wastewater in the last decades. To accomplish both of these obligations, utilization of wastewater should be done in such a manner so that the process used would treat the wastewater along with the production of some cherished products which can be reutilized further [4,5]. Use of waste water for energy generation is economic, as this does not require expensive phenomenon. While several emerging technologies contribute to the wastewater resource recovery challenge, biological approaches give the greatest promise to recover essential resources from effluent in an efficient manner. The article will generally concentrate on various methods of resource recovery from domestic and industrial wastewater [6]. The next generation of Domestic Wastewater Treatment Plants (DWWTP) targets energy efficiency and the complete use of wastewater for energy generation. There are also increasing concerns to extract useful products, especially renewable energy, from various forms of waste and wastewater from various industrial effluents [7]. Moreover, the fossil sources are very limited and may deplete in the coming future, so alternative sources of energy have to be developed. Therefore, the best approaches include the use of wastewater for production of energy products like bioethanol, biogas, biodiesel, etc., which further can be transformed into electricity [8]. Such energy recovery approaches may help mitigate wastewater sector electricity consumption and show promising areas for renewable energy policy implementation. Our analysis looks only at energy usage and future savings; while very significant, the economics of energy recovery mechanisms of wastewater treatment plants are reserved for a separate examination.

2. Characteristics of Wastewater

The characteristics of wastewater significantly affect the treatment approach to be pursued, as well as the reactor design selection process. For such characteristics, the most important are concentration for suspended solids, organic strength (BOD or COD), temperature, pH, and inhibitor presence [9]. Many reactor designs can damaged by suspended solids and accumulation of grits. For this purposes, liquid waste or wastewater is considered to have a concentration of suspended solids below 1000 mg/L with small quantities of grit (inorganic non-soluble solids), often removed by simple pretreatment. Defined as such, wastewater can be graded as low, i.e., below 1000 for industrial, agricultural (including flushed manures), and pulp and paper, medium, i.e., 1000–10,000 for food processing, canning, citrus processing, milk processing, juice processing, and brewery, and high, i.e., 10,000–200,000 for ethanol production, distillery, biodiesel production, petrochemical, and slaughter house concentration [10–12].

2.1. Sources of Wastewater

On the basis of the source from which the wastewater is being generated, there are various types of wastewater, some of them are listed below:

1. Domestic Wastewater (DW)
2. Sewage Sludge (SS)
3. Diary Wastewater (DWW)
4. Winery Wastewater (WWW)
5. Tannery Wastewater (TWW)
6. Textile Wastewater (TxWW)
7. Food Wastewater (FWW)
8. Phenolic Wastewater (PWW)
9. Carpet Mill Wastewater (CMWW)
10. Slaughter House Wastewater (SHWW)
11. Pharmaceuticals Wastewater (PhWW)
12. Beverage Wastewater (BWW)
13. Paper industry Wastewater (PWW)
14. Palm Oil Mill Wastewater (POMW)
15. Olive Oil Mill Wastewater (OOMW)

2.2. Features and Pollutnats of the Wastewater

Wastewater is generally characterized on the basis of physical (color, odor, and turbidity) and chemical (pH, alkalinity, biochemical oxygen demand (BOD), chemical oxygen demand (COD), dissolved oxygen (DO), total organic carbon (TOC), total dissolved solids (TDS), total suspended solids (TSS), conductivity, nitrogen, phosphorus, heavy metals, volatile solids (VS), oil, fats, grease and gases), etc. Different types of sources, along with their typical properties, are listed and discussed in Table 1.

Table 1. Characteristics and sources of some wastewater effluents.

Waste Water Type	Sources	pH	Chemical Oxygen Demand (COD) (mg/L)	Biochemical Oxygen Demand (BOD) * (mg/L)	Dissolved Oxygen (DO) (mg/L)	Total Solids (g/L)	Total Dissolved Solids (TDS) (g/L)	Total Suspended Solids (TSS) (g/L)	(VS) (g/L)	Alkalinity (mg/L)	References
DW	Toilets	-	740	350	-	-	-	450	320	1850	[13]
DWW	Dairy or Milk Industry	3.3	4705	1800	6.3	43.62	5.3	38.32	39.84	-	[14]
FWW	Petha Sweet Industry	11.9	5882	580	3.8	5.44	5.22	0.22	1.64	2400	[15]
OOMW	Olive Mills	4.8	132,300	-	-	41.8	-	-	36.8	-	[16]
POMWW	Palm Oil Mills	4.2	51,000	25,000	-	-	-	18,000	-	-	[17]
SHWW	Slaughter House	5.3–6.8	58,000–20,150	2200–9800	-	-	-	2.4–4.7	-	-	[18]

* BOD after 5 days.

Most of the wastewater contains a chemical, biological matter, and other objectionable matter that differ from source to source from which it generates. Industrial effluent includes a significant quantity of harmful chemicals and heavy metals, i.e., zinc, copper, nickel, lead, cadmium, arsenic, antimony, mercury, etc. [19], with lower biological content. Wastewater from households contains lower levels of chemicals comparatively to industrial wastewater, but high levels of organic matter, whereas agricultural wastewater includes high levels of chemicals in the form of pesticides, weedicides, fertilizers, etc., and biological substances like algae, fungi, bacteria, etc. [5,20]. Waste water consists of 70% organic compounds and 30% inorganic compounds, along with a variety of gases. Organic compounds are mainly carbohydrates, fats, and proteins, whereas inorganic matter consists of heavy metals, nitrogen, phosphorus, sulphur, and chloride, etc. Hydrogen sulfide, methane, ammonia, oxygen, nitrogen, and carbon-dioxide are commonly dissolved gases present in wastewater [21]. Biologically, wastewater consists of different types liverworts, seedy plants, ferns and mosses, bacteria, fungi, algae, and protozoans along with various types pathogens are also found in wastewater, which comes from the human beings suffering from various diseases [20,22].

3. Treatment Methods of Wastewater

Various types of pollutants present in the wastewater can be removed by using different strategies. Various treatment methods are used on the basis of source and location for wastewater treatment. Primary treatment can reduce BOD by 20–30% and suspended solids by as much as 60% [23]. This step includes reduction of oil, grease, fats, sand, and coarse solids. Secondary treatment will minimize BOD and total suspended solids by up to 85 percent. This step includes degradation of dissolved contents of the sewage within a biological degradation of system, as shown in Figure 1. The last step of secondary treatment is the removal of biological matter from the treated water with very low levels of organic material and suspended solids [24]. Microbes in the wastewater consume food in the form of organic matter, turning it into carbon dioxide, water, and electricity. Tertiary treatment can remove up to 99 percent of sewage impurities. Some operators add chlorine as a disinfectant before discharging the water. Sometimes nitrogen and phosphorus removal are done by tertiary treatment. Tertiary treatment uses advanced equipment and technologies to further eliminate or discard contaminants or particular pollutants [25].

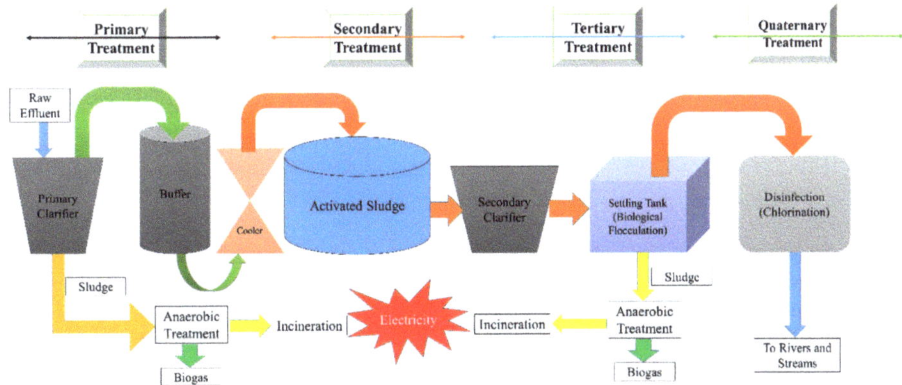

Figure 1. Schematic representation of waste treatment plant.

On the basis of type of matter present, ultrafiltration, sedimentation, sand filtration, etc., are physical methods used to treat the wastewater or industrial effluent as shown in Figure 2. In the chemical treatment method, chlorine is the most widely used chemical that acts as oxidizing agent to kill the bacteria that decompose the water. Another disinfecting oxidizing agent called ozone is used to purify wastewater [10]. Biological treatment methods use biological agents like plants and microorganisms in this way to remove the harmful pollutants. Biological wastewater treatment is done by oxidation bed or aerated systems, and post precipitation. It can be classified into various groups such as aerobic, anaerobic, and anoxic systems or suspended growth and attached growth according to the growth mechanism of microorganisms [26,27].

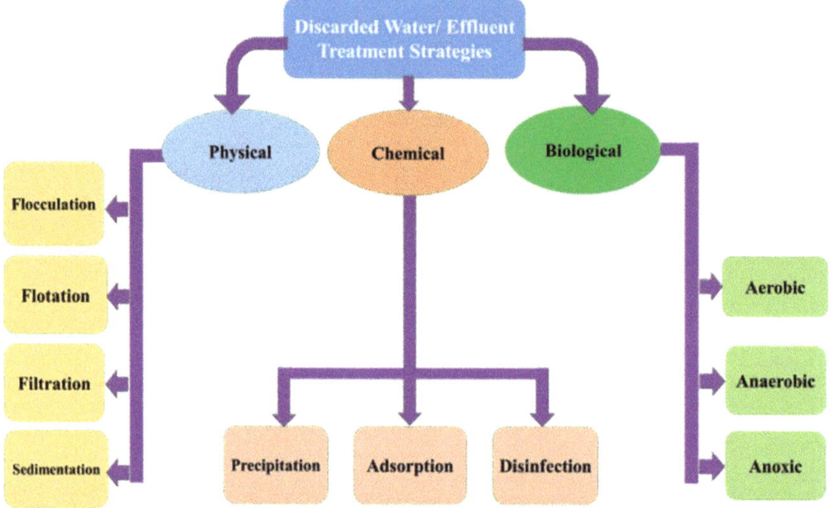

Figure 2. Different methods of wastewater treatment.

4. Wastewater as a Source of Renewable Energy

Wastewater produced by various sources is usually full of various minerals, nutrients, and organic matter, and these act as a wonderful source of metabolites for the growth and development of various microorganisms, algae, and plants which can be utilize to produce various renewable energy products. Organic matter rich wastewater, when decomposed in an oxygen-free environment, especially deep in a landfill, releases methane gas. Wastewater which is rich in organic matter when it decomposes, particularly deep in a landfill in an oxygen-free environment, releases methane gas [28]. This methane can be collected and used, instead of released into the atmosphere, to generate heat and electricity. Most wastewater treatment systems include different stages and eventually create solid sludge that is treated by thermal hydrolysis to increase the amount of methane it can generate. The processed waste then enters in an anaerobic digester, which ends up breaking it down. The resulting product is a methane-rich gas, or biogas, which can be used for on-site energy needs, or further refined and used instead of natural gas [29]. However, the solid wastewater residues produce a nutrient-rich "digestate" that can be used as a biofertilizer for soil conditioning and to improve plant production. Usually, the concentration of reducing matter in wastewater is expressed as the COD, which indicates how much oxygen is needed to oxidize the reducing matter [30]. A typical wastewater has a 0.5 kg/m^3 COD value and theoretically has the ability to generate 1.47 to 10^7 joules of energy per kg COD oxidized to CO_2 and H_2O, and the wastewater energy density is 0.74 to 10^7 J/m^3. Heidrich et al. [31] have recently statistically calculated the internal chemical energy of wastewater measured at 1.68×10^7 J/m^3 for wastewater combined with household wastewater and industrial wastewater, and 0.76×10^7 J/m^3 for pure household wastewater. Accordingly, a fair estimation of the theoretical energy density in wastewater is in the order of 10^7 J/m^3, which is five times the energy used to treat wastewater, and based on this data, the USA has an approximate capacity of 1.2×10^{15} J/day, 4.4×10^{17} J/year of wastewater renewable energy production [32–34].

4.1. Approaches to Determining Energy Potential from Wastewater

Process streams must be differentiated according to their capacity as energy sources according to the following input streams, process intermediates, and energy outputs in order to relate different technical options for energy recovery from wastewater.

4.1.1. Inputs

Carbonaceous material, especially the organic material (dissolved/suspended) present in wastewater, conducts its chemical energy potential. Exceptions are energy in the form of heat (for wastewater above ambient temperature) and the use of wastewater as growth media for species to absorb carbon dioxide using sunlight energy (photosynthesis) because of their inorganic constituents [35].

4.1.2. Intermediates

A number of intermediate compounds synthesized by the microorganisms, algae, and green plants that act as store houses of biochemical energy. These intermediary compounds are utilized by the microorganism as well animals to produce the energy to fulfill their energy requirement for day to day activities. But using some specific microorganisms, as well as advanced technological strategies, these compounds can also be utilize for the production of a number of energy products such as gaseous hydrogen or methane, liquid ethanol or biodiesel, or solid dry biomass [36].

4.1.3. Outputs

In particular, intermediate fuels provided by methane from wastewater can be used as energy products for heat, electricity generation, and even in propulsion vehicles. Currently available systems are much less effective and can transform only 25 to 35 percent of thermal energy into electrical energy, resulting in tremendous energy losses [8,37]. For higher recovery and better efficiencies, therefore, combined heat and power application is recommended. Therefore, technologies that convert inputs into intermediates by harnessing carbon-bound energy into biogas, bio-ethanol, and biodiesel, and then converting these intermediates into outputs via combustion/gasification into gaseous fuels, or converting inputs directly into outputs via heat recovery through heat pumps, and microbial fuel cells for electricity generation [38–40]. To maximize the potential for wastewater energy recovery, each of these technologies must first be understood in terms of physical, chemical, and biological concepts and constraints, and in terms of maturity and current penetration level.

4.2. Valuable Energy Products from Wastewater

Controlled treatment of wastewater can produce a variety of value-added items. Biological wastewater systems are commonly used to obtain valuable materials from wastewater. There may be either aerobic and anaerobic digestion, or effluent co-digestion. Wastewater treatment produces a large number of energy products that can be used in the form of biofuels, which are described in Table 2 as well as in the section below:

Table 2. Types of useful energy products recovered from different types of wastewater, and their characteristics and operational features.

Bioenergy Produced	Wastewater Stream (WWS)	Characteristic of WWS	Operating Conditions	Remarks	Reference
Biogas: Microorganisms involved- Acetogens and Methanogens.	Palm Oil Mill Effluent	High BOD and COD due to high organic content; pH = 3.4–5.2	Anerobic Digestion (AD) using Up-flow anaerobic sludge fixed film	71.9% CH_4	[41]
	Distillery Stillage	High COD and BOD due to presence of many organic compounds like polysaccharides, proteins, etc.	AD using Up-flow Anerobic Sludge Blanket (UASB)	90% COD; Biogas productivity = 0.43 m^3/kg.	[42]
	Textile printing and dyeing wastewater	High pH, high turbidity, poor biodegradability	AD using Anerobic Baffled Reactor; pH = 6.8–7.3; 30–35 °C; HRT = 4 days	83% CH_4; 71.5% COD removal	[43]
	Recycled Papermill Wastewater	Organic wastewater with high COD concentration	AD using UASB; 37 °C; OLR = 8.3 g COD/L/d; HRT = 15.14 h. d.	The highest biogas volume of 176.1 L/kg COD removed	[11]
	Cattle Slaughterhouse Wastewater	Rich in organic components and nutrient	AD using modified UASB; OLR of 10 g $L^{-1}d^{-1}$; HRT = 1 day	>90% COD removal; biogas (27 L/d); 89% CH_4	[44]
	Municipal Wastewater	High COD and BOD	AD using UASB and Dynamic Membrane Filter	Methane yield was 354 ± 37 mL $CH_4/gCOD_{utilized}$	[45]
	Molasses Wastewater	High amount of carbohydrates	AD using UASB	91.2% COD removal; 67.3%–78.9% CH_4	[46]
Bioethanol	Cassava Liquid Waste	Rich in glucose and starch	Acid/Enzyme pretreatment followed by SSF by *Zymomonas mobilis*	95% theoretical ethanol yield	[7]
	Municipal wastewater/Municipal Sewage sludge	Rich in organics	Saccharification by *Bacillus flexus* followed by fermentation by yeast 6% wt.; pH = 6.5; 30 °C; 10 days	Bioethanol yield are greater than 40 mL/L	[47]
	Agricultural Wastewater	Rich in organic nutrients	Algae cultivation followed by saccharification and fermentation	-	[39]
	Winery Wastewater	High concentration of organic and inorganic contaminants	Biorefinery concept	-	[48]
	Diary Wastewater	High COD and BOD due to presence of whey in liquid waste	Engineered *Lactobacillus lactis* can serve as novel cell factory; Fed Batch Strategy	71% Theoretical ethanol yield	[49]
	Olive mill wastewater	Rich in phenols and polyphenols	Biotreatment using: *Candida tropicalis*, *Pichia kudriavzevii*, *Pichia manshurica*, *Kluyveromyces marxianus*	Considerable amount of ethanol formed	[38]

Table 2. *Cont.*

Bioenergy Produced	Wastewater Stream (WWS)	Characteristic of WWS	Operating Conditions	Remarks	Reference
Biodiesel	Textile Wastewater	Contain variety of dyes, phosphates, nitrates, and auxiliary chemicals; high BOD, COD	Microalgal (*Chlorella* sp., *Scenedesmus* sp.) Bioremediation followed by harvesting, lipid extraction, and biodiesel production	Good yield of biodiesel obtained at lab scale with 60%–80% COD removal	[40]
	Diary Wastewater Sludge	Sludge contain 3–4% wt. total solids	Sludge dewatering and drying-two stage solvent extraction-Dewaxing-Drying-Transesterification-Washing	At optimum condition, 97.4% biodiesel yield from refined lipid	[50]
	Sago Processing Wastewater from Cassava based Industry	Rich in starch; high COD and BOD	Lipid production for biodiesel feedstock using *Candida tropicalis* ASY2	Innovative and ecologically sustainable technology with 84% COD removal, 92% BOD removal; high oleic acid content = 41.33%	[51]
	Tannery Wastewater	High oxygen demand	Microalgae Cultivation by *Chlorella* sp., *Scenedesmus* sp. Supplemented with Kelp waste extract	Good alternative for TWW treatment and biodiesel production	[52]
	Domestic Wastewater	Nutrient rich	Microalgae cultivation by *Nostoc* sp., *Chlorella* sp.	It is a suitable and non-expensive method for biodiesel production	[13]
Biohydrogen	Paper board Mill Wastewater	High organic and inorganic contaminants are present	Anerobic digestion by mixed culture bacteria (Hydrogen producer) using Continuous up flow anerobic reactor; HRT = 9.6 h	H_2 yield = 5.29 mmol/g COD (70%)	[53]
	Beverage Industry Wastewater (Alcohol distillery)	Rich in starch and glucose	Reactor- Anerobic Sequencing Batch Reactor; Inoculum- Sludge from Red Bull Distillery anerobic tank; HRT = 16 h	H_2 yield = 172 mL/g COD removed (34.7%)	[9]
	Sugary Wastewater	Rich in simple sugars	Reactor- Continuous Stirred Tank Reactor (CSTR); Inoculum- Municipal sewage treatment sludge; HRT = 1 h	H_2 yield = 1.37 mol/mol hexose (40%)	[54]
	Cheese Processing Wastewater	High COD and BOD due to whey	Reactor-CSTR; Inoculum-Anerobic digestor sludge; HRT = 24–84 h	H_2 yield = 5–22 mmol/g COD (45%)	[55]
	Rice mill wastewater	Rich in polysaccharides	After proper pretreatment and hydrolysis (Acid pretreatment with enzymatic hydrolysis); hydrogen produced using *Enterobacter aerogenes* RM 08 in Batch Reactor	H_2 yield = 1.97 mol H_2/mol sugar	[56]

Table 2. *Cont.*

Bioenergy Produced	Wastewater Stream (WWS)	Characteristic of WWS	Operating Conditions	Remarks	Reference
Bioelectricity	Kitchen Wastewater	High Organic and Protein content	Using Microbial Fuel Cell consisting photosynthetic microorganism as cathode catalyst (*Synecochococcus* sp. and *Chlorococcum* sp.); Anode Mixed culture; 1600 lx; CO_2 Supply	Power density = 41.48 mW/m^2 *Synecochococcus* sp.; 30.20 mW/m^2 *Chlorococcum* sp.	[37]
	Diary Wastewater	High COD and BOD	Dual chambered Microbial Fuel Cell is used with anolyte pH = 7	Aerobic metabolism gives Power density of 192 mW/m^2 with 91% COD removal while anerobic metabolism gives power density of 161 mW/m^2 with 90% COD removal	[12]
	Food processing wastewater	Rich polysaccharides, proteins; High BOD	Anode-Buffer solution; Cathode-Food processing wastewater; both are separated by proton exchange membrane in two compartment Microbial Fuel Cell (MFC) reactors; no catalyst and mediator	Power density = 230 mW/m^2 86% COD removal	[57]
	Sewage Sludge	Rich in organics	Anode-Graphite with neutral red (NR)/graphite with Mn^{4+}/platinum and polyanilineco-modified; Bacteria: *Escherichia coli; System configuration: Single Chamber*	Power density = 152/91/6000 mW/m^2 respectively	[58]
Microbial Fuel Cell	Swine Wastewater	8320 mg/L soluble COD	Single chambered MFC; Carbon paper electrode; cathode covered with platinum one side	261 mW/m^2	[59]
	Urban Wastewater	Low BOD	Salt bridge is present; graphite electrodes	25 mW/m^2	[60]
	Beer brewery Wastewater	2240 mg/L COD	Single Chamber, air cathode MFC, carbon cloth electrodes	205 mW/m^2	[61]
	Wastewater from Paper recycling industry	High organic and inorganic contaminants are present	Mixed culture of *Enterobacter* sp., U-tube MFC	5.5 mW/m^2	[62]
	Chocolate Industry Wastewater	1459 mg/L COD	Activated sludge from Municipal Wastewater treatment plant; dual chambered MFC; Graphite electrodes	1500 mW/m^2	[63]
	Starch Processing Wastewater	Rich in glucose and starches	Air cathode MFC; carbon paper anode	239.4 mW/m^2	[64,65]
	Sewage Sludge	12,110 mg/L Total COD	Two chambered MFC; graphite fiber brush electrodes	9.1 W/m^3	[66,67]

4.2.1. Biogas

Biogas is produced in an oxygen-free environment by anaerobic digestion of organic matter using microorganisms. Biogas processing requires several stages of a number of microorganisms: Microorganisms turn complex organic compounds into less complex organic compounds that are then converted into organic acids in initial hydrolysis reactions. Then methane forming microorganisms use these acids to form methane, the principal component of biogas [2,11]. Biogas is a mixture of gases containing typically 50–70% of methane. In addition, anaerobic digestion can produce hydrogen either as a component of the biogas, or as the major product. The latter requires that particular species, such as *Rhodobacter* or *Enterobacter* sp., to dominate the microbial population [41,44]. Current models indicate significantly greater recovery of energy from the digestion of biomass as methane. Hydrogen fermentation may become more attractive with the advancement of fuel-cell technology. Biogas can be used with little alteration in many applications (Stoves, Boilers). The gas needs substantial modifications for applications in combustion engines (generators, motor car engines) in order to eliminate non-methane components [2,29].

4.2.2. Bioethanol

Generally, bioethanol is produced from lignocellulosic biomass, but due to advancements in wastewater treatment technologies, it can also be produced from wastewater, and bio-electrolytic conversion is one of the innovative technologies in this direction [36]. A number of wastewater effluents (alone or with other wastes) had been utilized for ethanol production, i.e., OMW and olive pomace [16], apple pomace hydrolysate [68], etc.

4.2.3. Biodiesel

Algal growth in water bodies is an indication of water pollution. However, nutrient rich waste water can be utilized for the production of blue-green algae that accumulate the lipids in it and that in turn can be utilized for the production of biodiesel trough transesterification [69,70]. Algal biomass thus obtained can also be utilized as animal feed and also spread out as fertilizer. Certain species of algae, grown in wastewater is capable of produce oil up to 80% as its storage product and produce more oil, around 23 times more than the best oil-seed plant. In terms of oil produced per unit area, algal productivity exceeds palm oil by 10 bend, and jatropha, canola, and sunflower crops by more than 30 bend [20]. Oil transesterification to biodiesel, algal biomass, and glycerol can be utilized to produce energy products, and glycerol may be used as fuel for burning directly or can be converted to hydrogen and bioethanol by fermentation [71]. The viability of algal biomass for the production of biodiesel can be improved by instant wastewater treatment, use as animal feed, and the production of secondary energy products. Microalgae were also used in phytoremediation of wastewater sources [51].

4.2.4. Biohydrogen

Wastewater may be used through dark fermentation for the production of hydrogen. A microbial consortium contains a wide variety of bacteria, and some of them inhibit hydrogen production (i.e., hydrogenophilic-methanogenesis) by their consumption (homo-acetogens and methanogens) [54]. So, for optimum hydrogen production, the activity of inhibitory microbes gets suppressed or they are killed, either by heat pretreatment of inoculum or by increasing dilution rate and by decreasing the pH of reactors. Another reliable and finely honed process is catalytic methane to hydrogen conversion [72]. Different wastewater and effluent have been used for hydrogen production, these are TxW, PWW, rejected water and seed sludge, MW, OMW, BW, etc. [54,71,73–75].

4.2.5. Biomethane

Methane is another gas produced from anaerobic digestion of wastewater and effluent. Methane can be produced either by hydrogenophilic-methanogenesis (abiotically) or by fermentation of organic matter present in wastewater (biologically) [76]. Anaerobic digestion process is more proficient over aerobic process for production of methane due to low energy rations, high energy production in the form of methane, and low sludge production with high organic ejection rates. Various wastewater effluents, i.e., PWW, brown water, rejected water and seed sludge, sugarcane juice in the effluent discharged from sugarcane industry, etc., have been used for methane production [22,55,77].

4.2.6. Bioelectricity

Bio-electrochemical conversion process is involved in production of electricity from wastewater and effluent, in which catalytic commotion of microorganisms utilize the organic matter and produce the electrons that may be received by cathodes to generate the electricity [57,78]. Organic electron donors are catalyzed for oxidation in the anodic chamber by electrochemical bacteria, and electrons are supplied to the anode, which can be arrested in the form of bio-electricity [79]. In this process, exchange of protons, generated from anodic chamber to cathodic chamber, where these protons are involved in the production of cherished products [69]. Algal biomass, which is generated by photosynthesis, can be a direct organic source of electricity generation in Microbial Fuel Cells (MFC) and for producing biofuels [68].

4.2.7. Syngas

Syngas consists mainly of a mixture of carbon monoxide, carbon dioxide, and hydrogen that can be used as combustion fuel (heat energy value 8–14 MJ/kg or 10–20 MJ/Nm^3), or converted to liquid fuels using a biological or chemical process [80]. Syngas can be used to produce synthetic petroleum through the synthesis of Fischer–Tropsch, or through the gasoline methanol process. Conversely, the anaerobic bacteria can transform the syngas carbon monoxide into ethanol, with average yields of 340 L of ethanol per ton (municipal solid waste, agricultural waste, animal waste, etc.) [81]. The combustion of biomass (or syngas) in the presence of excess oxygen supply results in full oxidation and the production of hot flue gasses usually used to produce steam to drive electric turbines for electricity production with an output of approximately 30% [82].

5. Strategies and Mechanisms for Recovery of Renewable Energy Products from Waste Water

To produce the various valuable energy products from the wastewater, it has to be digested with the aid of a variety of microorganisms to produce the specific type of energy product. Even after digestion, the remaining sludge is treated again with various physical and chemical methods to obtain more energy from the wastewater and it's left over materials. A broad variety of energy products and value-added compounds can be extracted from wastewater effluents; energy in electricity form can somewhat minimize electricity scarcity [77]. There are various techniques involved in the production of energy products from wastewater namely, MFC, bio-electrochemical system (BES), biochemical, chemical, and biological (aerobic and anaerobic digestion of effluent) [58]. The key methods used to produce energy from the waste water are:

5.1. Anaerobic Digestion to Produce Biogas

This is a biological process that involves using microorganisms to transform the organic waste into valuable products. Anaerobic treatment of liquid waste or wastewater provides the ability to rapidly minimize the organic content of the waste while reducing the energy usage of the treatment process and the production of microbial biomass or sludge [29,71]. AD, as shown in Figure 3, is a complex process that involves and carries a variety of reactions (in absence of oxygen) such as hydrolysis, acidogenesis, acetogenesis, and methanogenesis [2]. AD is a very useful process which is applicable on a wide variety of waste effluents (sewage sludge, industrial wastewater, domestic wastewater, etc.) for their conversion to useful products especially into various energy forms, i.e., biohydrogen and methane [55]. The conversion of organic compounds into sludge in wastewater generates a by-product which needs further treatment or disposal. Reduction in sludge and energy consumption are the two attributes which make it economically attractive for municipal and industrial waste streams to consider direct anaerobic pretreatment of wastewater. AD is affected by various factors like temperature (25–35 °C), pH (~7), moisture, carbon source, nitrogen, and C/N ratio [2]. AD of sewage sludge used in treatment plants is very useful now because of lower disposal costs, and it is ecofriendly, too. Direct anaerobic treatment may also provide excess energy for relatively warm wastewaters which contain significant degradable organic compounds [47]. However, even with low-strength wastewaters, the energy savings that can be achieved by avoiding most of the aeration costs are significant. Anaerobic treatment effluents, however, are often not suitable for direct discharge into the receiving waters without further treatment that may require aerobic polishing. Nonetheless, this treatment scheme can be explained by the reduced aeration demand and the production of sludge in aerobic treatment following anaerobic pretreatment. The average ambient temperature of the wastewater has an effect on anaerobic treatment design quality [83]. Some wastewaters of low and medium strength are relatively cool (<20 °C), and the energy needed to heat them to mesophilic temperatures is significant and not economical. Wastewater with a temperature of 20 °C and a COD of 20 g/L, and biogas generation produces around the same amount of energy needed to increase the liquid's temperature to 35 °C. So, treatment at ambient temperatures is only feasible for wastewaters of low and medium strength. Successful anaerobic treatment of wastewater as low as 15 °C is feasible, but application of anaerobic digestion should not be taken into consideration below 12 °C [84]. On the other end, there are many industrial wastewater sources that are very warm and can be considered as mesophilic (food processing) use, and in some cases, thermophilic (distillery waste) anaerobic digestion.

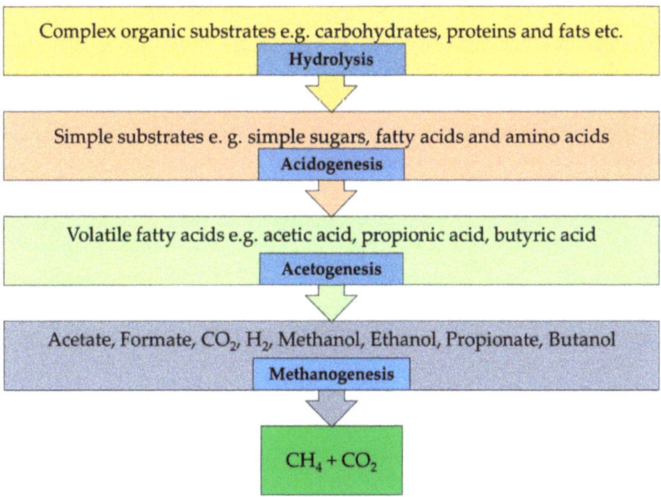

Figure 3. Showing stepwise anaerobic digestion process.

5.2. Fermentation to Bioethanol

It is well known that bioethanol is developed as a renewable liquid fuel. Bioethanol can be used alone or combined with traditional liquid fuels to form either Gasohol or Diesohol. Bioethanol is usually produced under anaerobic conditions through fermentation of simple sugars, such as glucose and fructose. Several yeasts, including, for example, *Saccharomyces* sp., and other bacteria including *Zymomonas* sp., undergo this fermentation [6,50]. A variety of industries, such as sugar, food processing, meat, and pulp and paper, have developed carbohydrate (glucose, fructose, lactose, etc.) rich wastewater that can be converted into bioethanol through fermentation. However, the current challenges are using waste streams in which organic carbon is not present as simple sugars through the use of chemical or biological pretreatment or novel microorganisms that ferment a wider range of organic substrates [53]. Currently, there is extensive work focusing on pretreatment procedures for cellulolytic. The small yields of ethanol obtained in fermentation (typically 10% (*v*/*v*)) currently need subsequent energy intensive distillation. Conventional ethanol plants will expend more than 30 per cent of bioethanol fuel's heat energy during the distillation process [49,85,86].

5.3. Microbial Fuel Cells

Fuel cells transform chemical energy into electrical energy. Microbial fuel cells work using bacteria that oxidize organic matter in wastewater to transfer electrons to an anode from where they pass to the cathode through a circuit to combine protons and oxygen to form water. Electricity is produced by the difference in potential coupled to electron flow [58]. Microbial fuel cells have become an emerging technology and a number of these have been successfully operated with both pure cultures and mixed cultures, which have either been enriched by sediment or activated sludge from wastewater treatment plants [87]. Wastewaters of very different characteristics can be used: Sanitary waste, wastewater for food production, dairy manure, swine wastewater, and corn stove. This technology may effectively use bacteria already present in wastewater as catalysts to produce electricity while treating wastewater simultaneously, but its advancement is hindered by low power output and high material costs. To date, microbial fuel cells have not been used in large-scale applications, but are used to produce energy for BOD sensors, robots, and small telemetry systems [59,88].

5.4. Combustion, Gasification, and Pyrolysis

The heating of sludge in the presence of a small supply of oxygen contributes to gasification and syngas output. In order to generate pyrolytic oil, biochar, and non-condensable gases, combustion and pyrolysis require a fully inert atmosphere at moderate to high temperatures (300–900 °C). Bio-oil can also be used as a liquid fuel or converted into a synthetic gas (CO and H_2), whereas biochar, non-condensable gases, and bio-oil can also be used through combustion to produce electricity and heat [80,89]. Gasification involves the thermochemical conversion of organic compounds through partial oxidation at high temperatures (650–1000 °C) to optimize gaseous products (CO, H_2, CO_2, and light hydrocarbons), particularly synthesis gas (CO and H_2) [81]. Depending on the gasifying agent and temperature, the energy content of the natural gas ranges from 4–28 MJ/Nm. Additionally, the biomass contained in wastewater, such as microbial and algal biomass, and other biomass, can also be gasified into energy products, including heat, steam, electricity, syngas and liquid fuels, and biogas [2,90]. If the heat energy is also captured and combined heat and power (CHP) are given, the output can be improved up to 50% and up to 80%. The viability of applying combustion or gasification has to do with moisture content, and the practical problems of tar formation, mineral content, over bed burning, and bed agglomeration. The feedstock must be relatively dry; with 40 to 50 percent maximum moisture content [82]. Co-pyrolysis of sewage sludge with other non-biodegradable waste such as polythene and plastic waste was also performed at 525 °C in a stirred batch reactor under N_2 atmosphere. This potential synergetic strategy resulted in better yield of H_2 and CH_4 as compared to individual pyrolysis of sludge, and that can be used as gaseous fuel. This combined feasible management provides an alternative for both residues to be generated in a better output [91]. Dryers may be used in the design, but there is a direct trade-off between the amount of energy in the feedstock and the amount of energy spent on drying. Since gaseous and solid phase contaminants are potentially generated from the application of these technologies, there are many possible negative environmental consequences, including heavy metals, dioxins, furans, and NOx gases [92]. However, evidence suggests that technological interventions can control these emissions and use of combined cycle gas turbine, the generated gas can be diverted to various end uses such as direct combustion for heat and electricity generation.

5.5. Incineration of Sludge for Energy Production

Sludge is one of the useful byproducts from wastewater treatment plants predominantly consists of 75% mud, and 25% solid matter, but for any further use, it must be treated to remove the pathogens. The residual sludge can also be used for the manufacture of different energy products from it after wastewater treatment [22,30]. Typically, bio-solids are comprise of huge quantities of water that can be collected by using dewatering machines for further use either as biofertilizer or by incineration for heat and electricity generation. This technique is used by various Municipal Solid Waste (MSW) organizations or waste management firms to dispose of the bio-solids and extract the energy from it to fund their operational expenses [44]. Thanks to its global effect on waste minimization, resource optimization, and renewable energy production, energy recovery from wastewater and sludge in contemporary to wastewater management remains assured. Figure 1 shows the focus energy conversion methods, which shows the sludge conversion pathways to biogas, heat, and electricity. Commercially incinerated bio-solids with several hearth furnaces (MHF) and fluid bed furnaces (FBF) [47,53]. MHF burns bio-solids and allows the hot air to dry incoming bio-solids, reduces moisture, and increases the efficiency of MHF for heat generation, while FBF is a modern and efficient technology that provides continuous operation without any multi-stage device that eventually increases the efficiency of the overall process and technology of incineration [93]. Residual gas pollution is a challenge in both of these incineration technologies, but can be addressed with the use of advanced scrubber systems to make these technologies more efficient and environmentally friendly [94].

6. Advancements and Integrated Technologies to Recover Energy Products from Waste Water

Anaerobic wastewater treatment is currently the most commonly used method for extracting energy from wastewater. The energy is harvested as methane production. Removal of the energy lost due to heat dissipation during energy conversion from different reducing matters to methane, energy consumption to sustain microbial activity, and residual reducing matters of wastewater after treatment, 80% of the chemical energy found in the original reducing matters can be transferred to methane [24,95]. In view of the fact that only about 35% of methane's chemical energy can be converted into electricity through the combustion cycle, the overall efficiency of energy recovery is about 28%. If more efficient CH_4-driven chemical fuel cells are created, this number will theoretically increase to 40%. Our future perspectives should be production of more and more energy products to minimize the cost of overall wastewater treatment processes [43,75]. While streams and technologies can be matched individually, the integration of technologies and waste streams holds the greatest promise in achieving long-term energy security while maximizing wastewater treatment. There are many examples where this worked. Thus, our aim should be to design an approach in such a way that wastes that we generate can be reconverted into pathogen free resources without any cost and without polluting our environment. Some of the steps which may be taken are as follows:

6.1. Scale up of MFCs

MFCs utilized now are designed for lab scale purposes, but to know the utility of MFC, these should be scale up to a practical level so the energy recovery can be enhanced and optimized [57]. The restriction in scaling of MFCs are high scaling cost, pH buffers, high internal resistance, high material cost, and low efficiency of mixed cultures on an electrode and these limitations may be overcome by reactor engineering, biological employment and material development [63].

6.2. Digitalization of Process

The parameters and techniques used in a process should be monitored to an extent to achieve a high yield of by-products and chemical compounds from the treatment of wastewater and effluents [12]. In case of electricity generation electro-chemical parameters, i.e., electric current, indicators, and electrode potential, are some useful parameters which can be monitored for optimum electricity production [37,68].

6.3. Statistical Modelling

More optimization of the process may be obtained by the use of mathematical models, especially two directional. Use of these models facilitate the complex process and make it easy to perform at a practical level rather than lab scale [64,95].

6.4. Multilateral Approach

In the hybrid approach, more than one useful product is produced in a single process, for example, integrated hydrogen and methane production, this leads to recover maximum energy from wastewater [96]. MFC coupled with anaerobic membrane bioreactor and integrated photo-bioreactor is a game changer for this energy recovery from the wastewater [66,69]. This technique is capable of treating high levels of wastewater along with polishing treatment for removal of specific pollutant types from the effluent.

6.5. Use of Genetically Modified Microorganisms

Energy generation by wastewater treatment can also be improved by selection of the best microbial species responsible for specific biofuel production and specific pollutant removal; this will lead to reducing the operation costs and energy consumption for the process [62,97].

6.6. Hydrothermal Carbonization of Sewage Sludge

Hydrothermal carbonization (HTC) is a thermochemical process that can be used as a solid fuel or soil conditioner to transform liquid biomass into so-called biocoal. HTC can be an alternative to anaerobic digestion or a supplement in sewage sludge treatment. In the latter case, digested sludge serves as feedstock to HTC [98]. In 2010, the first industrial plant in Germany (Karlsruhe) was built in a WWTP. HTC's benefit is almost complete recycling of organic matter, very strong dewaterability of the resulting sludge, and an increased energy balance [99]. HTC is a recent, on the market technology, although the process has been known for over a hundred years. It is conceivable that HTC will develop into a standard sludge treatment technology in the future.

6.7. Integrated Algal Biodiesel

Historically, algal biodiesel has been suggested to be financially feasible only with concomitant wastewater treatment or animal feed production, valuable secondary products, or supplemental energy products. More recent analysis shows that recovery of algal biodiesel from wastewaters can be beneficial after 2 to 4 years with a fair breakeven [40,100]. There has been a lot of speculative interest in algal biodiesel recently, which had been fueled largely by the increased price of diesel before the reversal in 2008. Many new companies have been set up to use algae to develop biofuels and to obtain certified emission reductions (CERs) by reducing CO_2 (IGV GmbH, Nuthetal, Germany, undated) [35]. One example is Aquaflow Bionomic (Nelson, New Zealand), which reported harvesting crude oil to refine it into paraffinic kerosene for use as jet fuel, from wild algae grown on oxidation ponds used in the domestic and agro-industrial waste stream treatment trains. Within a 60 ha facility, this plant treats 5 billion liters of water per year [101]. More innovations will show whether these ventures are financially viable and will be introduced not for demonstration purposes, but for growth.

6.8. Advanced Integrated Wastewater Pond Systems

These consist of an anaerobic digester and algal pond with high concentrations. For nine years, a facility in Grahamstown, South Africa, has been tested for effectiveness in wastewater treatment. Nutrient and organic removal levels were reached comparable with traditional wastewater treatment works and negligible *E. coli* counts [24,50]. Anaerobic digestion biogas provides energy, and the algae may be used as fertilizer or fuel (e.g., biodiesel). Despite these and many other international examples of wastewater energy projects, in many countries there is no overall view of the potential or a plan for harnessing this renewable energy source. Anaerobic digesters are used by many urban wastewater treatment plants as part of the wastewater treatment process [76,102]. Although some use heat internally to control digester temperatures and to heat building space, however, the majority vent or flare the gas. This shows the weak integration of energy usage and the potential for reducing greenhouse gas emissions have not been understood [103]. Wastewater treatment plants must be integrated to use biogas to dry and pellet the wastewater sludge, thus reducing the on-site disposal costs and environmental burdens while providing a potential source of energy. The pellets have an energy content of ~16.6 MJ/kg and were used as additional fuel in their kilns by a local cement factory [99].

7. Benefits of Using Wastewater for Energy

Examples of where energy is extracted from wastewater to produce a range of energy products at varying scales (from small rural to large industrial operations) exist worldwide. A limited range of examples to demonstrate the energy potential from wastewater technologies is presented.

7.1. Domestic Biogas

In 1975, the Chinese government began a household biogas mass implementation strategy. Units were being installed within a few years at a rate of 1.6 million per annum. The technology continued to be developed and implemented, and in 2005, China had 17 million digesters generating 6.5 billion m^3 of biogas per annum [11,27]. Importantly, one fifth of households in rural areas get electricity from biogas. In India, Nepal, Vietnam, and Sri Lanka, this trend of rapid installation of biogas units has repeated itself. There are currently more than 2 million family-sized units in service in India, and over 200,000 families move from the traditional fireplace to cooking and heating biogas per year [2].

7.2. Biogas for Agricultural Use

Meili village (province of Zhejiang, China) slaughters 28,000 pigs, 10,000 ducks, 1 million ducklings, and 100,000 chickens each year, and the wastewater is fed to an anaerobic digester that generates enough biogas for more than 300 households and 7200 tons of organic fertilizer each year (ISIS, 2006) [2]. A similar process is used in Linköping (Sweden), where the biogas is upgraded to vehicle fuel standard for all public transit vehicles in the city (>60 buses), converted to run on biogas in 2005 (IEA Biogas, undated) [104]. Throughout Ireland, wastewater from farms in Ballytobin and the food processing industry produce electrical and heat energy by anaerobic digesters for the local farming population. With gas turbines and combined heat and power, this plant produces an additional 150,000 kWh of electricity and 500,000 kWh of heat energy per annum (IEA Bioenergy, undated) [91].

7.3. Bioethanol

A few examples of wastewater and wastewater sludge use for bioethanol production have been published. Finland's VTT Technical Research Center has developed technology for distributed ethanol production by fermentation of food processing wastewater. This technology allows production even on a small scale and is estimated to be capable of meeting 2% of the total volume of petrol sold in Finland, and is currently being marketed by St1 Biofuels Oy [38,48].

7.4. Energy Production

To sustain rising populations and expanding cities, the planet needs more resources. The use of waste for energy for many cities is an inexpensive, sustainable, and readily available source of energy. Because sewage treatment plants can use biogas generated from their own sludge to power their operations, they can be self-sufficient in energy. It means that the primary purpose of a sewage plant-eliminating toxins and disease-causing pathogens is not disrupted by power outages [9,15,31].

7.5. Emissions Reductions

Methane accounts for 16 per cent of global greenhouse gas emissions, and it is very strong—about 30 times more powerful than carbon dioxide, a greenhouse gas. Sludge-to-energy systems harness this methane for energy instead of allowing it to escape into the atmosphere where climate change will be fueling. Although methane releases carbon dioxide when harnessed for energy, if methane-rich biogas is used instead of fossil fuels, the net emissions are negligible [35].

7.6. Waste Management

Many developing countries lack the infrastructure necessary to manage solid waste and sludge properly. Such toxic, foul-smelling waste is frequently dumped directly onto land or surrounding waterways in these areas, where it can threaten public health. For example, in China, more than 70 per cent of urban solid waste and sludge are landfilled or dumped—sometimes illegally. One alternative is a sludge-to-energy strategy [91,105].

7.7. Economic Benefits

Sludge-to-energy systems reduce the need for more costly and polluting power sources, such as fossil fuels. However, those who run waste-to-energy operations will directly benefit from selling the gas and solid digestate [54].

8. Barriers in Recovery for Renewable Energy from Wastewater

The interrelationship between energy and organic wastewater content should facilitate energy recovery operations from different sources, including wastewater treatment plants wastewater [106]. Combining the anaerobic digesters with biosolid incineration for electricity generation from wastewater utilities will reduce energy consumption through 5 to 85 per cent [83]. But still there are many challenges that must be took in consideration for efficient utilization of wastewater to transform it into energy related products and other valuable products. Recovery of energy from the wastewater depends upon its organic content, and any deviation in it further adds to uncertainties in its utilization in the reactors for energy production [105]. Low temperature is another crucial challenge in the operation of anaerobic digesters because most of the microorganisms work in ambient temperatures, i.e., 15–35 °C and if there is any change from this range, the organism will not work efficiently and the kinetics of overall process falls down and reduce the production of energy components, i.e., CH_4 and H_2 [65,100]. Additionally, it will make it difficult to perform the anaerobic digestion to produce the valuable energy products from the wastewater. Wastewater is mainly composed of different organic materials and nitrogenous wastes. Almost all biodegradable matter in wastewater is converted into methane, but there is the chance of forming nitrous gas (potent greenhouse gas) during the partial nitration process from the nitrogenous wastes, and it may also reduce pH and oxygen levels that are important for survival of various microorganism inside the digester [2,107]. Sometimes phenolic substances are inhibitory to microbes, and great care should be taken in the selection of microbes for production of desired energy products. Moreover, the complexity of wastewater due to the introduction of new chemicals and substances from various anthropogenic activities is another growing challenge. Such activities not only change the uniformity of wastewater, but also make energy recovery from wastewater an uphill task [108,109]. Potential barriers to hydropower generation at wastewater treatment plants include a lack of excess heat, flow rate variations, turbine failure due to blockages, or particulate matter present in wastewater, especially in raw sewage. A resource recovery process is not cost effective due to excessive operational cost. While various technologies have been explored in the academic arena for the recovery of water, electricity, fertilizers, and other wastewater products, none of these have ever been implemented on a large scale because of technological immaturity and/or non-technical bottlenecks [110]. In order to treat wastewater more effectively and recover the essential energy-related products, all these problems and concerns have to be tackled in a very comprehensive way in order to solve all the wastewater-related issues besides developing more efficient and ecofriendly technologies.

9. Conclusions

Although domestic wastewater cannot completely satisfy the industrialized society's energy demands, it is a valuable resource that should be widely exploited and used in the future. Wastewater these days is no longer wastewater, however, most of agencies look at it as a resource rich in organic content and minerals. These all components could be utilized for the production of renewable energy products or to grow plants and algae that later on can be converted to various types of biofuels as per the requirement. The complex nature of effluent and wastewater make it difficult for the existing technology to remove all the pollutants effectively. So, there is an urgent need to develop a combined strategy using biological, chemical, and physical processes to treat the wastewater as well as to recover valuable products from this untapped resource. Using some of the advanced technologies like membrane bioreactors can be useful, but these have higher operation and maintenance costs compared with conventional processes. Moreover continuous efforts should be there on the development of suitable consortia that not only work on variable conditions, but also help to recover the valuable products from the wastewater. Despite all these facts, on one hand, there is a need for strict discharging standards to be implemented, and on the other hand, researchers and the scientific community need to develop the cutting-edge technologies to recover all the desired products in a sustainable manner.

Author Contributions: Conceptualization, R.K.B.; formal analysis, D.S. and S.M.; data curation, D.S. and S.M.; writing—original draft preparation, R.K.B.; writing—review and editing, A.W. All authors have read and agreed to the published version of the manuscript.

Funding: There is no any funding source for this study.

Acknowledgments: The authors would like to acknowledge all the facilities provided by the Department of Biotechnology Himachal Pradesh University, Summer Hill Shimla; Department of Microbiology, College of Basic Sciences, CSKHPKV Palampur, are duly acknowledged.

Conflicts of Interest: The authors declare that there is no conflict of interests regarding the publication of this paper.

References

1. Puyol, D.; Batstone, D.J.; Hülsen, T.; Astals, S.; Peces, M.; Krömer, J.O. Resource Recovery from Wastewater by Biological Technologies: Opportunities, Challenges, and Prospects. *Front. Microbiol.* **2017**, *7*, 2106. [CrossRef] [PubMed]
2. Bhatia, R.K.; Ramadoss, G.; Jain, A.K.; Dhiman, R.K.; Bhatia, S.K.; Bhatt, A.K. Conversion of Waste Biomass into Gaseous Fuel: Present Status and Challenges in India. *Bioenerg. Res.* **2020**. [CrossRef]
3. Shah, M.P. Bioremediation-Waste Water Treatment. *J. Bioremediat. Biodegrad.* **2018**, *9*, 10. [CrossRef]
4. Ijaz, A.; Iqbal, Z.; Afzal, M. Remediation of sewage and industrial effluent using bacterially assisted floating treatment wetlands vegetated with *Typha domingensis*. *Water Sci. Technol.* **2016**, *74*, 2192–2201. [CrossRef]
5. Reungsang, A.; Sittijunda, S.; Sreela-or, C. Methane production from acidic effluent discharged after the hydrogen fermentation of sugarcane juice using batch fermentation and UASB reactor. *Renew. Energy* **2016**, *86*, 1224–1231. [CrossRef]
6. De Amorim, E.L.C.; Dos Amorim, S.N.C.; Macêdo, W.V.; Batista, E.A. Biohydrogen and Bioethanol Production from Agro-Industrial Wastewater. In *Organic Pollutants in Wastewater II*; Ahamed, M.I., Hasan, S.W., Eds.; Materials Research Foundations: Millersville, PA, USA, 2018; pp. 122–137.
7. Akponah, E.; Akpomie, O.O.; Ubogu, M. Bio-ethanol production from cassava effluent using *Zymomonas mobilis* and *Saccharomyces cerevisiae* isolated from rafia palm (*Elaesis guineesi*) SAP. *Eur. J. Exp. Biol.* **2013**, *3*, 247–253.
8. Kassongo, J.; Togo, C.A. Evaluation of full-strength paper mill effluent for electricity generation in mediator-less microbial fuel cells. *Afr. J. Biotechnol.* **2011**, *10*, 15564–15570. [CrossRef]

9. Prakash, J.; Sharma, R.; Ray, S.; Koul, S.; Kalia, V.C. Wastewater: A Potential Bioenergy Resource. *Indian J. Microbiol.* **2018**, *58*, 127–137. [CrossRef]
10. Bhatnagar, A.; Sillanpää, M. Utilization of agro-industrial and municipal waste materials as potential adsorbents for water treatment—A review. *Chem. Eng. J.* **2010**, *157*, 277–296. [CrossRef]
11. Bakraoui, M.; Karouach, F.; Ouhammou, B.; Aggour, M.; Essamri, A.; el Bari, H. Biogas production from recycled paper mill wastewater by UASB digester: Optimal and mesophilic conditions. *Biotechnol. Rep.* **2020**, *25*, e00402. [CrossRef]
12. Elakkiya, E.; Matheswaran, M. Comparison of anodic metabolisms in bioelectricity production during treatment of dairy wastewater in Microbial Fuel Cell. *Bioresour. Technol.* **2013**, *136*, 407–412. [CrossRef] [PubMed]
13. Mostafa, S.S.M.; Shalaby, E.A.; Mahmoud, G.I. Cultivating Microalgae in Domestic Wastewater for Biodiesel Production. *Not. Sci. Biol.* **2012**, *4*, 56–65. [CrossRef]
14. Shah, N.; Patel, A. *Indian Food Industry Mag*; Association of Food Scientists & Technologists (India): Mysore, India, 2013; pp. 45–49.
15. Singhal, Y.; Singh, R. Energy recovery from petha industrial wastewater by anaerobic digestion. *Int. J. Sci. Eng.* **2015**, *3*, 146–151.
16. Battista, F.; Mancini, G.; Ruggeri, B.; Fino, D. Selection of the best pretreatment for hydrogen and bioethanol production from olive oil waste products. *Renew. Energy* **2016**, *88*, 401–407. [CrossRef]
17. Kaman, S.P.D.; Tan, I.A.W.; Lim, L.L.P. Palm oil mill effluent treatment using coconut shell—Based activated carbon: Adsorption equilibrium and isotherm. In *Proceedings of the MATEC Web of Conferences, Kuching, Sarawak, Malaysia, 26–28 October 2016*; Hasan, A., Khan, A.A., Mannan, M.A., Hipolito, C.N., Mohamed Sutan, N., Othman, A.-K.H., Kabit, M.R., Abdul Wahab, N., Eds.; EDP Sciences: Les Ulis, France, 2017; Volume 87.
18. Sivaramakrishna, D.; Sreekanth, D.; Sivaramakrishnan, M.; Sathish Kumar, B.; Himabindu, V.; Narasu, M.L. Effect of system optimizing conditions on biohydrogen production from herbal wastewater by slaughterhouse sludge. *Int. J. Hydrog. Energy* **2014**, *39*, 7526–7533. [CrossRef]
19. Kong, Q.; Li, L.; Martinez, B.; Chen, P.; Ruan, R. Culture of Microalgae *Chlamydomonas reinhardtii* in Wastewater for Biomass Feedstock Production. *Appl. Biochem. Biotechnol.* **2010**, *160*, 9–18. [CrossRef]
20. Steinbusch, K.J.J.; Hamelers, H.V.M.; Plugge, C.M.; Buisman, C.J.N. Biological formation of caproate and caprylate from acetate: Fuel and chemical production from low grade biomass. *Energy Environ. Sci.* **2011**, *4*, 216–224. [CrossRef]
21. Li, M.; Cheng, X.; Guo, H. Heavy metal removal by biomineralization of urease producing bacteria isolated from soil. *Int. Biodeterior. Biodegrad.* **2013**, *76*, 81–85. [CrossRef]
22. Ge, H.; Jensen, P.D.; Batstone, D.J. Pre-treatment mechanisms during thermophilic–mesophilic temperature phased anaerobic digestion of primary sludge. *Water Res.* **2010**, *44*, 123–130. [CrossRef]
23. Hennebel, T.; Boon, N.; Maes, S.; Lenz, M. Biotechnologies for critical raw material recovery from primary and secondary sources: R&D priorities and future perspectives. *New Biotechnol.* **2015**, *32*, 121–127. [CrossRef]
24. Sharma, S.K.; Sanghi, R. *Advances in Water Treatment and Pollution Prevention*, 1st ed.; Sharma, S.K., Sanghi, R., Eds.; Springer: Dordrecht, The Netherlands, 2012; ISBN 978-94-007-4203-1.
25. Leal, L.H. Removal of Micropollutants from Grey Water-Combining Biological and Physical/Chemical Processes. Ph.D. Thesis, Wageningen University, Wageningen, The Netherlands, 2010.
26. Rathour, R.K.; Sharma, V.; Rana, N.; Bhatia, R.K.; Bhatt, A.K. Bioremediation of simulated textile effluent by an efficient bio-catalyst purified from a novel Pseudomonas fluorescence LiP-RL5. *Curr. Chem. Biol.* **2020**, *14*, 1–12. [CrossRef]
27. Abioye, O.P.; Mustapha, O.T.; Aransiola, S.A. Biological Treatment of Textile Effluent Using *Candida zeylanoides* and *Saccharomyces cerevisiae* Isolated from Soil. *Adv. Biol.* **2014**, *2014*, 670394. [CrossRef]
28. Koch, K.; Helmreich, B.; Drewes, J.E. Co-digestion of food waste in municipal wastewater treatment plants: Effect of different mixtures on methane yield and hydrolysis rate constant. *Appl. Energy* **2015**, *137*, 250–255. [CrossRef]
29. Maragkaki, A.E.; Fountoulakis, M.; Gypakis, A.; Kyriakou, A.; Lasaridi, K.; Manios, T. Pilot-scale anaerobic co-digestion of sewage sludge with agro-industrial by-products for increased biogas production of existing digesters at wastewater treatment plants. *Waste Manag.* **2017**, *59*, 362–370. [CrossRef] [PubMed]

30. Tontti, T.; Poutiainen, H.; Heinonen-Tanski, H. Efficiently Treated Sewage Sludge Supplemented with Nitrogen and Potassium Is a Good Fertilizer for Cereals. *Land Degrad. Dev.* **2016**, *28*, 742–751. [CrossRef]
31. Heidrich, E.S.; Curtis, T.P.; Dolfing, J. Determination of the Internal Chemical Energy of Wastewater. *Environ. Sci. Technol.* **2011**, *45*, 827–832. [CrossRef]
32. Rao, P.; Kostecki, R.; Dale, L.; Gadgil, A. Technology and Engineering of the Water-Energy Nexus. *Annu. Rev. Environ. Resour.* **2017**, *42*, 407–437. [CrossRef]
33. Stillwell, A.; Hoppock, D.; Webber, M. Energy Recovery from Wastewater Treatment Plants in the United States: A Case Study of the Energy-Water Nexus. *Sustainability* **2010**, *2*, 945–962. [CrossRef]
34. Bhatia, S.K.; Joo, H.-S.; Yang, Y.-H. Biowaste-to-bioenergy using biological methods—A mini-review. *Energy Convers. Manag.* **2018**, *177*, 640–660. [CrossRef]
35. Bhatia, S.K.; Bhatia, R.K.; Jeon, J.-M.; Kumar, G.; Yang, Y.-H. Carbon dioxide capture and bioenergy production using biological system—A review. *Renew. Sustain. Energ. Rev.* **2019**, *110*, 143–158. [CrossRef]
36. Evcan, E.; Tari, C. Production of bioethanol from apple pomace by using cocultures: Conversion of agro-industrial waste to value added product. *Energy* **2015**, *88*, 775–782. [CrossRef]
37. Naina Mohamed, S.; Ajit Hiraman, P.; Muthukumar, K.; Jayabalan, T. Bioelectricity production from kitchen wastewater using microbial fuel cell with photosynthetic algal cathode. *Bioresour. Technol.* **2020**, *295*, 122226. [CrossRef] [PubMed]
38. Jamai, L.; Ettayebi, M. Production of bioethanol during the bioremediation of olive mill wastewater at high temperatures. In Proceedings of the 2015 3rd International Renewable and Sustainable Energy Conference (IRSEC), Marrakech, Morocco, 10–13 December 2015; pp. 1–6.
39. Ungureanu, N.; Vladut, V.; Biris, S.-S. Capitalization of wastewater-grown algae in bioethanol production. In Proceedings of the 19th International Scientific Conference Engineering for Rural Development, Latvia University of Life Sciences and Technologies, Jelgava, Latvia, 20–22 March 2020; pp. 1859–1864.
40. Fazal, T.; Mushtaq, A.; Rehman, F.; Ullah Khan, A.; Rashid, N.; Farooq, W.; Rehman, M.S.U.; Xu, J. Bioremediation of textile wastewater and successive biodiesel production using microalgae. *Renew. Sustain. Energ. Rev.* **2018**, *82*, 3107–3126. [CrossRef]
41. Foong, S.Z.Y.; Chong, M.F.; Ng, D.K.S. Strategies to Promote Biogas Generation and Utilisation from Palm Oil Mill Effluent. *Process. Integr. Optim. Sustain.* **2020**, 1–17. [CrossRef]
42. Mikucka, W.; Zielińska, M. Distillery Stillage: Characteristics, Treatment, and Valorization. *Appl. Biochem. Biotechnol.* **2020**, 1–24. [CrossRef]
43. Xu, H.; Yang, B.; Liu, Y.; Li, F.; Shen, C.; Ma, C.; Tian, Q.; Song, X.; Sand, W. Recent advances in anaerobic biological processes for textile printing and dyeing wastewater treatment: A mini-review. *World J. Microbiol. Biotechnol.* **2018**, *34*, 165. [CrossRef]
44. Musa, M.A.; Idrus, S.; Harun, M.R.; Marzuki, T.F.T.M.; Wahab, A.M.A. A comparative study of biogas production from cattle slaughterhouse wastewater using conventional and modified upflow anaerobic sludge blanket (UASB) reactors. *Int. J. Environ. Res. Public Health* **2020**, *17*, 283. [CrossRef]
45. Quek, P.J.; Yeap, T.S.; Ng, H.Y. Applicability of upflow anaerobic sludge blanket and dynamic membrane-coupled process for the treatment of municipal wastewater. *Appl. Microbiol. Biotechnol.* **2017**, *101*, 6531–6540. [CrossRef]
46. Yun, J.; Lee, S.D.; Cho, K.-S. Biomethane production and microbial community response according to influent concentration of molasses wastewater in a UASB reactor. *Appl. Microbiol. Biotechnol.* **2016**, *100*, 4675–4683. [CrossRef]
47. Manyuchi, M.M.; Chiutsi, P.; Mbohwa, C.; Muzenda, E.; Mutusva, T. Bio ethanol from sewage sludge: A bio fuel alternative. *S. Afr. J. Chem. Eng.* **2018**, *25*, 123–127. [CrossRef]
48. Interreg ADRION/OIS-AIR Utilization of Winery Wastewater for Bioethanol Production Applying Biorefinery Concept. Available online: https://www.oisair.net/technology/view/53/utilization-of-winery-wastewater-for-bioethanol-production-applying-biorefinery-concept (accessed on 22 August 2020).
49. Liu, J.; Dantoft, S.H.; Würtz, A.; Jensen, P.R.; Solem, C. A novel cell factory for efficient production of ethanol from dairy waste. *Biotechnol. Biofuels* **2016**, *9*, 33. [CrossRef] [PubMed]
50. Leandro, M.J.; Marques, S.; Ribeiro, B.; Santos, H.; Fonseca, C. Integrated process for bioenergy production and water recycling in the dairy industry: Selection of *Kluyveromyces* strains for direct conversion of concentrated lactose-rich streams into bioethanol. *Microorganisms* **2019**, *7*, 545. [CrossRef] [PubMed]

51. Thangavelu, K.; Sundararaju, P.; Srinivasan, N.; Muniraj, I.; Uthandi, S. Simultaneous lipid production for biodiesel feedstock and decontamination of sago processing wastewater using Candida tropicalis ASY2. *Biotechnol. Biofuels* **2020**, *13*, 35. [CrossRef] [PubMed]
52. Nagi, M.; He, M.; Li, D.; Gebreluel, T.; Cheng, B.; Wang, C. Utilization of tannery wastewater for biofuel production: New insights on microalgae growth and biomass production. *Sci. Rep.* **2020**, *10*, 1530. [CrossRef]
53. Farghaly, A.; Tawfik, A.; Danial, A. Inoculation of paperboard mill sludge versus mixed culture bacteria for hydrogen production from paperboard mill wastewater. *Environ. Sci. Pollut. Res.* **2016**, *23*, 3834–3846. [CrossRef]
54. Li, Y.-C.; Liu, Y.-F.; Chu, C.-Y.; Chang, P.-L.; Hsu, C.-W.; Lin, P.-J.; Wu, S.-Y. Techno-economic evaluation of biohydrogen production from wastewater and agricultural waste. *Int. J. Hydrog. Energy* **2012**, *37*, 15704–15710. [CrossRef]
55. Parihar, R.K.; Upadhyay, K. Production of biohydrogen gas from dairy industry wastewater by anaerobic fermentation process. *Int. J. Appl. Res.* **2016**, *2*, 512–515.
56. Balasubramanian, R.; Sircar, A.; Sivakumar, P.; Anbarasu, K. Production of biodiesel from dairy wastewater sludge: A laboratory and pilot scale study. *Egypt. J. Pet.* **2018**, *27*, 939–943. [CrossRef]
57. Lefebvre, O.; Neculita, C.M.; Yue, X.; Ng, H.Y. Bioelectrochemical treatment of acid mine drainage dominated with iron. *J. Hazard. Mater.* **2012**, *241*, 411–417. [CrossRef]
58. Rahimnejad, M.; Adhami, A.; Darvari, S.; Zirepour, A.; Oh, S.-E. Microbial fuel cell as new technology for bioelectricity generation: A review. *Alex. Eng. J.* **2015**, *54*, 745–756. [CrossRef]
59. Kiran, V.; Gaur, B. Microbial fuel cell: Technology for harvesting energy from biomass. *Rev. Chem. Eng.* **2013**, *29*, 189–203. [CrossRef]
60. Slate, A.J.; Whitehead, K.A.; Brownson, D.A.C.; Banks, C.E. Microbial fuel cells: An overview of current technology. *Renew. Sustain. Energy Rev.* **2019**, *101*, 60–81. [CrossRef]
61. Wagner, R.C.; Porter-Gill, S.; Logan, B.E. Immobilization of anode-attached microbes in a microbial fuel cell. *Amb Expr.* **2012**, *2*, 2. [CrossRef] [PubMed]
62. Xiao, L.; He, Z. Applications and perspectives of phototrophic microorganisms for electricity generation from organic compounds in microbial fuel cells. *Renew. Sustain. Energy Rev.* **2014**, *37*, 550–559. [CrossRef]
63. Eom, H.; Chung, K.; Kim, I.; Han, J.-I. Development of a hybrid microbial fuel cell (MFC) and fuel cell (FC) system for improved cathodic efficiency and sustainability: The M2FC reactor. *Chemosphere* **2011**, *85*, 672–676. [CrossRef]
64. Lv, Z.; Xie, D.; Li, F.; Hu, Y.; Wei, C.; Feng, C. Microbial fuel cell as a biocapacitor by using pseudo-capacitive anode materials. *J. Power Sources* **2014**, *246*, 642–649. [CrossRef]
65. Malaeb, L.; Katuri, K.P.; Logan, B.E.; Maab, H.; Nunes, S.P.; Saikaly, P.E. A Hybrid Microbial Fuel Cell Membrane Bioreactor with a Conductive Ultrafiltration Membrane Biocathode for Wastewater Treatment. *Environ. Sci. Technol.* **2013**, *47*, 11821–11828. [CrossRef]
66. Liang, M.; Tao, H.; Li, S.-F.; Li, W.; Zhang, L.-J.; Ni, J.-R. Treatment of Cu^{2+}-containing wastewater by microbial fuel cell with excess sludge as anodic substrate. *Environ. Sci. Technol.* **2011**, *32*, 179–185.
67. Bhatia, S.K.; Mehariya, S.; Bhatia, R.K.; Kumar, M.; Pugazhendhi, A.; Awasthi, M.K.; Atabani, A.E.; Kumar, G.; Kim, W.; Seo, S.-O.; et al. Wastewater based microalgal biorefinery for bioenergy production: Progress and challenges. *Sci. Total Environ.* **2020**, 141599. [CrossRef]
68. Pittman, J.K.; Dean, A.P.; Osundeko, O. The potential of sustainable algal biofuel production using wastewater resources. *Bioresour. Technol.* **2011**, *102*, 17–25. [CrossRef]
69. Xiao, L.; Young, E.B.; Berges, J.A.; He, Z. Integrated Photo-Bioelectrochemical System for Contaminants Removal and Bioenergy Production. *Environ. Sci. Technol.* **2012**, *46*, 11459–11466. [CrossRef] [PubMed]
70. Lundquist, T.J.; Woertz, I.C.; Quinn, N.W.T.; Benemann, J.R. *A Realistic Technology and Engineering Assessment of Algae Biofuel Production*; Energy Biosciences Institute: Berkeley, CA, USA, 2010.
71. Cavinato, C.; Bolzonella, D.; Fatone, F.; Cecchi, F.; Pavan, P. Optimization of two-phase thermophilic anaerobic digestion of biowaste for hydrogen and methane production through reject water recirculation. *Bioresour. Technol.* **2011**, *102*, 8605–8611. [CrossRef] [PubMed]
72. Jeihanipour, A.; Bashiri, R. Perspective of Biofuels from Wastes BT—Lignocellulose-Based Bioproducts. In *Lignocellulose-Based Bioproducts*; Karimi, K., Ed.; Springer International Publishing: Cham, Switzerland, 2015; pp. 37–83. ISBN 978-3-319-14033-9.

73. Ghimire, A.; Sposito, F.; Frunzo, L.; Trably, E.; Escudié, R.; Pirozzi, F.; Lens, P.N.L.; Esposito, G. Effects of operational parameters on dark fermentative hydrogen production from biodegradable complex waste biomass. *Waste Manag.* **2016**, *50*, 55–64. [CrossRef]
74. Villano, M.; Aulenta, F.; Ciucci, C.; Ferri, T.; Giuliano, A.; Majone, M. Bioelectrochemical reduction of CO_2 to CH_4 via direct and indirect extracellular electron transfer by a hydrogenophilic methanogenic culture. *Bioresour. Technol.* **2010**, *101*, 3085–3090. [CrossRef]
75. Chaudhary, P.; Chhokar, V.; Kumar, A.; Beniwal, V. Bioremediation of Tannery Wastewater. In *Advances in Environmental Biotechnology*; Kumar, R., Sharma, A.K., Ahluwalia, S.S., Eds.; Springer Singapore: Singapore, 2017; pp. 125–144. ISBN 978-981-10-4041-2.
76. Lim, J.W. Anaerobic Co-digestion of Brown Water and Food Waste for Energy Recovery. In Proceedings of the 11th Edition of the World Wide Workshop for Young Environmental Scientists (WWW-YES-2011)-Urban Waters: Resource or Risks? Arcueil, France, 5–10 June 2011.
77. Khalid, A.; Arshad, M.; Anjum, M.; Mahmood, T.; Dawson, L. The anaerobic digestion of solid organic waste. *Waste Manag.* **2011**, *31*, 1737–1744. [CrossRef]
78. Mansoorian, H.J.; Mahvi, A.H.; Jafari, A.J.; Amin, M.M.; Rajabizadeh, A.; Khanjani, N. Bioelectricity generation using two chamber microbial fuel cell treating wastewater from food processing. *Enzym. Microb. Technol.* **2013**, *52*, 352–357. [CrossRef]
79. Jadhav, D.A.; Ghosh Ray, S.; Ghangrekar, M.M. Third generation in bio-electrochemical system research—A systematic review on mechanisms for recovery of valuable by-products from wastewater. *Renew. Sustain. Energy Rev.* **2017**, *76*, 1022–1031. [CrossRef]
80. Calì, G.; Deiana, P.; Bassano, C.; Meloni, S.; Maggio, E.; Mascia, M.; Pettinau, A. Syngas Production, Clean-Up and Wastewater Management in a Demo-Scale Fixed-Bed Updraft Biomass Gasification Unit. *Energies* **2020**, *13*, 2594. [CrossRef]
81. Ren, J.; Cao, J.-P.; Zhao, X.-Y.; Yang, F.-L.; Wei, X.-Y. Recent advances in syngas production from biomass catalytic gasification: A critical review on reactors, catalysts, catalytic mechanisms and mathematical models. *Renew. Sustain. Energy Rev.* **2019**, *116*, 109426. [CrossRef]
82. de Filippis, P.; Scarsella, M.; de Caprariis, B.; Uccellari, R. Biomass Gasification Plant and Syngas Clean-up System. *Energy Procedia* **2015**, *75*, 240–245. [CrossRef]
83. Turovskiy, I.S. Biosolids or Sludge? The Semantics of Terminology. Available online: https://www.wwdmag.com/biosolids-or-sludge-semantics-terminology (accessed on 15 July 2020).
84. Bhatia, S.K.; Kim, S.-H.; Yoon, J.-J.; Yang, Y.-H. Current status and strategies for second generation biofuel production using microbial systems. *Energy Convers. Manag.* **2017**, *148*, 1142–1156. [CrossRef]
85. Jayakody, L.N.; Liu, J.-J.; Yun, E.J.; Turner, T.L.; Oh, E.J.; Jin, Y.-S. Direct conversion of cellulose into ethanol and ethyl-β-d-glucoside via engineered *Saccharomyces cerevisiae*. *Biotechnol. Bioeng.* **2018**, *115*, 2859–2868. [CrossRef] [PubMed]
86. Snyder & Associates Reusing Municipal Wastewater for Ethanol Production. Available online: https://www.snyder-associates.com/projects/reusing-municipal-wastewater-for-ethanol-production/ (accessed on 22 August 2020).
87. Forss, J.; Lindh, M.V.; Pinhassi, J.; Welander, U. Microbial Biotreatment of Actual Textile Wastewater in a Continuous Sequential Rice Husk Biofilter and the Microbial Community Involved. *PLoS ONE* **2017**, *12*, 16. [CrossRef] [PubMed]
88. Gude, V.G. Wastewater treatment in microbial fuel cells—An overview. *J. Clean. Prod.* **2016**, *122*, 287–307. [CrossRef]
89. Speight, J.G. Upgrading by Gasification. In *Heavy Oil Recovery and Upgrading*; Gulf Professional Publishing: Houston, TX, USA, 2019; pp. 559–614. ISBN 978-0-12-813025-4.
90. Bhatt, A.K.; Bhatia, R.K.; Thakur, S.; Rana, N.; Sharma, V.; Rathour, R.K. Fuel from Waste: A Review on Scientific Solution for Waste Management and Environment Conservation BT—Prospects of Alternative Transportation Fuels. In *Prospects of Alternative Transportation Fuels. Energy, Environment, and Sustainability*; Springer: Berlin, Germany, 2018; pp. 205–233.
91. Uzoejinwa, B.B.; He, X.; Wang, S.; Abomohra, A.E.; Hu, Y.; He, Z.; Wang, Q. Co-pyrolysis of seaweeds with waste plastics: Modeling and simulation of effects of co-pyrolysis parameters on yields, and optimization studies for maximum yield of enhanced biofuels. *Energy Sources Part A Recover. Util. Environ. Eff.* **2019**, *42*, 954–978. [CrossRef]

92. Bhatia, S.K.; Jagtap, S.S.; Bedekar, A.A.; Bhatia, R.K.; Patel, A.K.; Pant, D.; Rajesh Banu, J.; Rao, C.V.; Kim, Y.-G.; Yang, Y.-H. Recent developments in pretreatment technologies on lignocellulosic biomass: Effect of key parameters, technological improvements, and challenges. *Bioresour. Technol.* **2020**, *300*, 122724. [CrossRef]
93. Bafghi, M.F.; Yousefi, N. Role of *Nocardia* in Activated Sludge. *Malays. J. Med. Sci.* **2016**, *23*, 86–88.
94. Fischer, F.; Bastian, C.; Happe, M.; Mabillard, E.; Schmidt, N. Microbial fuel cell enables phosphate recovery from digested sewage sludge as struvite. *Bioresour. Technol.* **2011**, *102*, 5824–5830. [CrossRef]
95. Mani, D.; Kumar, C. Biotechnological advances in bioremediation of heavy metals contaminated ecosystems: An overview with special reference to phytoremediation. *Int. J. Environ. Sci. Technol.* **2014**, *11*, 843–872. [CrossRef]
96. Gan, Z.; Sun, H.; Wang, R.; Hu, H.; Zhang, P.; Ren, X. Transformation of acesulfame in water under natural sunlight: Joint effect of photolysis and biodegradation. *Water Res.* **2014**, *64*, 113–122. [CrossRef]
97. Bhatia, S.K.; Bhatia, R.K.; Yang, Y.-H. An overview of microdiesel—A sustainable future source of renewable energy. *Renew. Sustain. Energy Rev.* **2017**, *79*, 1078–1090. [CrossRef]
98. Deng, C.; Kang, X.; Lin, R.; Murphy, J.D. Microwave assisted low-temperature hydrothermal treatment of solid anaerobic digestate for optimising hydrochar and energy recovery. *Chem. Eng. J.* **2020**, *395*, 124999. [CrossRef]
99. Seiple, T.E.; Skaggs, R.L.; Fillmore, L.; Coleman, A.M. Municipal wastewater sludge as a renewable, cost-effective feedstock for transportation biofuels using hydrothermal liquefaction. *J. Environ. Manag.* **2020**, *270*, 110852. [CrossRef] [PubMed]
100. Bhatia, S.K.; Yang, Y.-H. Microbial production of volatile fatty acids: Current status and future perspectives. *Rev. Environ. Sci. Biotechnol.* **2017**, *16*, 327–345. [CrossRef]
101. Rawat, I.; Ranjith Kumar, R.; Mutanda, T.; Bux, F. Dual role of microalgae: Phycoremediation of domestic wastewater and biomass production for sustainable biofuels production. *Appl. Energy* **2011**, *88*, 3411–3424. [CrossRef]
102. Riaño, B.; Molinuevo, B.; García-González, M.C. Potential for methane production from anaerobic co-digestion of swine manure with winery wastewater. *Bioresour. Technol.* **2011**, *102*, 4131–4136. [CrossRef]
103. Intanoo, P.; Rangsanvigit, P.; Malakul, P.; Chavadej, S. Optimization of separate hydrogen and methane production from cassava wastewater using two-stage upflow anaerobic sludge blanket reactor (UASB) system under thermophilic operation. *Bioresour. Technol.* **2014**, *173*, 256–265. [CrossRef] [PubMed]
104. Buitrón, G.; Martínez-Valdez, F.J.; Ojeda, F. Biogas Production from a Highly Organic Loaded Winery Effluent Through a Two-Stage Process. *Bioenerg. Res.* **2019**, *12*, 714–721. [CrossRef]
105. Roy, M.M.; Dutta, A.; Corscadden, K.; Havard, P.; Dickie, L. Review of biosolids management options and co-incineration of a biosolid-derived fuel. *Waste Manag.* **2011**, *31*, 2228–2235. [CrossRef]
106. Ramprakash, B.; Muthukumar, K. Comparative study on the performance of various pretreatment and hydrolysis methods for the production of biohydrogen using *Enterobacter aerogenes* RM 08 from rice mill wastewater. *Int. J. Hydrog. Energy* **2015**, *40*, 9106–9112. [CrossRef]
107. Conrad, R. Importance of hydrogenotrophic, aceticlastic and methylotrophic methanogenesis for methane production in terrestrial, aquatic and other anoxic environments: A mini review. *Pedosphere* **2020**, *30*, 25–39. [CrossRef]
108. Dadrasnia, A.; Usman, M.; Kang, T.; Velappan, R.; Shahsavari, N.; Vejan, P.; Mahmud, A.; Ismail, S. Microbial Aspects in Wastewater Treatment—A Technical Review. *Environ. Technol.* **2017**, *2*, 75–84. [CrossRef]
109. Seow, T.W.; Lim, C.K.; Nor, M.H.M.; Mubarak, M.F.M.; Lam, C.Y.; Yahya, A.; Ibrahim, Z. Review on Wastewater Treatment Technologies. *Int. J. Appl. Environ. Sci.* **2016**, *11*, 111–126.
110. Liu, Y.-J.; Gu, J.; Liu, Y. Energy self-sufficient biological municipal wastewater reclamation: Present status, challenges and solutions forward. *Bioresour. Technol.* **2018**, *269*, 513–519. [CrossRef] [PubMed]

Article

Wastewater Treatment Costs: A Research Overview through Bibliometric Analysis

Leticia Gallego-Valero [1,*], Encarnación Moral-Parajes [1] and Isabel María Román-Sánchez [2]

[1] Department of Economy, University of Jaén, 23071 Jaén, Spain; Emoral@ujaen.es
[2] Department of Economics and Business, University of Almería, 04120 Almería, Spain; Iroman@ual.es
* Correspondence: Lgallego@ujaen.es; Tel.: +34-651888733

Abstract: Given the problem of water scarcity and the importance of this resource for the sustainability of the planet, wastewater treatment and its costs have become a key issue for proper water management. Using bibliometric analysis of publications in the Web of Science database, this study presents an overview of the research on wastewater treatment costs in the period 1950–2020. The worldwide search returned 22,788 articles for wastewater treatment costs, which compares poorly to the results for research on wastewater treatment, accounting for only 12.34% of the total output on wastewater treatment. The findings of this study reveal the leading countries in this field of research (China, USA, India, Spain and the UK), with the articles being published in a wide range of high impact journals. Similarly, there are very few results on UV and chlorination costs, despite the importance of these two treatments for wastewater disinfection and reuse. This study is aimed at researchers in this field, helping them to identify recent trends, and at the main institutions in the scientific community working on this subject.

Keywords: cost; database; treatment; wastewater; water; Web of Science

Citation: Gallego-Valero, L.; Moral-Parajes, E.; Román-Sánchez, I.M. Wastewater Treatment Costs: A Research Overview through Bibliometric Analysis. *Sustainability* **2021**, *13*, 5066. https://doi.org/10.3390/su13095066

Academic Editor: Andreas N. Angelakis

Received: 31 March 2021
Accepted: 27 April 2021
Published: 30 April 2021

1. Introduction

Given the importance of water resources, appropriate water management is needed as well as more sustainable exploitation of this resource [1–4]. At a global level, water scarcity is an economic, sanitation and even security issue [5]. Moreover, the problem is expected to become more acute in the future, with this resource playing a fundamental role in the sustainability of the planet [6,7]. The United Nations (UN) considers clean water and sanitation to be a priority objective, and one of its goals is to ensure universal access to safe and affordable drinking water. European Union policy is also aimed at protecting this resource, with the implementation of the Water Framework Directive (2000/60/EC) and the Urban Waste Water Treatment Directive (91/271/EEC). Factors such as the population growth in many urban areas, agricultural productivity, the economic development of different countries, industrialization, energy production, improvements in health and sanitation systems, and the expansion of irrigation systems in arid regions, have underscored the fact that conventional resources alone cannot meet the constantly growing demand [8–14].

Wastewater costs suppose a great concern given the need for a growing resource [15]. In response to the problem of scarcity, which has become hugely important in countries with high levels of water stress [16], hydric resources should be managed more efficiently [17]. Appropriate water management aimed at increasing the supply of water necessarily involves the use of wastewater treatment. [18–20]. Wastewater management is expensive and poses problems regarding how to finance it and how to reduce treatment costs [21]. Adequate wastewater management is necessary to finance the investment in wastewater treatment plants (WWTPs) and the costs of treatment technologies, and to improve the environmental quality of water resources [22,23]. Treatment methods improve the quality of the water and, when the treated water is reused, increase the quantity of the resource [24,25].

Water reuse is a process with few adverse environmental impacts when compared with desalination or water transfers and offers economic and social benefits [17]. This process gives rise to a resource, the reused wastewater, that can help to improve the quality and quantity of the planet's water supplies [26–28]. Treating wastewater prior to its discharge helps ensure the good status of water resources [29], facilitating the use of reclaimed water as an additional source of water supply that is safe and economical [30]. In view of the growing demand, sanitation and purification treatments constitute an indispensable tool for cleaning the water that is returned to ecosystems and increasing the quantity of the available resource, regardless of climate conditions [31–34]. This way, reused wastewater can be seen as a source of irrigation supply that is both economical and safe in terms of human health and the environment [35–37], helping to tackle the problem of water scarcity [38–40], boosting supply and decreasing the dependence on groundwater and surface water resources [41].

The analysis and study of the costs of the different treatments is crucial in order to boost their efficiency, cut costs and help ensure the widespread use of such treatments [42]. There is a need for cheaper, more robust and more effective processes for wastewater decontamination and disinfection, always bearing in mind the need to protect human health and the environment [17]. It is increasingly important to adopt appropriate measures to bring down operating costs, which entails an evaluation of the efficiency of WWTPs. By doing so, it is possible to identify WWTPs that make better use of their economic resources without reducing the quality of the treated water. This information can then be used to determine the appropriate operational practices to be applied in other WWTPs in order to reduce operating costs. In addition, this cost-cutting is beneficial to society as a whole, since it is the citizens who bear these costs through the payment of water tariffs [43]. Economic evaluation is also a useful tool in the implementation of efficient and effective water management strategies and policies, thus supporting various institutions' policy decisions [44–46].

The present study is carried out through bibliometrics, a technique that uses statistical methods to analyse the scientific output published and which contains sub-fields such as structural, dynamic, evaluative and predictive scientometrics. Bibliometric analysis has been applied to almost all scientific fields, and all types of literature can be studied in this way, identifying features such as topics, authors, publication dates, reference literature, content, etc. [47,48]. The use of the internet as a data collection tool is accepted by the scientific community [49]. In this regard, Web of Science (WoS), published by Thomson Reuters, is a hugely relevant database for evaluating research [50].

The main aim of this work is the quantitative and qualitative analysis of the dynamics of global research on the costs of wastewater treatments since 1950, as well as an analysis of the research on the costs of chlorination and UV disinfection treatments. These treatments enable the reclamation of water for reuse, which contributes to an efficient management of the resources used, keeping them circulating in the economic system for as long as possible, and thereby generating less waste and avoiding the unnecessary use of new resources. They therefore help to reduce environmental impacts, as well as contributing to the restoration and regeneration of natural capital, in line with the tenets of the circular economy [51]. The application of these treatments contributes to sustainability by allowing the value of resources to remain in the economy for as long as possible and reducing waste generation to a minimum. To achieve this objective, bibliometric techniques are used to identify, organize and analyse the main elements of the topics in question, using the WoS database and statistical processing tools. The results obtained are useful for the scientific community to gain an understanding of the current environment and upcoming trends in the lines of research on these subjects, and to make decisions before embarking on research.

2. Materials and Methods

2.1. Bibliometric Analysis

The article performs a bibliometric analysis of wastewater treatment costs. This method makes it possible to identify, organize and evaluate the constituent elements of a specific area of study, and is, thus, an important tool for literature reviews [52–54]. Bibliometric analysis involves various different types of indicators relating to quantity (productivity), quality (impact of publications) and structure (analysing connections) [55]. This article uses the WoS database from Thomson Reuters to conduct the bibliometric analysis. WoS has high visibility in the different areas of knowledge, a selection filter for prestigious publications, and is also widely used to carry out bibliometric studies [56,57]. The bibliometric analysis technique has been used to study areas such as the use of water or wastewater [58,59], wastewater treatment by advanced oxidation processes [17], infectious diseases and microbiology [60], or renewable energies, sustainability and the environment [61].

2.2. Data Selection and Processing

The sample of documents analysed in this study was obtained by conducting a search of the entire WoS database with the term "wastewater treatment cost" in the option "topic":

- The period analysed was 1950–2020. The analysis yielded a final sample of 22,788 results on wastewater treatment costs. The sample selection was conducted in January 2021.
- The following variables were analysed: evolution, areas of study, main countries, main journals, and main institutions.
- An additional analysis section was introduced to examine differences in scientific research on treatments for water intended for reuse, ultraviolet (UV) and chlorination, as these treatments are the most commonly used options for water disinfection, for various economic and environmental reasons.

Bibliometric studies distinguish between three types of indicators [55]: quantity indicators, which refer to productivity; quality indicators, which refer to the impact of publications; and structural indicators, which measure the connections established between the different agents. In this study, quality and quantity indicators were analysed. In addition to the measure of productivity of the countries and institutions, the following indicators were used to evaluate the quality of the journals in which the documents were published:

- SCImago Journal Rank (SJR): measure of the scientific influence of scholarly journals that accounts for both the number of citations received by a journal and the importance or prestige of the journals where the citations where made [62].
- Quartile in which the journal is positioned.
- Number of citations.
- Journal Impact Factor (JIF): measure of the frequency with which the average article in a journal has been cited in a particular year.
- Total publications.

After selecting the sample, the information available in the WoS database was downloaded and prepared for analysis by eliminating duplications, correcting mistakes and adding non-complete information [54]. First, the evolution of the field over the period 1950–2020 was analysed. Secondly, the main areas of study in wastewater research classified by WoS were identified, before reviewing the leading countries in this research. In the next step, the main journals were identified along with their SJR, JIF and total citations index for the year 2019 (to evaluate the impact of the journals). The main institutions are shown below. Lastly, the results of the search for research on the costs of UV and chlorination treatments were studied.

3. Results

3.1. Evolution of the Research on Wastewater Treatment Costs

The scientific community has shown very little interest in analysing wastewater treatment costs, although there has been some growth in the theoretical and applied literature on the subject over the last decade. Figure 1 presents the evolution of articles published in this field from 1950 to 2020, revealed by the bibliometric analysis using WoS (2020) as the main database. The search for "wastewater treatment cost" returns 22,788 results, a very small number compared to those returned by the search "wastewater treatment", with 184,697 articles published. In percentage terms, wastewater treatment cost research comprises only 12.34% of the total for wastewater treatment. The research on wastewater treatment costs shows an increasing trend between 1950 and 2020, with a marked rise from 2010, albeit with some fluctuations. It can be seen that, in both cases, the majority of articles are concentrated from 2010 onwards (16,204 studies, 71.11% of the total for "wastewater treatment cost" and 125,235 articles, 67.81% of the total for "wastewater treatment"). The period between 1950 and 1980 yields very few results, although the number grows over time. Looking at the results since 1980, 20 articles were published for "wastewater treatment cost" and 235 for "wastewater treatment" in 1980, while the corresponding figures for 2020 were 2528 and 16,708, respectively. The growth in the research on wastewater treatment costs is slower than that for wastewater treatment: although both show a rising trend during the period 1950–2020, research on "wastewater treatment" grows at a much faster rate, especially after the year 2000.

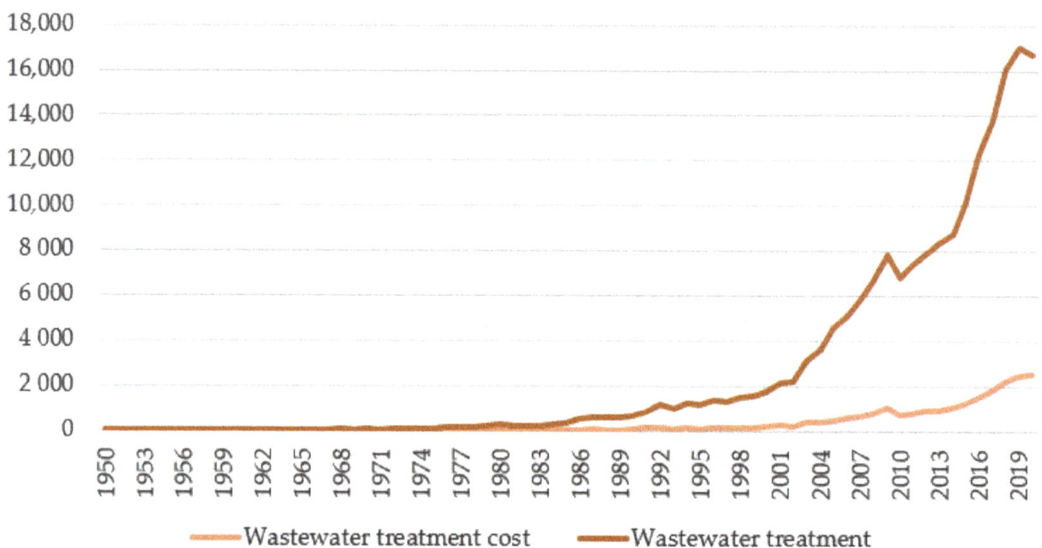

Figure 1. Trends in the research on wastewater treatment costs and wastewater treatment (number of articles). Source: own elaboration from WoS (2021).

3.2. Main Areas of Study in Wastewater Treatment Costs Research

The results of the bibliometric analysis make it possible to distinguish between the different disciplines to which the analysed scientific articles belong. It should be noted that an article can belong to more than one category; for this reason, the results are analysed in percentages. Figure 2 shows the main areas of study in wastewater treatment costs in the period under study. Among the many areas of research, the most important are Environmental Sciences and Ecology, accounting for 14% of the total, Engineering (11%) and Water Resources (10%). These areas are followed by Public Environmental Occupational Health (8%), Energy and Fuels (6%), Business Economics (6%) and Chemistry (6%). The item labelled "other" (39% of the total) includes a wide and diverse range of areas, none of them with a percentage higher than 5%, such as Materials Science, Toxicology, Biochemistry Molecular Biology, Mathematics, Biodiversity Conservation, Physics, Marine Freshwater Biology, Plant Sciences, Microbiology, Mathematical Computational Biology, Computer Science, Food Science Technology, Polymer Science, Meteorology Atmospheric Sciences, Science Technology (other topics), Biotechnology Applied Microbiology, Instruments Instrumentation and Agriculture.

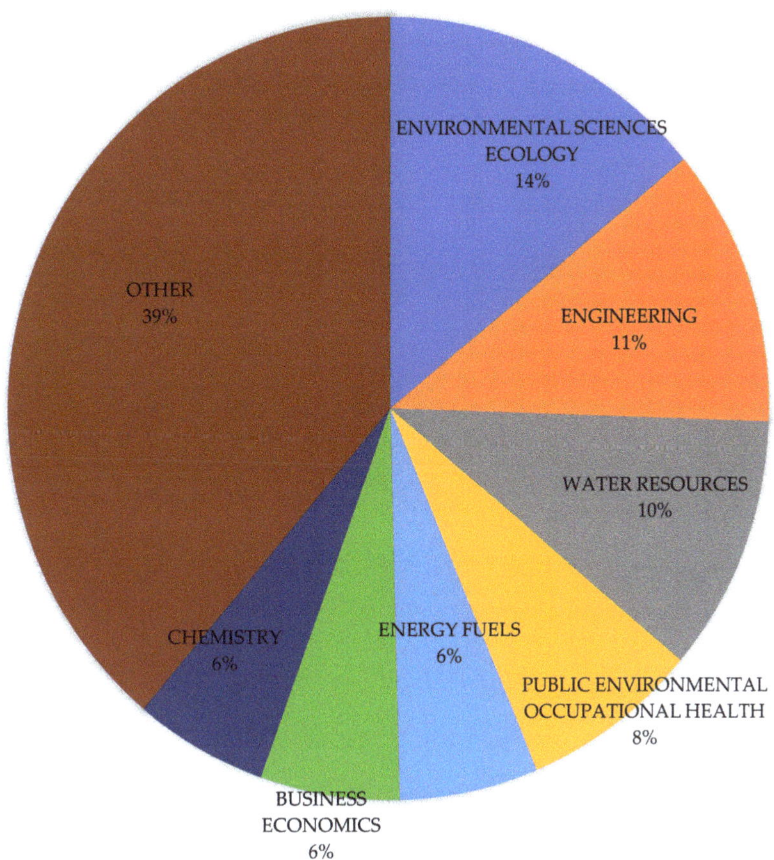

Figure 2. Main areas of study in wastewater treatment costs (percentage). Source: own elaboration from WoS (2021).

3.3. Relevant Countries in Wastewater Treatment Costs Research

The articles published on wastewater treatment costs in the period 1950–2020 come from a total of 162 countries. Figure 3 shows the map of the countries with results on wastewater treatment cost research, although the majority of studies in this field come from a relatively small number of main countries. In order of quantity of results, they are China, the USA, India, Spain, the UK, Australia, Brazil, Canada, Turkey and Iran. Altogether, there are countries from almost all continents, and with a wide diversity of economic and sociocultural characteristics.

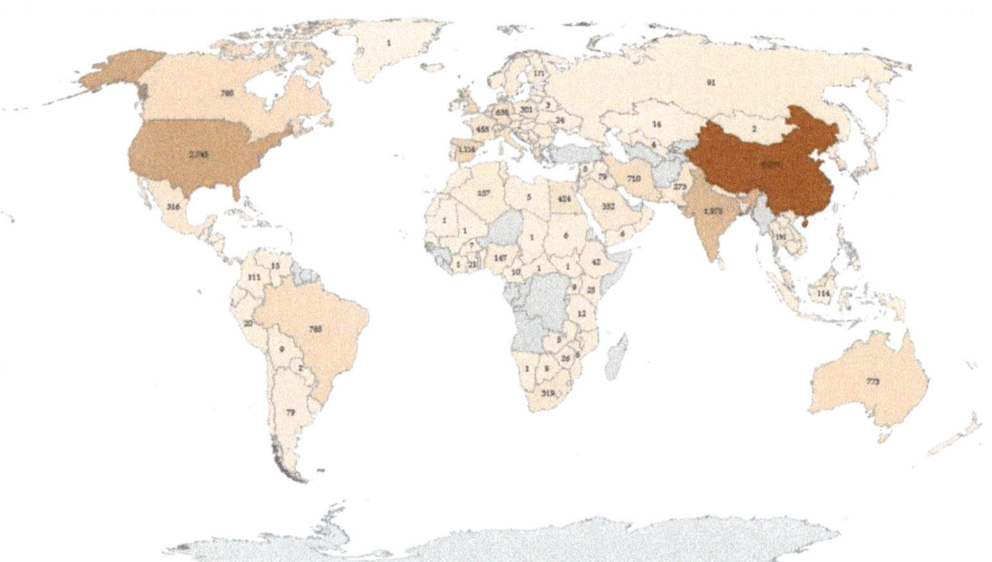

Figure 3. Map of the research on wastewater treatment costs (number of articles). Source: own elaboration from WoS (2021).

Figure 4 presents the top countries in the field. China is the most prolific country, representing 28.84% of the total, with 6571 results. The second place is occupied by the USA (12.05% of the total, 2745 articles). Third is India, with 1975 results and 8.67%. These countries are followed by Spain (1116 results, 4.90%), the UK (1112 results, 4.88%), Australia, Brazil, Canada, Turkey, Iran, Italy, Germany, Malaysia, South Korea, France, Japan, Egypt and the Netherlands.

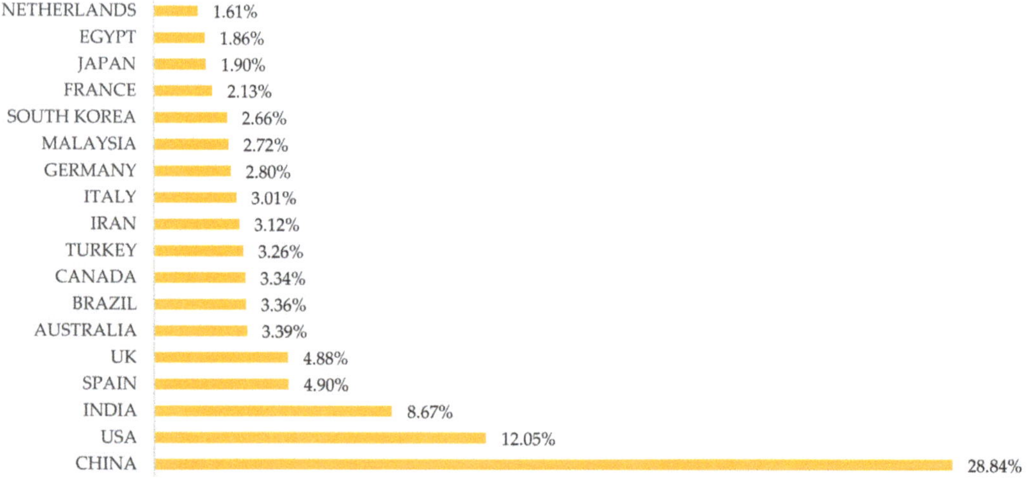

Figure 4. Top countries in wastewater treatment cost research (percentage). Source: own elaboration from WoS (2021).

3.4. Journals in Wastewater Treatment Cost Research

This part presents the most relevant journals publishing articles on wastewater treatment cost research and analyses their main indexes (Table 1). The main journals by number of articles are Water Science and Technology (1471 articles), Chemical Engineering Journal (709) and Desalination and Water Treatment (651). Taken together, these three journals published 12.42% of the total papers on this research subject. Among the list of journals, there are diverse nationalities, with the United Kingdom, the Netherlands, Switzerland and the USA being the most prolific countries.

Table 1. Journals and relevant indexes in wastewater treatment cost research. Source: own elaboration based on WoS (2021).

Journal	Articles	SJR (2019)	Country	JIF (2019)	Total Citations (2019)
Water Science and Technology	1471	0.47 (Q2)	United Kingdom	1.638	20,937
Chemical Engineering Journal	709	2.32 (Q1)	Switzerland	10.652	129,806
Desalination and Water Treatment	651	0.33 (Q2)	Italy	0.854	14,535
Journal of Hazardous Materials	552	2.01 (Q1)	Netherlands	9.038	110,068
Water Research	523	2.93 (Q1)	United Kingdom	9.130	99,442
Journal of Cleaner Production	492	1.89 (Q1)	Netherlands	7.246	104,138
Bioresource Technology	468	2.43 (Q1)	Netherlands	7.539	131,781
Journal of Environmental Management	452	1.32 (Q1)	USA	5.647	44,264
Environmental Science and Pollution Research	415	0.79 (Q2)	Germany	3.056	46,033
Desalination	320	1.81 (Q1)	Netherlands	7.098	44,845
Environmental Technology	296	0.49 (Q2)	United Kingdom	2.213	7947
Journal of Environmental Chemical Engineering	281	0.93 (Q1)	United Kingdom	4.300	13,023
Science of the Total Environment	254	1.66 (Q1)	Netherlands	6.551	134,962
Chemosphere	240	1.53 (Q1)	United Kingdom	5.778	94,799
Environmental Science & Technology	221	2.7 (Q1)	USA	7.864	187,995
Journal of Chemical Technology and Biotechnology	202	0.66 (Q1)	United Kingdom	2.750	12,232
Water	176	0.66 (Q1)	Switzerland	2.544	13,460
International Journal of Environmental Science and Technology	174	0.52 (Q2)	USA	2.540	6522
Water Environment Research	173	0.3 (Q3)	USA	1.369	3120
Water, Air and Soil Pollution	169	0.54 (Q2)	Switzerland	1.900	15,219

It is important to note that there is a broad range of journals publishing articles on wastewater treatment costs. The journal with the largest number of articles is Water Science and Technology, with 6.46% of the total sample. This journal has an SJR in 2019 of 0.47 (quartile Q2), a JIF of 1.638 and a total of 20,937 citations in 2019. The second is Chemical Engineering Journal, with 3.11% of the total sample. The SJR in 2019 for this journal is 2.32 (Q1), with a JIF of 10.652 in 2019 and 129,806 total citations in 2019. In third place is Desalination and Water Treatment, with 2.86% of the total sample and an SJR (2019) of 0.33 (Q2), a JIF index of 0.854 and 14,535 total citations. Below these three results, there is a wide variety of journals, almost all of which present high index scores—most are in the first and second quartiles. These 20 journals published 36.15% of the articles, most of them included in the first two quartiles of the SJR. They include Journal of Hazardous Materials, Water Research, Journal of Cleaner Production, Bioresource Technology, Journal of Environmental Management, Environmental Science and Pollution Research, Desalination, Environmental Technology, Journal of Environmental Chemical Engineering, Science of the Total Environment, Chemosphere, Environmental Science & Technology, Journal of Chemical Technology and Biotechnology, Water, International

Journal of Environmental Science and Technology, Water Environment Research, and Water, Air and Soil Pollution.

3.5. *Leader Institutions in Wastewater Treatment Cost Research*

Figure 5 presents the main institutions focusing on wastewater treatment costs research. This large number of institutions (62) accounts for 31.10% of the total results, which indicates a low concentration index at the institutional level. The leading institution in number of related articles is the Chinese Academy of Sciences (China), with 561 articles (2.46% of the total), followed by the Indian Institute of Technology System (India), with 445 articles (1.95% of the total). Institutions from China, India and the USA predominate, with those from the USA having a lower concentration by institution. It is worth noting the wide dispersion of articles among different institutions.

3.6. *Differences in Scientific Research on Treatments for Wastewater Intended for Reuse: UV and Chlorination*

This section analyses the results on disinfection treatments for wastewater intended for reuse; namely, UV and chlorination. These are the most commonly applied treatments for wastewater disinfection for various economic and environmental reasons [63]. The classification as UV or chlorination reveals a difference, with 1118 articles for UV and 140 for chlorination, representing 4.91% and 0.61%, respectively, of the total research on wastewater treatment costs. This is a very small number of results considering the importance of these treatments for the reuse of wastewater, and the fact that they are the most commonly used options. The analysis of the number of publications over the period 1950–2020 shows a growing amount of research on wastewater treatment costs relating to UV and chlorination over the years. Figure 6 shows the evolution of the studies published on these two treatments throughout the period analysed, clearly depicting the scant research up until the year 2000. From then on, we see a rising trend in both cases, with a particularly notable increase in studies on UV, especially from 2014 onwards. In the last 10 years, we see a rise in publications on UV from 29 in 2010 to 147 in 2020, representing a growth rate of 36.99%. In parallel, those addressing chlorination register an increase of 27.27% in the last decade, with two results in 2010 and eight in 2020.

Figure 7 shows the leading countries in research on wastewater treatment costs related to UV and chlorination: China is the leader on UV treatment, with 256 results (22.90% of total results for UV), and the USA for chlorination treatment, with 31 articles (22.14% of total results for chlorination). In the research on the costs of tertiary treatment with UV, China is followed by India, the USA, Spain, Iran, Brazil, Italy, Canada, Germany and Egypt, in that order. In the case of chlorination cost research, after the USA come China, Brazil, India, Australia, Canada, France, Italy, Spain and the United Kingdom. It can be seen that 8 of the 10 leading countries are the same for the two treatments, although they hold different positions according to their numbers of results.

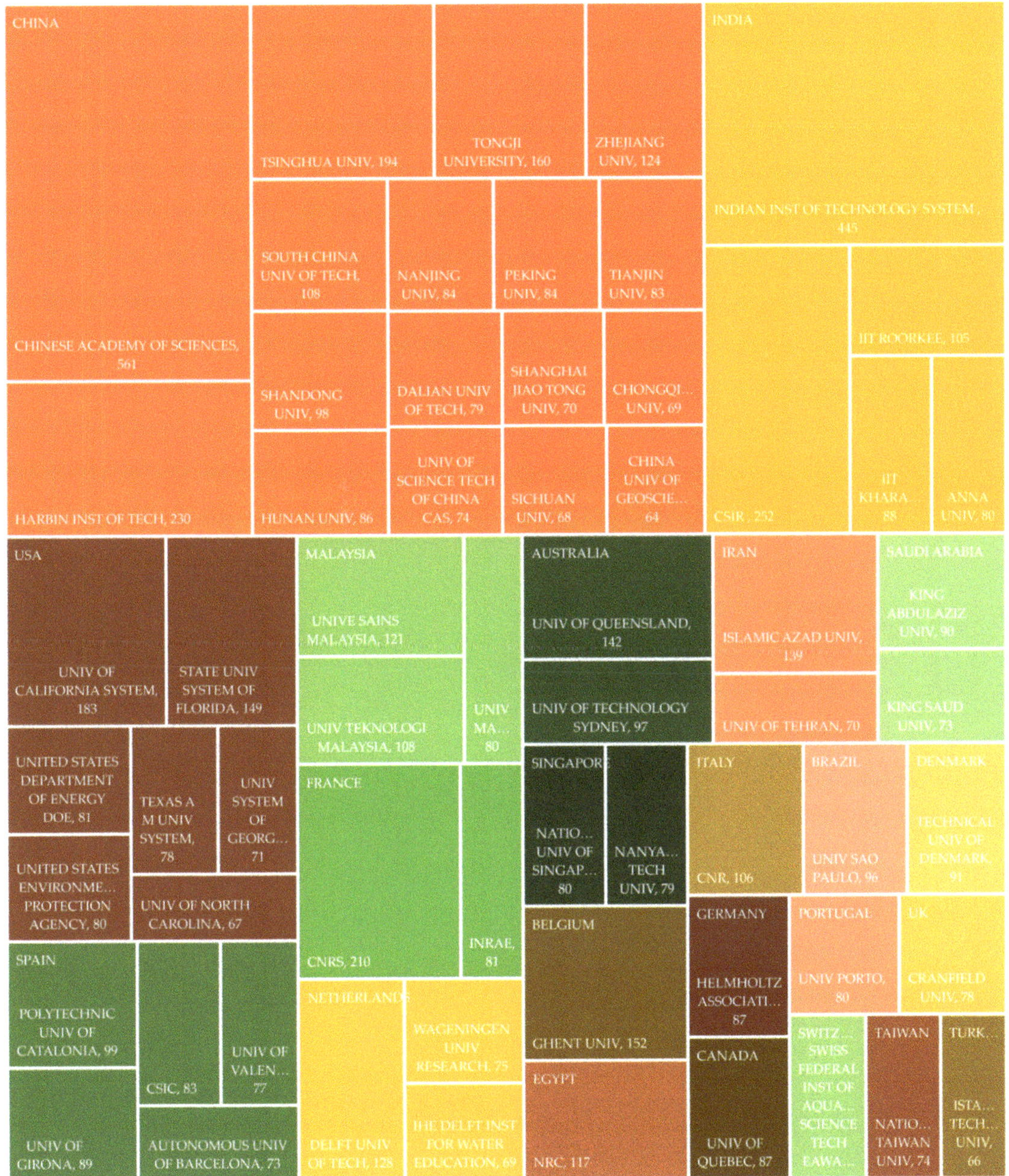

Figure 5. Main institutions in wastewater treatment cost research. Source: own elaboration from WoS (2021).

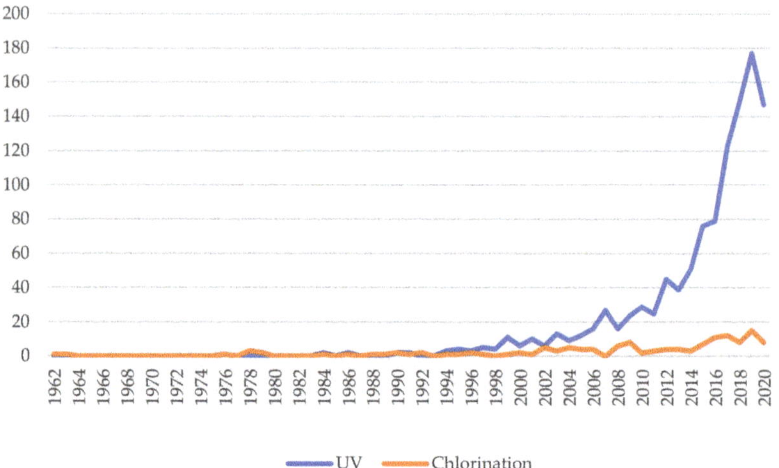

Figure 6. Evolution of articles on UV and chlorination treatments for wastewater reuse from 1950 to 2020 (number of articles). Source: own elaboration from WoS (2021).

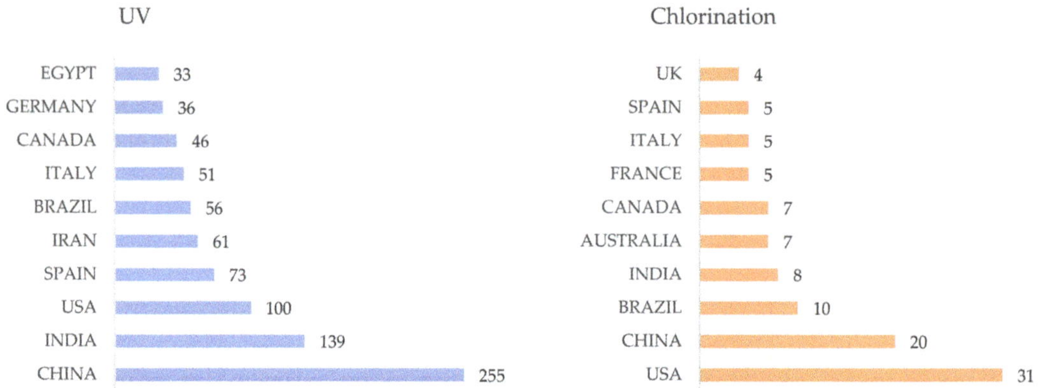

Figure 7. Leading countries in the research on wastewater treatment costs related to UV and chlorination (number of articles). Source: own elaboration from WoS (2021).

4. Discussion and Conclusions

4.1. Summary of Findings

The objective of this work was to show the current status and evolution of research on the wastewater treatment costs during the period 1950–2020. To achieve this goal, the main drivers of the subject, the main lines of research, the trends over several years, and the gaps in research were analyzed in depth. In addition, two disinfection treatments for water reuse have been considered: chlorination and UV. Bibliometric techniques have been used to carry out this study, with the WoS database and statistical processing tools. Globally, research on wastewater treatment costs has yielded 22,788 results: an increasing trend of wastewater treatment costs research was observed. Nevertheless, this topic remains largely understudied compared to the results for the search "wastewater treatments", representing only 12.34% of the total output for research on wastewater treatments. For the sake of easy understanding, the findings of this study are summarized as follows:

(1) The evolution of research suggests that the study of wastewater treatment costs has experienced a steady increase from 2010. This changing trend is related to the growing social concern for the environment and its resources. Reflecting the rise in "wastewater treatment" research, there has been a rise in research on wastewater treatment costs since 1950, showing strong growth since 2010, with 71.11% of the articles being published from this date onwards.

(2) The main areas of research on this topic are very diverse, with the most important being Environmental Sciences and Ecology (14% of the total), Engineering (11%) and Water Resources (10%).

(3) As for the countries of origin of the scientific output in this field during the period 1950–2020, it is worth noting the high dispersion; all together, 162 countries were responsible for the results in wastewater treatment costs research. The main countries are China (28.84% of the total), the USA (12.05%), India (8.67%), Spain (4.90%) and the UK (4.88%).

(4) Concerning to the leader institutions, there is great diversity from which the research on this subject comes, as reflected in the Low Concentration Index at the institutional level; some that stand out are the Chinese Academy of Sciences (China), with 2.46% of the total, and the Indian Institute of Technology System (India), with 1.95% of the total.

(5) Journal analysis revealed that the articles relating to the analysed subject are published in international journals with a high impact factor. There is a wide range of journals that publish articles on the costs of wastewater treatment. Researchers' preferred journals are Water Science and Technology, Chemical Engineering Journal, and Desalination and Water Treatment, with high impact factors of 1.638, 10.652 and 0.854 (JIF), respectively. Taken together with the number of publications, these are an indicator of the high level of institutional scientific quality. These three journals together have published 12.42% of the total research in this field.

(6) Regarding to UV and chlorination costs, very few results were returned in the search for articles (4.91% for UV and 0.61% for chlorination) in comparison with the total results for wastewater treatment costs. It is remarkable that these topics have registered an increase over the period analysed, with a particularly notable rise from the year 2000. The leading countries in this research are the following:

- For UV, China (22.90%), India, the USA, Spain, Iran, Brazil, Italy, Canada, Germany and Egypt.
- For chlorination, the USA (22.14%), China, Brazil, India, Australia, Canada, France, Italy, Spain and the United Kingdom.

4.2. *Implications and Limitations*

The bibliometric analysis conducted in this study shows the trends in wastewater treatment costs research. It is very scarce in comparison with the research on wastewater treatment, highlighting the existence of a research field yet to be explored. The findings of this study provide valuable information for researchers and institutions, helping them explore much-needed management pathways. In practical terms, this study focuses on the importance of the improvements in the management of wastewater treatment.

Despite its contributions, this study has limitations. It is relevant to remark the fact that the findings of this study might not fully reflect the complete research on wastewater treatment costs, given that the information is derived only from the WoS database. There are other databases, such as Scopus, with quality publications, that are not taken into account.

4.3. *Future Research Opportunities*

Using a bibliometric analysis and the WoS, this study identifies the contributions to the topic of wastewater treatment costs, with China, the USA, India, Spain and the UK being the leading countries in this regard. From 2010, we see greater concern for sustainability, which has been promoted by the UN since 2000 when it declared its Millennium Development Goals. Indeed, one of those goals was to ensure environmental sustainability, which was followed by the inclusion of water in the Sustainable Development Goals in 2015. Nevertheless, the bibliometric analysis carried out reveals a topic that calls for further research, given the scarcity of scientific publications in the area of wastewater treatment costs. In addition, the few results obtained for UV and chlorination costs contrast to the prominence of these two treatments in the disinfection and reuse of wastewater; indeed, they are the most commonly used options for this purpose. The scientific contribution in this field is minimal compared to that on wastewater treatments, revealing a segment that has yet to be explored. This study serves as a guide for researchers, pointing to new trends, and also informs the scientific community and institutions, while the growing interest in the subject enables improvements in the management of wastewater treatment. In addition, society should be more aware of the need to fund wastewater treatments given the relevant role they play in securing additional water resources.

Wastewater reclamation has the disadvantage of entailing high costs, above the average of those associated with naturally-occurring resources. The study of wastewater treatment costs is essential for the proper management of water resources. The decision-making process regarding the planning and management of water resources requires the constant generation of information to achieve high levels of efficiency. Increasingly, the use of reclaimed water has strategic value in that it takes the pressure off water resources, while minimizing health risks for downstream users and helping to maintain the quality of ecosystems. Wastewater treatment and reclamation processes are an essential element of the efficient management of the water cycle, becoming hugely important in countries with high levels of water stress.

Author Contributions: Conceptualization, L.G.-V., E.M.-P. and I.M.R.-S.; methodology, L.G.-V. and I.M.R.-S.; validation, E.M.-P. and I.M.R.-S.; formal analysis, L.G.-V.; investigation, L.G.-V.; resources, L.G.-V.; data curation, L.G.-V.; writing—original draft preparation, L.G.-V.; writing—review and editing, E.M.-P. and I.M.R.-S.; visualization, L.G.-V.; supervision, E.M.-P.; project administration, E.M.-P. All authors have read and agreed to the published version of the manuscript.

Funding: This research received no external funding.

Institutional Review Board Statement: Not applicable.

Informed Consent Statement: Not applicable.

Data Availability Statement: The data presented in this study are available on request from the corresponding author.

Conflicts of Interest: The authors declare no conflict of interest.

References

1. Vázquez, V.L.; Rodríguez, G.; Daddi, T.; De Giacomo, M.R.; Polders, C.; Dils, E. Policy challenges in transferring the integrated pollution Prevention and control approach to Southern Mediterranean countries: A case study. *J. Clean. Prod.* **2015**, *107*, 486–497. [CrossRef]
2. Martin-Carrasco, F.; Garrote, L.; Iglesias, A.; Mediero, L. Diagnosing Causes of Water Scarcity in Complex Water Resources Systems and Identifying Risk Management Actions. *Water Resour. Manag.* **2012**, *27*, 1693–1705. [CrossRef]
3. Beamonte, E.; Casino, A.; Veres, E.J. Medición de la calidad del agua mediante indicadores. Relación entre éstos y las tarifas de abastecimiento. *Estud. Econ. Apl.* **2010**, *28*, 357–374.
4. Burak, S.; Margat, J. Water Management in the Mediterranean Region: Concepts and Policies. *Water Resour. Manag.* **2016**, *30*, 5779–5797. [CrossRef]
5. Katz, D. Water use and economic growth: Reconsidering the Environmental Kuznets Curve relationship. *J. Clean. Prod.* **2015**, *88*, 205–213. [CrossRef]
6. Su, M.-H.; Huang, C.-H.; Li, W.-Y.; Tso, C.-T.; Lur, H.-S. Water footprint analysis of bioethanol energy crops in Taiwan. *J. Clean. Prod.* **2015**, *88*, 132–138. [CrossRef]
7. Xu, Y.; Huang, K.; Yu, Y.; Wang, X. Changes in water footprint of crop production in Beijing from 1978 to 2012: A logarithmic mean Divisia index decomposition analysis. *J. Clean. Prod.* **2015**, *87*, 180–187. [CrossRef]
8. Uche, J.; Martínez-Gracia, A.; Círez, F.; Carmona, U. Environmental impact of water supply and water use in a Mediterranean water stressed region. *J. Clean. Prod.* **2015**, *88*, 196–204. [CrossRef]

9. Vörösmarty, C.J.; Green, P.; Salisbury, J.; Lammers, R.B. Global water resources: Vulnerability from climate change and pop-ulation growth. *Science* **2000**, *289*, 284–288. [CrossRef]
10. Gleick, P.H. *The World's Water*; Island Press: Washington, DC, USA, 2011.
11. Böhmelt, T.; Bernauer, T.; Buhaug, H.; Gleditsch, N.P.; Tribaldos, T.; Wischnath, G. Demand, supply, and restraint: Determi-nants of domestic water conflict and cooperation. *Global Environ. Chang.* **2014**, *29*, 337–348. [CrossRef]
12. Gizelis, T.-I.; Wooden, A.E. Water resources, institutions, & intrastate conflict. *Polit. Geogr.* **2010**, *29*, 444–453. [CrossRef]
13. De Graaf, I.; van Beek, L.; Wada, Y.; Bierkens, M. Dynamic attribution of global water demand to surface water and groundwater resources: Effects of abstractions and return flows on river discharges. *Adv. Water Resour.* **2014**, *64*, 21–33. [CrossRef]
14. Morote, Á.-F.; Hernández, M. Urban sprawl and its effects on water demand: A case study of Alicante, Spain. *Land Use Policy* **2016**, *50*, 352–362. [CrossRef]
15. Valero, L.G.; Pajares, E.M.; Sánchez, I.M.R.; Pérez, J.A.S. Analysis of Environmental Taxes to Finance Wastewater Treatment in Spain: An Opportunity for Regeneration? *Water* **2018**, *10*, 226. [CrossRef]
16. Organization for Economic Co-Operation and Development (OECD). *Water Pricing: Current Practices and Recent Trends*; Organization for Economic Co-Operation and Development: Paris, France, 1999.
17. Garrido-Cardenas, J.A.; Esteban-García, B.; Agüera, A.; Sánchez-Pérez, J.A.; Manzano-Agugliaro, F. Wastewater Treatment by Advanced Oxidation Process and Their Worldwide Research Trends. *Int. J. Environ. Res. Public Health* **2019**, *17*, 170. [CrossRef]
18. Fytianos, G.; Tziolas, E.; Papastergiadis, E.; Samaras, P. Least Cost Analysis for Biocorrosion Mitigation Strategies in Concrete Sewers. *Sustainability* **2020**, *12*, 4578. [CrossRef]
19. Caballero, J.A.; Labarta, J.A.; Quirante, N.; Carrero-Parreño, A.; Grossmann, I.E. Environmental and Economic Water Management in Shale Gas Extraction. *Sustainability* **2020**, *12*, 1686. [CrossRef]
20. Dang, H.T.; Dinh, C.V.; Nguyen, K.M.; Tran, N.T.; Pham, T.T.; Narbaitz, R.M. Loofah Sponges as Bio-Carriers in a Pilot-Scale Integrated Fixed-Film Activated Sludge System for Municipal Wastewater Treatment. *Sustainability* **2020**, *12*, 4758. [CrossRef]
21. Pajares, E.M.; Valero, L.G.; Sánchez, I.M.R. Cost of Urban Wastewater Treatment and Ecotaxes: Evidence from Municipalities in Southern Europe. *Water* **2019**, *11*, 423. [CrossRef]
22. Juszczyk, S.; Bałlina, R.; Juszczyk, J. Environment protection costs in polish dairy cooperatives. *Ann. Pol. Assoc. Agric. Agribus. Econ.* **2020**, *XXII*, 49–59. [CrossRef]
23. Perez, J.A.S.; Sanchez, I.M.R.; Carra, I. Promoting environmental technology using sanitary tax: The case of agro-food industrial wastewater in Spain. *Environ. Eng. Manag. J.* **2014**, *13*, 961–969. [CrossRef]
24. Ling, J.; Germain, E.; Murphy, R.; Saroj, D. Designing a Sustainability Assessment Tool for Selecting Sustainable Wastewater Treatment Technologies in Corporate Asset Decisions. *Sustainability* **2021**, *13*, 3831. [CrossRef]
25. Arden, S.; Morelli, B.; Schoen, M.; Cashman, S.; Jahne, M.; Ma, X.; Garland, J. Human Health, Economic and Environmental Assessment of Onsite Non-Potable Water Reuse Systems for a Large, Mixed-Use Urban Building. *Sustainability* **2020**, *12*, 5459. [CrossRef] [PubMed]
26. Hardisty, P.E.; Sivapalan, M.; Humphries, R. Determining a sustainable and economically optimal wastewater treatment and discharge strategy. *J. Environ. Manag.* **2013**, *114*, 285–292. [CrossRef]
27. Bonoli, A.; Pennellini, S.; Eshtawi, T.; Qrenawi, L.; Solaroli, M.; Al-Zaanin, N.; Abu Saada, G.; Barilli, E. Appropriate solutions for wastewater reuse in agriculture: A pilot plant in Rafah, Gaza strip. *Environ. Eng. Manag. J.* **2020**, *19*, 1683–1692. [CrossRef]
28. Perulli, G.D.; Gaggia, F.; Sorrenti, G.; Donati, I.; Boini, A.; Bresilla, K.; Manfrini, L.; Baffoni, L.; Di Gioia, D.; Grappadelli, L.C.; et al. Treated wastewater as irrigation source: A microbiological and chemical evaluation in apple and nectarine trees. *Agric. Water Manag.* **2021**, *244*, 106403. [CrossRef]
29. Molinos-Senante, M.; Sancho, F.H.; Garrido, R.S. Estado actual y evolución del saneamiento y la depuración de aguas residuales en el contexto nacional e internacional. *Anales Geografía Universidad Complutense* **2012**, *32*, 69–89. [CrossRef]
30. Melgarejo, J. *Efectos Ambientales y Económicos de la Reutilización del Agua en España*; Universidad de Alicante: Alicante, Spain, 2009.
31. Chen, Z.; Ngo, H.H.; Guo, W. A critical review on sustainability assessment of recycled water schemes. *Sci. Total. Environ.* **2012**, *426*, 13–31. [CrossRef]
32. Goodwin, D.; Raffin, M.; Jeffrey, P.; Smith, H.M. Applying the water safety plan to water reuse: Towards a conceptual risk management framework. *Environ. Sci. Water Res. Technol.* **2015**, *1*, 709–722. [CrossRef]
33. Ruiz-Rosa, I.; García-Rodríguez, F.J.; Mendoza-Jiménez, J. Development and application of a cost management model for wastewater treatment and reuse processes. *J. Clean. Prod.* **2016**, *113*, 299–310. [CrossRef]
34. Gharbia, S.S.; Aish, A.; Abushbak, T.; Qishawi, G.; Shawa, I.A.-; Gharbia, A.S.; Zelenakova, M.; Gill, L.; Pilla, F. Evaluation of wastewater post-treatment options for reuse purposes in the agricultural sector under rural development conditions. *J. Water Process. Eng.* **2016**, *9*, 111–122. [CrossRef]
35. Fito, J.; Van Hulle, S.W.H. Wastewater reclamation and reuse potentials in agriculture: Towards environmental sustainability. *Environ. Dev. Sustain.* **2021**, *23*, 2949–2972. [CrossRef]
36. Leonel, L.P.; Tonetti, A.L. Wastewater reuse for crop irrigation: Crop yield, soil and human health implications based on giardiasis epidemiology. *Sci. Total. Environ.* **2021**, *775*, 145833. [CrossRef]
37. Ait-Mouheb, N.; Mayaux, P.L.; Mateo-Sagasta, J.; Hartani, T.; Molle, B. Water reuse: A resource for Mediterranean agriculture. In *Water Resources in the Mediterranean Region*; Elsevier: Amsterdam, The Netherlands, 2020; pp. 107–136.

38. Hochstrat, R.; Wintgens, T.; Kazner, C.; Jeffrey, P.; Jefferson, B.; Melin, T. Managed aquifer recharge with reclaimed water: Approaches to a European guidance framework. *Water Sci. Technol.* **2010**, *62*, 1265–1273. [CrossRef]
39. Garcia, X.; Pargament, D. Reusing wastewater to cope with water scarcity: Economic, social and environmental considerations for decision-making. *Resour. Conserv. Recycl.* **2015**, *101*, 154–166. [CrossRef]
40. Wilcox, J.; Nasiri, F.; Bell, S.; Rahaman, S. Urban water reuse: A triple bottom line assessment framework and review. *Sustain. Cities Soc.* **2016**, *27*, 448–456. [CrossRef]
41. Gallego-Valero, L.; Moral-Pajares, E.; Román-Sánchez, I.M. Crop production and irrigation: Deciding factors of wastewater reuse in Spain? *Desalin. Water Treat.* **2019**, *150*, 91–98. [CrossRef]
42. San Juan, J.L.; Caligan, C.J.; Garcia, M.M.; Mitra, J.; Mayol, A.P.; Sy, C.; Culaba, A. Multi-objective optimization of an inte-grated algal and sludge-based bioenergy park and wastewater treatment system. *Sustainability* **2020**, *12*, 7793. [CrossRef]
43. Castellet, L.; Molinos-Senante, M. Efficiency assessment of wastewater treatment plants: A data envelopment analysis ap-proach integrating technical, economic, and environmental issues. *J. Environ. Manag.* **2016**, *167*, 160–166. [CrossRef]
44. Berbeka, K.; Czajkowski, M.; Markowska, A. Municipal wastewater treatment in Poland—Efficiency, costs and returns to scale. *Water Sci. Technol.* **2012**, *66*, 394–401. [CrossRef]
45. Djukic, M.; Jovanoski, I.; Ivanovic, O.M.; Lazic, M.; Bodroza, D. Cost-benefit analysis of an infrastructure project and a cost-reflective tariff: A case study for investment in wastewater treatment plant in Serbia. *Renew. Sustain. Energy Rev.* **2016**, *59*, 1419–1425. [CrossRef]
46. Moran, D.; Dann, S. The economic value of water use: Implications for implementing the Water Framework Directive in Scotland. *J. Environ. Manag.* **2008**, *87*, 484–496. [CrossRef]
47. Fu, X.; Niu, Z.; Yeh, M.-K. Research trends in sustainable operation: A bibliographic coupling clustering analysis from 1988 to 2016. *Clust. Comput.* **2016**, *19*, 2211–2223. [CrossRef]
48. Pérez, A.T.E.; Camargo, M.; Rincón, P.C.N.; Marchant, M.A. Key challenges and requirements for sustainable and industrialized biorefinery supply chain design and management: A bibliographic analysis. *Renew. Sustain. Energy Rev.* **2017**, *69*, 350–359. [CrossRef]
49. Bryman, A.; Bell, E. *Business Research Methods*; Oxford University Press: Evans Road Cary, NC, USA, 2015.
50. Archambault, É.; Campbell, D.; Gingras, Y.; Larivière, V. Comparing bibliometric statistics obtained from the Web of Science and Scopus. *J. Am. Soc. Inf. Sci. Technol.* **2009**, *60*, 1320–1326. [CrossRef]
51. Prieto-Sandoval, V.; Jaca, C.; Ormazabal, M. Towards a consensus on the circular economy. *J. Clean. Prod.* **2018**, *179*, 605–615. [CrossRef]
52. Huang, L.; Zhang, Y.; Guo, Y.; Zhu, D.; Porter, A.L. Four dimensional Science and Technology planning: A new approach based on bibliometrics and technology roadmapping. *Technol. Forecast. Soc. Chang.* **2014**, *81*, 39–48. [CrossRef]
53. Aznar-Sánchez, J.A.; García-Gómez, J.J.; Velasco-Muñoz, J.F.; Carretero-Gómez, A. Mining Waste and Its Sustainable Management: Advances in Worldwide Research. *Minerals* **2018**, *8*, 284. [CrossRef]
54. Albort-Morant, G.; Henseler, J.; Leal-Millán, A.; Cepeda-Carrión, G. Mapping the field: A bibliometric analysis of green in-novation. *Sustainability* **2017**, *9*, 1011. [CrossRef]
55. Durieux, V.; Gevenois, P.A. Bibliometric Indicators: Quality Measurements of Scientific Publication. *Radiology* **2010**, *255*, 342–351. [CrossRef]
56. Pauly, D.; Stergiou, K. Equivalence of results from two citation analyses: Thomson ISI's Citation Index and Google's Scholar service. *Ethics Sci. Environ. Polit.* **2005**, *9*, 33–35. [CrossRef]
57. Prigle, J. Trends in the use of ISI citation databases for evaluation. *Learn. Publ.* **2008**, *21*, 85–91. [CrossRef]
58. Montoya, F.G.; Baños, R.; Meroño, J.E.; Manzano-Agugliaro, F. The research of water use in Spain. *J. Clean. Prod.* **2016**, *112*, 4719–4732. [CrossRef]
59. López-Serrano, M.J.; Velasco-Muñoz, J.F.; Aznar-Sánchez, J.A.; Román-Sánchez, I.M. Sustainable Use of Wastewater in Agriculture: A Bibliometric Analysis of Worldwide Research. *Sustainability* **2020**, *12*, 8948. [CrossRef]
60. Ramos, J.M.; González-Alcaide, G.; Gutiérrez, F. Bibliometric analysis of the Spanish scientific production in Infectious Diseases and Microbiology. *Enferm. Infecc. Microbiol. Clin.* **2016**, *34*, 166–176. [CrossRef]
61. Romo-Fernandez, L.M.; Guerrero-Bote, V.P.; Moya-Anegon, F. Analysis of the Spanish scientific production in Renewable Energy, Sustainability and the Environment (Scopus, 2003–2009) in the global context. *Investig. Bibliotecol.* **2013**, *27*, 125–151.
62. Trinh, L.T.; Duong, C.C.; Van Der Steen, P.; Lens, P.N. Exploring the potential for wastewater reuse in agriculture as a climate change adaptation measure for Can Tho City, Vietnam. *Agric. Water Manag.* **2013**, *128*, 43–54. [CrossRef]
63. Gómez-López, M.; Bayo, J.; García-Cascales, M.; Angosto, J. Decision support in disinfection technologies for treated wastewater reuse. *J. Clean. Prod.* **2009**, *17*, 1504–1511. [CrossRef]

MDPI

Review

Microalgal Production of Biofuels Integrated with Wastewater Treatment

Merrylin Jayaseelan [1], Mohamed Usman [2], Adishkumar Somanathan [3], Sivashanmugam Palani [4], Gunasekaran Muniappan [5] and Rajesh Banu Jeyakumar [6,*]

1 Department of Food Science and Nutrition, Sarah Tucker College, Tirunelveli 627007, Tamil Nadu, India; merrylin_86@yahoo.co.in
2 Department of Civil Engineering, PET Engineering College, Vallioor 627117, Tamil Nadu, India; mdus@live.in
3 Department of Civil Engineering, Anna University Regional Campus, Tirunelveli 600025, Tamil Nadu, India; adishk2002@yahoo.co.in
4 Department of Chemical Engineering, National Institute of Technology, Tiruchirappalli 110040, Tamil Nadu, India; psiva@nitt.edu
5 Department of Physics, Anna University Regional Campus, Tirunelveli 600025, Tamil Nadu, India; mgunat75@gmail.com
6 Department of Life Sciences, Central University of Tamil Nadu, Thiruvarur 610001, Tamil Nadu, India
* Correspondence: rajeshces@gmail.com

Abstract: Human civilization will need to reduce its impacts on air and water quality and reduce its use of fossil fuels in order to advance towards a more sustainable future. Using microalgae to treat wastewater as well as simultaneously produce biofuels is one of the approaches for a sustainable future. The manufacture of biofuels from microalgae is one of the next-generation biofuel solutions that has recently received a lot of interest, as it can remove nutrients from the wastewater whilst capturing carbon dioxide from the atmosphere. The resulting biomass are employed to generate biofuels, which can run fuel cell vehicles of zero emission, power combustion engines and power plants. By cultivating microalgae in wastewater, eutrophication can be prevented, thereby enhancing the quality of the effluent. Thus, by combining wastewater treatment and biofuel production, the cost of the biofuels, as well as the environmental hazards, can be minimized, as there is a supply of free and already available nutrients and water. In this article, the steps involved to generate the various biofuels through microalgae are detailed.

Keywords: microalgae; wastewater treatment; biofuel; nutrient removal

Citation: Jayaseelan, M.; Usman, M.; Somanathan, A.; Palani, S.; Muniappan, G.; Jeyakumar, R.B. Microalgal Production of Biofuels Integrated with Wastewater Treatment. *Sustainability* **2021**, *13*, 8797. https://doi.org/10.3390/su13168797

Academic Editor: Alessio Siciliano

Received: 14 June 2021
Accepted: 20 July 2021
Published: 6 August 2021

1. Introduction

From sea ice in the Arctic to microbiotic crusts in deserts, the term "algae" refers to a wide group of (mainly) autotrophic, aquatic creatures found all over the world [1,2]. Algae are usually one of two types such as macro and micro algae. Macroalgae are generally considered terrestrial plants that returned to a damp environment, according to evolutionary theory. They are classified into red, brown, and green algae and are diverse forms of multicellular eukaryotes, each with a respective evolution pathway. They have leaves and branches and may be fixed firmly [3]. Microalgae, on the other hand, are unicellular and range in size from nanometers to millimetres. Microalgae is defined by phycologists as a creature with chlorophyll and a body (thallus) that is not divided into roots, leaves (thallophytes), and stems [4]. They comprise both the prokaryotes and eukaryotes. Microalgae fix carbon dioxide more efficiently than terrestrial plants, and are widely known for capturing both atmospheric and industrial pollutants [5].

According to the research conducted earlier, production of algae biomass are 5–10 times higher than terrestrial plants, indicating a considerable chance of increased biomass production [6]. Microalgae are considered as the most successful feedstock for biodiesel synthesis due to their high photosynthetic activity, effective capture of the emitted carbon dioxide, and remarkable environmental adaptation, including high algal production [7–9]. Thus, algae utilize carbon dioxide, along with sunlight and water, to produce sugars through photosynthesis. The products thus obtained are broken down into carbohydrates, proteins, lipids, nucleic acids, etc. The fatty acid content varies with each type of algae. Fatty acids can account for up to 40–50% of the entire mass of some algae species. Table 1 details the general composition of various microalgae.

Table 1. General composition of various microalgae on percentage dry matter basis [10].

Microalgae Species	Protein	Carbohydrates	Fatty Acids
Anabaena cylindrica	43–56	25–30	4–7
Aphanizomenonflos-aquae	62	23	3
Chlamydomonas rheinhardii	48	17	21
Chlorella pyrenoidosa	57	26	2
Chlorella vulgaris	51–58	12–17	14–22
Dunaliella salina	57	32	6
Euglena gracilis	39–61	14–18	14–20
Porphyridiumcruentum	28–39	40–57	9–14
Scenedesmus obliquus	50–56	10–17	12–14
Spirogyra sp.	6–20	33–64	11–21
Arthrospira maxima	60–71	13–16	6–7
Spirulina platensis	46–63	8–14	4–9
Synechococcus sp.	63	15	11

Current biofuel production from microalgal biomass is limited by a lack of dependable and cost-effective technologies for producing and harvesting algal feedstocks [11]. Muchrecent research has proposed that algal biomass production be combined with wastewater treatment and recycling to equalize the expense of fertilisers and freshwater necessary for microalgae growing [12–15]. This combination of algal biomass generation and wastewater treatment also helps to purify wastewater [16–18]. Microalgal cells′ nutrient absorption ability can lower the nitrogen and phosphorus content of wastewater to a relatively low level, allowing it to fulfil the increasingly stricter nutrient discharge standards [19,20]. Most nutrients in the wastewater are eliminated, suggesting the possibility of integrating wastewater treatment with algal biomass production. In addition, the nutrients in the wastewater are not only eliminated from the wastewater, but they are also caught by the microalgae and returned to the environment as agricultural fertilisers. Another benefit of microalgae-based wastewater treatment is that the algal cells fix the greenhouse gas (carbon dioxide) through photosynthesis. Microalgae cultivations have been widely employed for wastewater treatment so far, demonstrating the capacity to remove nutrients from a variety of wastewater types, including wastewater from industries [21,22], municipal wastewater [23,24], cattle wastewater [14,25], and aquaculture wastewater [18]. Microalgae may utilize the nutrients in wastewater to flourish, and the wastewater may be treated at the same time. Large amounts of freshwater and nutrients necessary for algal development might be conserved by utilising wastewater. Using microalgae to remediate wastewater for biofuel production is a more sustainable and ecologically beneficial option. Algal growth is primarily reliant on carbon, nitrogen, and phosphorus. Various wastewater sources, such as urban wastewater, agricultural run-off, animal wastewaters, and industrial water, often provide adequate carbon, nitrogen, and phosphorus for algae to thrive efficiently. Algae may develop in a variety of wastewaters as long as there is enough carbon (both organic or inorganic), nitrogen (in the form of urea, ammonium, or nitrate), and other trace components [26]. Due to its efficiency and requirement, widespread production of algal biomass for biofuel and other purposes utilising wastewaters is attracting increased interest for study.Nutrient removal efficiency of microalga in different wastewater is tabulated in Table 2.

Table 2. Microalgal nutrient removal efficiencies from various wastewaters.

Wastewater	Strain	Removal of Total Nitrogen (%)	Removal of Total Phosphorus (%)	Reference
Municipal sewage water	*Chlorella*	97.81	89.39	[27]
Pharmaceutical wastewater	*Chlorella sorokiniana*	70	89	[28]
Landfill Leachate	*Chlorella vulgaris*	69	100	[29]
Aquaculture wastewater	*Chlorella vulgaris and Scenedesmus obliquus*	86.1	82.7	[18]
Textile wastewater	Mixed consortia of Microalga	95	70	[30]
Edible Oil Refinery wastewater	*Desmodesmussp*	96	53	[31]
Dairy wastewater	*Chlorella vulgaris*	85.47	65.96	[32]

Algae may be harvested from the treatment facility on a regular basis and used to make biofuel. In comparison to traditional wastewater treatment technologies, simultaneous wastewater treatment and algae culture can give a more cost-effective and environmentally friendly wastewater treatment. It has been demonstrated that it is a more cost-effective method of removing biochemical oxygen demand, pathogens, phosphate, and nitrogen than activated sludge [33]. This review paper discusses the various biofuels obtained through the nutrient removal in wastewater using microalgae (as shown in Figure 1). The fuels generated are biodiesel, biomethanol, biomethane, biohydrogen, bioprocessed esters and fatty acids, and synthestic fuels such as Fischer Tropsch Diesel, Dimethyl ether, and methanol.

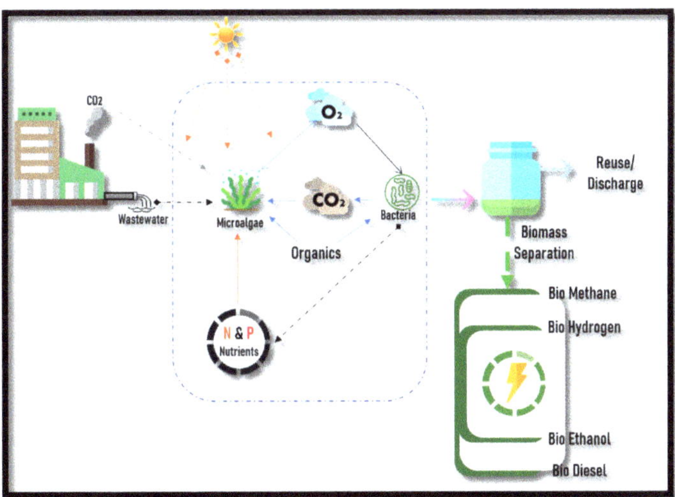

Figure 1. Microalgal wastewater treatment and the application of the produced biomass.

2. Biofuel Production Pathways

Biofuels using microalgae can be produced through various pathways such as (i) physicochemical pathway, which involves transesterification or esterification,(ii) biochemical conversion, which involves alcoholic fermentation, dark fermentation, anaerobic digestion, andbiophotosynthesis, and (iii) thermochemical conversions such as pyrolysis, gasification, hydrothermal processes, or hydro-processing. Biomethane is produced by anaerobic digestion [34,35], biohydrogen is generated by biophotosynthetic process [36,37], bioethanol is produced by fermentation [38,39], liquid fats by thermal liquefaction [40–43], and biodiesel by transesterification or esterification process [40,44,45].

2.1. Biodiesel

The triglyceride transesterification technique has been used in biodiesel production for more than 50 years [46]. During transesterification, fatty acid esters are formed when the triglycerides react with alcohol, and when the reaction is sped up by a catalyst.As the chemical processes involved in the manufacture of biodiesel are rather slow, catalysts are utilised to speed them up. Biodiesel manufacturing methods can be chemical or biotechnological, depending on the kind of catalyst used in the process. Biodiesel may be made from algal biomass in a variety of ways (including oil extraction from algal biomass) via esterification and direct transesterification of microalgae [47]. Fatty acid methyl esters (FAMEs), the chemical component of biodiesel, are usually generated in algal biodiesel processes by transesterification of algal oil with the alcohol (methanol) utilizing 98% concentrated sulfuric acid as a catalyst and n–hexane as a solvent. Extraction of oil from the microalgae without breaking their cells is a novel way in using nano catalysts for biodiesel synthesis from microalgae [48]. In situ transesterification is a promising method for avoiding oil extraction and directly converting lipids within microalgae cells to biodiesel in a single step, which might simplify biodiesel manufacturing procedures while also producing more biodiesel [49].

Generally, microalgae oil contains more free fatty acids compared to the oil from terrestrial plants; therefore, during biodiesel production from microalgae oil, free fatty acids should be esterified using acid catalysts prior to usual transesterification. In a research conducted by Ashokkumar, the fatty acids in the microalgae *Botryococcus braunii* were esterified using 1.5% concentrated sulphuric acid at a temperature of 55 °C for about 20 min prior to transesterification. The initial fatty acid content was 9.7% in the microalgae, and it decreased to 0.55 percent after esterification [50].

Unlike terrestrial oily plants (such as palm, soybean, or canola seed), algal oil may be recovered readily from the algal cell by compressing it, followed by solvent extraction. Mechanical crushing of algal biomass is another option, although extracting algal oil with the current technology is quite challenging. The majority of extraction methods are based on Bligh and Dryer′s 1959 approach. [51]. Cold pressing is the least costly method of extraction of algal oil [47]. This method may extract up to 70% of the oil contained inside the algae [52]. The addition of an organic solvent can improve the extraction level to 99%, but this comes at a cost in terms of processing [53].

When assessing the potential of different microalgal strains for biodiesel generation, it′s crucial to keep in mind that all microalgal oils are not suitable for the production of biodiesel. Biodiesel made from microalgae is similar to biodiesel made from other sources, such as oleaginous plants, in that it has limited oxidative stability due to a high degree of unsaturation.

Blending biodiesel with fossil fuel and/or adding chemical stabilisers can help in improving it [54–56]. The temperature-related features of diesel, such as cloud point and cold filter plugging point, become critical in cold climates for optimal fuel performance. [57]. Other important factors to be considered are the energy content which provides the intrinsic value of the fuel, acidic, and water content in the fuel which determines its corrosive nature, and the viscosity of the fuel to determine the proficient operation of the engine [54].

2.2. Biomethane

Biogas is one of the most promising biofuels, with the ability to alleviate some of the rising worries about fossil fuels, such as the energy calamity and change in the weather [58,59]. Application of microalgae have been shown to be efficient, practical, and cost-effective in biogas generation [60–62]. Microalgae are particularly well suited for combined nutrient removal through wastewater treatment and carbon dioxide sequestration, due to their ability to assimilate large amounts of carbon dioxide and the possibility of blending microalgal cultivation with flue gas emissions or biogas upgrading, which involves removing carbondioxide (as biogas) to increase methane percentage [63–65]. Microalgae cultivation at a wastewater treatment facility offers a free source of water and nutrients, while also contributing to the wastewater treatment process and allowing the recycling of vital nutrients that would otherwise be lost to the environment. The resulting microalgal biomass can subsequently be processed to extract nutrients for fertiliser production or oils for biodiesel generation. Biogas can also be produced through anaerobic digestion of residual biomass [66–68]. Biomethane is generated via biochemical conversion of biomass, followed by gas upgrading, or by thermochemical conversion of solid biomass through gasification, followed by gas cleaning, methanation as the process of synthesis, and biogas upgrading of the product.

Anaerobic digestion by microalgae was first detailed by Golueke and his coworkers in 1957 by means of *Chlorella* sp. and *Scenedesmus* sp.,whichgenerated nearly 0.17–0.32 L CH_4/gVS [69]. However, for two reasons, microalgae biomass is not currently considered a viable substrate for biogas production in the anaerobic digestion process: (i) the rigid cell wall of the microalgae confers a high level of resistance to microbial degradation, and (ii) due to the high protein content, the biomass has very low carbon-to-nitrogen (C/N) ratio, which is unfavourable [69,70]. Three different microalgae, belonging to the genus *Scenedesmus*, *Chlorella*, and *Chlamydomonas*, displayed higher carbon/nitrogen ratio in the biomass (24–26 on the basis of weight) when those microalgae were starved. Consecutively, they also lost their capacity to resist degradation by bacteria, thus leading to higher production of biomethane [71], with their conversion rates nearly equal to the theoretical threshold value [70].

By weakening or breaking the microalgae cell wall structure via pretreatment, we can improve the anaerobic biodegradability of microalgae; co-digestion also enhances the biogas yield by increasing the organic loading rate while controlling the concentration of ammonia. For a better understanding and optimization of the performance of the process, mathematical models and reactor design and operation techniques must be carefully studied. Finally, the biogas generated during the anaerobic digestion of microalgae should be improved before being burned on-site or injected into natural gas systems or utilized as liquefied petroleum gas.

To aid compression, to improve calorific value, and to avoid metal component corrosion, all hydrogen sulphide and other trace gases must be eliminated before biogas can be utilised in generators or fed to national gas networks. These steps can be through various chemical scrubbing methods [72,73]. However, the systems involved are huge and difficult to scale down, and can produce very hazardous by-products. Therefore, biotrickling and bioscrubbing have been developed as ecologically viable biological processes.

The use of the entire algal cell and the ability to employ low-quality algae sources, such as wastewater treatment or blooms, are the advantages of the biogas generation in general.

2.3. Biohydrogen

The term "biohydrogen" refers to hydrogen created biologically, most typically by algae, bacteria, and archaea, either through cultivating them or from organic waste sources [74]. Hydrogen is considered as yet another sustainable energy source generated by photosynthetic organisms, with a higher energy content of about 122 kJ/g, which is nearly 2.75 times greater than that of hydrocarbon fuels [75]; due to this reason, it has been considered a viable alternative to fossil fuels and as an carrier of energy. A number of microalgae species, such as *Anabaena* sp. [76], *Chlorella vulgaris* [77], *Nannochoropsis* sp. [78], *Chlamydomonas reinhardtii* [79], *Spirulina maxima* [80], and *Scenedesmus obliquus* [77,81], are capable of generating molecular hydrogen through the photofermentative metabolism. Among the various species, *Chlorella vulgaris* is the most commonly used untreated substrate for hydrogen generation. Hydrogen yield acquired from various species ranged from 0.37 to 19 mL of hydrogen/g VS, and highest hydrogen yield was achieved from *Chlorella vulgaris* [82] and the *Scenedesmus* sp. [83].

Microalgal hydrogen production is facilitated by the solar light, or by fermentation processes and various thermochemical techniques used to convert the biomass. Production of molecular hydrogen by microalgae is carried out in the presence of the enzyme hydrogenase, which utilizes water as the only electron donor. [84].However, as the hydrogenase enzyme that produces biohydrogen is particularly sensitive to oxygen and becomes inactive even at a partial pressure of 2% of oxygen, [85], the ability of microalgae to stimulate hydrogen production shows only after acclimatization to anaerobic environment and is vanished in the presence of even small amount of oxygen [86]. Thus, due to its incompatibility to oxygen, the ability of microalgae lasts only for a short period of time.

In 1939, Gaffron and Rubin conducted the first scientific experiment of generating hydrogen using microalgae. In their study, it was reported that *Scenedesmus obliquus* was able to generate hydrogen at low rates under two conditions, such as in the dark environment and by replacing the culture's atmosphere with nitrogen gas [87]. Thus, it can be stated that the production of hydrogen depends on the mechanism of algae to adapt to the transition phase during which the dark anaerobic condition is switched to oxygenic condition to perform photosynthesis. The electron transport pathway is re-oxidized only through this transition [88].

Due to the low rate of biohydrogen production, the microalgae should be pretreated to convert the complex carbohydrates into simple sugars, thereby increasing the surface area (through the disruption of cell and disintegration of the cell wall) for the action of microbes. To increase the production of biohydrogen, physical, chemical, and biological pretreatments are employed. Physical pretreatment includes mechanical, thermal, and ultrasonication. Chemical pretreatment is carried out by the addition of acid or base or by ozonation.Microbes and enzymes are used in biological pretreatment to disrupt the microalgal biomass and liberate intracellular components, thus increasing the biohydrogen production rate. Enzymes are selected based on the composition of the microalgal cell wall. Another method to increase the hydrogen yield is through metabolic engineering and screening for mutant varieties. The photosynthetic stages within the cell are changed to increase hydrogen production through metabolic engineering.

Hydrogen can be produced inthree different ways such as direct biophotolysis, indirect biophotolysis (as shown in Figure 2), and hydrogen production driven by ATP. In direct photolysis biohydrogen is produced by converting water to hydrogen using solar energy through photosynthesis, and is further used as a substrate for anaerobic bacteria during dark fermentation [76]. Such fermentative reactions are typically faster and produce more hydrogen [89].

In indirect photolysis, the microalga produces hydrogen in two steps. In step 1, carbon dioxide is captured through photosynthesis in the presence of solar light. In other terms, microalgae produce oxygen and build up carbon within the cells. In step 2, production of hydrogen takes place through the degradation of the accumulated carbon through anaerobic fermentation which occurs in the absence of oxygen and involves a series of complex biochemical events involving multi-enzyme systems [36,75]. Hydrogenase enzyme plays an important role in this method. As discussed earlier, it is more sensitive to oxygen, so various research is being carried out to develop hydrogenase enzyme which is not sensitive to oxygen. Closed Photobioreactors can be employed for indirect photolysis (Figure 3). The most widely used photobioreactors have atubular design which consists of numerous transparent tubes. The tubes are usually designed to have a diameter less than 10 cm to maximize penetration of solar rays. The microalgal broth is pumped through the tubes, where it is exposed to sunlight for photosynthesis, and then recycled back to a reservoir. Using either a mechanical pump or an airlift pump, the algal biomass is kept from settling by maintaining a very turbulent flow within the reactor [46]. A fraction of the algae is usually collected from the solar tubes. Thus, in this way microalgae can be harvested continuously.

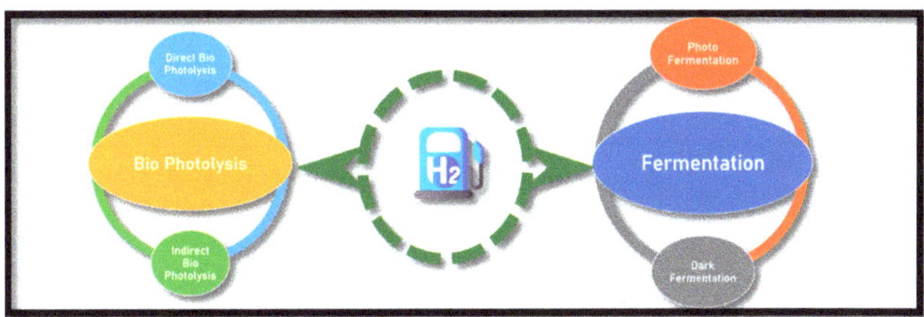

Figure 2. Biohydrogen Production through photolysis and fermentation.

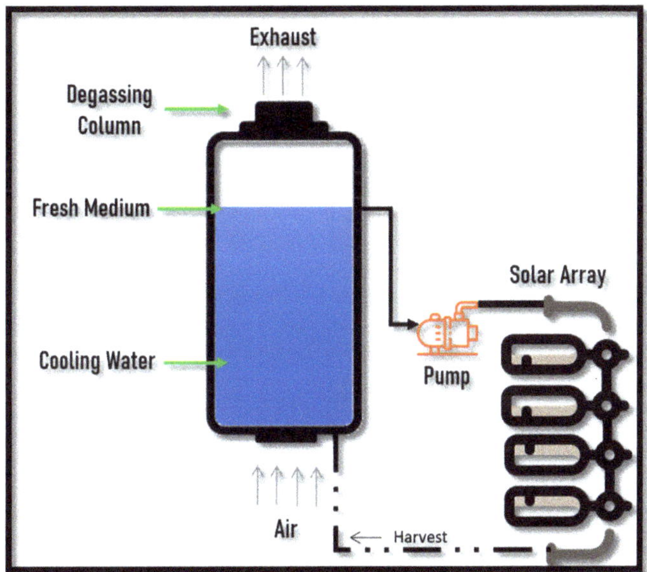

Figure 3. Closed Photobioreactors.

Various literatures have reported that usage of immobilized microalgal cells in the photobioreactors is more advantageous compared to the free cells. Immobilized cells provide increased cell retention time within the bioreactors and enhanced metabolic activity compared to the free cells [90]. Several strategies for improving hydrogen synthesis have been implemented, including varying light intensity, carbon supply, pH, temperature, and sulphur starvation [91,92].

Furthermore, the process's environmental and financial benefits are enhanced by the simultaneous treatment of wastewater and generation of valuable algal biomass [93]. According to Brennan and Owende (2010), in the immediate future, the combination of these processes will be the most reasonable commercial application, and it may be one of the most sustainable methods to generate biofuels [5]. Though the concept of generating hydrogen through fermentative metabolism was discovered years ago by *Chlamydomonas moewusii* [94]; the challenges faced by construction often hinder the application of algal hydrogen generation in a wastewater environment in terms of renewable energy generation and performance characteristics. Moreover, the action of volatile acids such as acetic acid and butyric acid (released during the anaerobic digestion in the wastewater treatment plant) on depletion of oxygen and consequent generation of biohydrogen and its continuity in producing hydrogen has to be studied [95].

2.4. *Bioethanol*

Bioethanol is considered a substitute for conventional petroleum, as they both have the same chemical and physical properties [96–98]. Microalgae biomass, in particular, has lately received a lot of interest as a viable renewable source for the production of biofuels. Third-generation bioethanol made from microalgae biomass could also be an environmentally beneficial fuel. As discussed earlier, microalgae are rich in lipids, enabling it to produce biodiesel. Similarly, some species of microalgae can store large amounts of carbohydrates, such as triacylglycerol and starch, within their cells. These carbohydrates can be used as a carbon source or substrate during fermentation to generate bioethanol [99,100]. Proteins can also be accumulated within the cells, along with carbohydrates and lipids under restriction of nitrogen or starvation [101]. Microalgae breakdown the complex nitrogen molecules into protein. Variation in salinity, light intensity, and temperature can also accumulate carbohydrates. Microalgae also lack lignin, and have low hemicellulose levels, making hydrolysis and fermentation yields more efficient [102].

There are three different routes to produce bioethanol from microalgae:(i) The first route is the conventional method, in which the biomass is pretreated, hydrolyzed enzymatically, and fermented using yeast [103]. (ii) The second route operates in the dark condition, and uses metabolic pathways to redirect photosynthesis to create hydrogen, acids, and ethanol [104]. (iii)The third method is to use photofermentation, which is impossible in nature [105]. (iv) The last route necessitates the use of genetic engineering to reroute microalgae′s pre-existing metabolic pathways for more subjective and efficient bioethanol synthesis. Bioethanol production from microalgae and cyanobacteria is a viable technical advancement, as they have shown to be more productive than crops such assugarcane and corn. Light is used as an energy source by genetically engineered strains to produce bioethanol from carbon dioxide and water in a single process [106].

2.4.1. Bioethanol by Hydrolysis and Fermentation

This route is entirely based on the production of microalgae biomass in photobioreactors, achieved by way of pretreatment steps which involve the hydrolysis of the biomass and breakdown of the cell walls. The biomass is preferably pretreated using enzymes. These treated biomass are further fermented with *Sacchromyces cerevisae* or bacteria to yield bioethanol.

2.4.2. Bioethanol by Dark Fermentation

In this route of dark fermentation, the organic biomass is converted into biohydrogen. Fermentative and hydrolytic microorganisms hydrolyze complex natural polymers into monomers which are, in the end, converted into a combination of low molecular weight organic acids and alcohols such as acetic acid and ethanol.

In the absence of light, certain microalgae and cyanobacteria are capable of expelling ethanol through the cell wall via an intracellular process. Some of the species include *Chlamydomonas moewusii*, *C.vulgaris*, *C.reinhardtii*, *Oscillatoria limnetica*, *O. limosa*, *Chlorococcum littorale*, etc. [102,107]

2.4.3. Bioethanol by Photofermentation

Algae can generate bioethanol directly through the photosynthetic process, which is referred to as thephotofermentative or photanol method [105,107]. Photofermentation is a technology whichis gaining popularity, especially after the announcement of industrial operations that will use modified algae to create bioethanol directly [108]. The photofermentative approach is a natural technique to transform sunlight into fermentation products via a highly efficient metabolic pathway [109].

There has been a recent surge in the number of compounds such as ethanol produced by modified cyanobacteria's photofermentation metabolism. This has been made possible by the information gained from traditional fermentation which is carried out by *Escherichia coli* and *Sacchromyces cerevisiae*, which have been genetically modified to generate biofuels such as bioethanol. Algae is a varied group of creatures with a lot of unwrapped genetic potential. Although microalgae are morphologically identical single-celled photosynthetic organisms, the functional genetic diversity is relatively high, as evidenced by the number of unique genes found among distinct species. This genetic diversity is being used to create novel algae strains for the production of biofuels [110].

The use of carbon dioxide and sunlight for carbon uptake and conversion into organic molecules is the most fundamental feature of photosynthesis. Thus, it can be understood that carbon dioxide and light are the two most important variables in increasing production of bioethanol. The photosynthetic efficiency of microalgae ranges from 6–10% of the incident light, whereas in higher plants, the efficiency is just 1 or 2 percent [111,112]. The refinement of geometry, the study of optimal growing circumstances, and the capacity of the photosynthetic machinery to absorb light all contribute to the desired result. In general, microalgae are suitable for the generation of third generation biofuels such as ethanol. Similar to biogas from microalgae, the ethanol production is determined by the pretreatment and the algal strain used [113].

3. Conclusions

With the rapid pace of economic development and energy consumption, as well as the limited supply of fossil fuels and the growing need for environmental protection, more attention is being paid to the development of ecologically friendly fuels such as biofuels to resolve the conflict. Microalgae-based biofuels are one of the most promising feedstocks for the next generation of biofuels due to their capacity to produce a high amount of lipids and minimum negative environmental consequence. Algal biofuels can be used in combination with a reduction in carbon dioxide in flue gas and wastewater treatment, as well as the generation of byproducts of high value. The markets for algal biofuel already exist, and are developing, but the markets' growth is constrained due high capital and operational costs, and also due to underdeveloped production technology. Thus, the cost of algal biofuel is increasing. Therefore, more research is to be conducted to improve the technologies used to convert biomass to biofuels, and also produce better harvesting technologies to make algal based biofuels more promising.

Author Contributions: Conceptualization, M.J.; software, M.U.; validation, G.M.; formal analysis, A.S.; investigation, S.P.; writing—original draft preparation, M.J.; and writing—review and editing, R.B.J. All authors have read and agreed to the published version of the manuscript.

Funding: This research received no external funding.

Institutional Review Board Statement: Not applicable.

Informed Consent Statement: Not applicable.

Conflicts of Interest: The authors declare no conflict of interest.

References

1. Steinman, A.D.; Vymazal, J. Algae and element cycling in wetlands. *J. N. Am. Benthol. Soc.* **1996**, *15*, 138–140. [CrossRef]
2. Larkum, A. *Photosynthesis in Algae*; Kluwer Academic Publisher: Dordrecht, The Netherlands, 2003.
3. Cock, J.M. *Introduction to Marine Genomics*; Springer: Berlin, Germany, 2010.
4. Lee, R.E. *Phycology*, 4th ed.; Cambridge University Press: Cambridge, UK, 2008.
5. Brennan, L.; Owende, P. Biofuels from microalgae—A review of technologies for production, processing, and extractions of biofuels and co-products. *Renew. Sustain. Energy Rev.* **2010**, *14*, 557–577. [CrossRef]

6. Feinberg, D.A.; Hill, A.M. Fuel from microalgae lipid products. *Algae* **1984**, *4*, 1–17.
7. Packer, M. Algal capture of carbon dioxide; biomass generation as a tool for greenhouse gas mitigation with reference to New Zealand energy strategy and policy. *Energy Policy* **2009**, *37*, 3428–3437. [CrossRef]
8. Ren, H.-Y.; Liu, B.-F.; Kong, F.; Zhao, L.; Ma, J.; Ren, N.-Q. Favorable energy conversion efficiency of coupling dark fermentation and microalgae production from food wastes. *Energy Convers. Manag.* **2018**, *166*, 156–162. [CrossRef]
9. Wendt, L.M.; Kinchin, C.; Wahlen, B.D.; Davis, R.; Dempster, T.A.; Gerken, H. Assessing the stability and techno-economic implications for wet storage of harvested microalgae to manage seasonal variability. *Biotechnol. Biofuels* **2019**, *12*, 80. [CrossRef] [PubMed]
10. Pritchard, H.N. Microalgae: Biotechnology and microbiology. *Q. Rev. Biol.* **1995**, *70*, 350–351. [CrossRef]
11. Christenson, L.; Sims, R. Production and harvesting of microalgae for wastewater treatment, biofuels, and bioproducts. *Biotechnol. Adv.* **2011**, *29*, 686–702. [CrossRef] [PubMed]
12. Pittman, J.K.; Dean, A.; Osundeko, O. The potential of sustainable algal biofuel production using wastewater resources. *Bioresour. Technol.* **2011**, *102*, 17–25. [CrossRef] [PubMed]
13. de Alva, M.S.; Luna-Pabello, V.M.; Cadena, E.; Ortíz, E. Green microalga Scenedesmusacutus grown on municipal wastewater to couple nutrient removal with lipid accumulation for biodiesel production. *Bioresour. Technol.* **2013**, *146*, 744–748. [CrossRef] [PubMed]
14. Prajapati, S.K.; Kaushik, P.; Malik, A.; Vijay, V.K. Phycoremediation coupled production of algal biomass, harvesting and anaerobic digestion: Possibilities and challenges. *Biotechnol. Adv.* **2013**, *31*, 1408–1425. [CrossRef] [PubMed]
15. Nayak, M.; Karemore, A.; Sen, R. Performance evaluation of microalgae for concomitant wastewater bioremediation, CO_2 biofixation and lipid biosynthesis for biodiesel application. *Algal Res.* **2016**, *16*, 216–223. [CrossRef]
16. Prandini, J.M.; da Silva, M.L.B.; Mezzari, M.P.; Pirolli, M.; Michelon, W.; Soares, H.M. Enhancement of nutrient removal from swine wastewater digestate coupled to biogas purification by microalgae *Scenedesmus* spp. *Bioresour. Technol.* **2016**, *202*, 67–75. [CrossRef] [PubMed]
17. Van Wagenen, J.; Pape, M.L.; Angelidaki, I. Characterization of nutrient removal and microalgal biomass production on an industrial waste-stream by application of the deceleration-stat technique. *Water Res.* **2015**, *75*, 301–311. [CrossRef] [PubMed]
18. Gao, F.; Li, C.; Yang, Z.-H.; Zeng, G.-M.; Feng, L.; Liu, J.-Z.; Liu, M.; Cai, H.-W. Continuous microalgae cultivation in aquaculture wastewater by a membrane photobioreactor for biomass production and nutrients removal. *Ecol. Eng.* **2016**, *92*, 55–61. [CrossRef]
19. Caporgno, M.; Taleb, A.; Olkiewicz, M.; Font, J.; Pruvost, J.; Legrand, J.; Bengoa, C. Microalgae cultivation in urban wastewater: Nutrient removal and biomass production for biodiesel and methane. *Algal Res.* **2015**, *10*, 232–239. [CrossRef]
20. Choudhary, P.; Prajapati, S.K.; Malik, A. Screening native microalgal consortia for biomass production and nutrient removal from rural wastewaters for bioenergy applications. *Ecol. Eng.* **2016**, *91*, 221–230. [CrossRef]
21. Chinnasamy, S.; Bhatnagar, A.; Hunt, R.W.; Das, K. Microalgae cultivation in a wastewater dominated by carpet mill effluents for biofuel applications. *Bioresour. Technol.* **2010**, *101*, 3097–3105. [CrossRef]
22. Wu, L.F.; Chen, P.C.; Huang, A.P.; Lee, C.M. The feasibility of biodiesel production by microalgae using industrial wastewater. *Bioresour. Technol.* **2012**, *113*, 14–18. [CrossRef]
23. Wang, L.; Min, M.; Li, Y.; Chen, P.; Chen, Y.; Liu, Y.; Wang, Y.; Ruan, R. Cultivation of green algae *Chlorella* sp. in different wastewaters from municipal wastewater treatment plant. *Appl. Biochem. Biotechnol.* **2009**, *162*, 1174–1186. [CrossRef]
24. Gao, F.; Li, C.; Yang, Z.-H.; Zeng, G.-M.; Mu, J.; Liu, M.; Cui, W. Removal of nutrients, organic matter, and metal from domestic secondary effluent through microalgae cultivation in a membrane photobioreactor. *J. Chem. Technol. Biotechnol.* **2016**, *91*, 2713–2719. [CrossRef]
25. Prajapati, S.K.; Choudhary, P.; Malik, A.; Vijay, V.K. Algae mediated treatment and bioenergy generation process for handling liquid and solid waste from dairy cattle farm. *Bioresour. Technol.* **2014**, *167*, 260–268. [CrossRef]
26. Mobin, S.; Alam, F. Biofuel production from algae utilizing wastewater. In Proceedings of the 19th Australasian Fluid Mechanics Conference, Melbourne, Australia, 8–11 December 2014.
27. Kiran, B.; Pathak, K.; Kumar, R.; Deshmukh, D. Cultivation of *Chlorella* sp. IM-01 in municipal wastewater for simultaneous nutrient removal and energy feedstock production. *Ecol. Eng.* **2014**, *73*, 326–330. [CrossRef]
28. Escapa, C.; Coimbra, R.; Paniagua, S.; García, A.; Otero, M. Nutrients and pharmaceuticals removal from wastewater by culture and harvesting of *Chlorella sorokiniana*. *Bioresour. Technol.* **2015**, *185*, 276–284. [CrossRef]
29. Khanzada, Z.T. Phosphorus removal from landfill leachate by microalgae. *Biotechnol. Rep.* **2020**, *25*, 419. [CrossRef]
30. Kumar, G.; Huy, M.; Bakonyi, P.; Bélafi-Bakó, K.; Kim, S.-H. Evaluation of gradual adaptation of mixed microalgae consortia cultivation using textile wastewater via fed batch operation. *Biotechnol. Rep.* **2018**, *20*, 289. [CrossRef]
31. Mar, C.C.; Fan, Y.; Li, F.-L.; Hu, G.-R. Bioremediation of wastewater from edible oil refinery factory using oleaginous microalga *Desmodesmus* sp. S1. *Int. J. Phytoremediat.* **2016**, *18*, 1195–1201. [CrossRef]
32. Choi, H.-J. Dairy wastewater treatment using microalgae for potential biodiesel application. *Environ. Eng. Res.* **2016**, *21*, 393–400. [CrossRef]
33. Green, F.B.; Bernstone, L.S.; Lundquist, T.J.; Oswald, W.J. Advanced integrated wastewater pond systems for nitrogen removal. *Water Sci. Technol.* **1996**, *33*. [CrossRef]
34. Spolaore, P.; Joannis-Cassan, C.; Duran, E.; Isambert, A. Commercial applications of microalgae. *J. Biosci. Bioeng.* **2006**, *101*, 87–96. [CrossRef]

35. Sialve, B.; Bernet, N.; Bernard, O. Anaerobic digestion of microalgae as a necessary step to make microalgal biodiesel sustainable. *Biotechnol. Adv.* **2009**, *27*, 409–416. [CrossRef] [PubMed]
36. Kapdan, I.K.; Kargi, F. Bio-hydrogen production from waste materials. *Enzym. Microb. Technol.* **2006**, *38*, 569–582. [CrossRef]
37. Fedorov, A.; Kosourov, S.; Ghirardi, M.L.; Seibert, M. Continuous hydrogen photoproduction by *Chlamydomonas reinhardtii*: Using a novel two-stage, sulfate-limited chemostat system. *Appl. Biochem. Biotechnol.* **2005**, *121*, 0403–0412. [CrossRef]
38. Dexter, J.; Fu, P. Metabolic engineering of cyanobacteria for ethanol production. *Energy Environ. Sci.* **2009**, *2*, 857–864. [CrossRef]
39. Choi, S.P.; Nguyen, M.-T.; Sim, S.J. Enzymatic pretreatment of *Chlamydomonas reinhardtii* biomass for ethanol production. *Bioresour. Technol.* **2010**, *101*, 5330–5336. [CrossRef] [PubMed]
40. Xu, H.; Miao, X.; Wu, Q. High quality biodiesel production from a microalga *Chlorella protothecoides* by heterotrophic growth in fermenters. *J. Biotechnol.* **2006**, *126*, 499–507. [CrossRef]
41. Banerjee, A.; Sharma, R.; Chisti, Y.; Banerjee, U.C. *Botryococcus braunii*: A renewable source of hydrocarbons and other chemicals. *Crit. Rev. Biotechnol.* **2002**, *22*, 245–279. [CrossRef] [PubMed]
42. Miao, X.; Wu, Q.; Yang, C. Fast pyrolysis of microalgae to produce renewable fuels. *J. Anal. Appl. Pyrolysis* **2004**, *71*, 855–863. [CrossRef]
43. Sawayama, S.; Inoue, S.; Dote, Y.; Yokoyama, S.-Y. CO_2 fixation and oil production through microalga. *Energy Convers. Manag.* **1995**, *36*, 729–731. [CrossRef]
44. Nagle, N.; Lemke, P. Production of methyl ester fuel from microalgae. *Appl. Biochem. Biotechnol.* **1990**, *24–25*, 355–361. [CrossRef]
45. Koberg, M.; Cohen, M.; Ben-Amotz, A.; Gedanken, A. Bio-diesel production directly from the microalgae biomass of Nannochloropsis by microwave and ultrasound radiation. *Bioresour. Technol.* **2011**, *102*, 4265–4269. [CrossRef] [PubMed]
46. Chisti, Y. Biodiesel from microalgae. *Biotechnol. Adv.* **2007**, *25*, 294–306. [CrossRef] [PubMed]
47. Johnson, M.B.; Wen, Z. Production of biodiesel fuel from the microalga *Schizochytrium limacinum* by direct transesterification of algal biomass. *Energy Fuels* **2009**, *23*, 5179–5183. [CrossRef]
48. Trindade, S.C. Nanotech biofuels and fuel additives. In *Biofuel's Engineering Process Technology*; IntechOpen: London, UK, 2011.
49. Wahlen, B.; Willis, R.M.; Seefeldt, L. Biodiesel production by simultaneous extraction and conversion of total lipids from microalgae, cyanobacteria, and wild mixed-cultures. *Bioresour. Technol.* **2011**, *102*, 2724–2730. [CrossRef]
50. AshokKumar, V.; Agila, E.; Sivakumar, P.; Salam, Z.; Rengasamy, R.; Ani, F.N. Optimization and characterization of biodiesel production from microalgae *Botryococcus* grown at semi-continuous system. *Energy Convers. Manag.* **2014**, *88*, 936–946. [CrossRef]
51. Lewis, T.; Nichols, P.D.; McMeekin, T.A. Evaluation of extraction methods for recovery of fatty acids from lipid-producing microheterotrophs. *J. Microbiol. Methods* **2000**, *43*, 107–116. [CrossRef]
52. Danielo, O. *An Algae-Based Fuel*; Biofuture: Venelles, France, 2005.
53. Metzger, P.; Largeau, C. *Botryococcus braunii*: A rich source for hydrocarbons and related ether lipids. *Appl. Microbiol. Biotechnol.* **2004**, *66*, 486–496. [CrossRef]
54. Knothe, G. *The Biodiesel Handbook*; AOCS Press: Urbana, IL, USA, 2005.
55. Kumar, M.; Sharma, M.P. Selection of potential oils for biodiesel production. *Renew. Sustain. Energy Rev.* **2016**, *56*, 1129–1138. [CrossRef]
56. Ribeiro, N.M.; Pinto, A.C.; Quintella, C.; da Rocha, G.O.; Teixeira, L.; Guarieiro, L.L.N.; Rangel, M.D.C.; Veloso, M.C.C.; Rezende, M.J.C.; da Cruz, R.S.; et al. The role of additives for diesel and diesel blended (ethanol or biodiesel) fuels: A review. *Energy Fuels* **2007**, *21*, 2433–2445. [CrossRef]
57. Bhale, P.V.; Deshpande, N.; Thombre, S.B. Improving the low temperature properties of biodiesel fuel. *Renew. Energy* **2009**, *34*, 794–800. [CrossRef]
58. Zabed, H.M.; Boyce, A.N.; Sahu, J.N.; Faruq, G. Evaluation of the quality of dried distiller's grains with solubles for normal and high sugary corn genotypes during dry–grind ethanol production. *J. Clean. Prod.* **2017**, *142*, 4282–4293. [CrossRef]
59. Ayala-Parra, P.; Liu, Y.; Field, J.; Sierra-Alvarez, R. Nutrient recovery and biogas generation from the anaerobic digestion of waste biomass from algal biofuel production. *Renew. Energy* **2017**, *108*, 410–416. [CrossRef]
60. Zamalloa, C.; Boon, N.; Verstraete, W. Anaerobic digestibility of *Scenedesmus obliquus* and *Phaeodactylum tricornutum* under mesophilic and thermophilic conditions. *Appl. Energy* **2012**, *92*, 733–738. [CrossRef]
61. Saratale, R.G.; Kumar, G.; Banu, R.; Xia, A.; Periyasamy, S.; Saratale, G.D. A critical review on anaerobic digestion of microalgae and macroalgae and co-digestion of biomass for enhanced methane generation. *Bioresour. Technol.* **2018**, *262*, 319–332. [CrossRef]
62. Ward, A.; Lewis, D.; Green, F. Anaerobic digestion of algae biomass: A review. *Algal Res.* **2014**, *5*, 204–214. [CrossRef]
63. Park, J.; Craggs, R.; Shilton, A. Wastewater treatment high rate algal ponds for biofuel production. *Bioresour. Technol.* **2011**, *102*, 35–42. [CrossRef] [PubMed]
64. Sivakumar, G.; Xu, J.; Thompson, R.W.; Yang, Y.; Randol-Smith, P.; Weathers, P.J. Integrated green algal technology for bioremediation and biofuel. *Bioresour. Technol.* **2012**, *107*, 1–9. [CrossRef] [PubMed]
65. Xu, J.; Zhao, Y.; Zhao, G.; Zhang, H. Nutrient removal and biogas upgrading by integrating freshwater algae cultivation with piggery anaerobic digestate liquid treatment. *Appl. Microbiol. Biotechnol.* **2015**, *99*, 6493–6501. [CrossRef]
66. Ehimen, E.; Sun, Z.; Carrington, C.; Birch, E.; Eaton-Rye, J. Anaerobic digestion of microalgae residues resulting from the biodiesel production process. *Appl. Energy* **2011**, *88*, 3454–3463. [CrossRef]
67. Sforza, E.; Barbera, E.; Girotto, F.; Cossu, R.; Bertucco, A. Anaerobic digestion of lipid-extracted microalgae: Enhancing nutrient recovery towards a closed loop recycling. *Biochem. Eng. J.* **2017**, *121*, 139–146. [CrossRef]

68. Gonzalez, L.M.; Correa, D.F.; Ryan, S.; Jensen, P.D.; Pratt, S.; Schenk, P.M. Integrated biodiesel and biogas production from microalgae: Towards a sustainable closed loop through nutrient recycling. *Renew. Sustain. Energy Rev.* **2018**, *82*, 1137–1148. [CrossRef]
69. Golueke, C.G.; Oswald, W.J.; Gotaas, H.B. Anaerobic digestion of algae. *Appl. Microbiol.* **1957**, *5*, 47–55. [CrossRef]
70. Klassen, V.; Blifernez-Klassen, O.; Wobbe, L.; Schlüter, A.; Kruse, O.; Mussgnug, J.H. Efficiency and biotechnological aspects of biogas production from microalgal substrates. *J. Biotechnol.* **2016**, *234*, 7–26. [CrossRef]
71. Klassen, V.; Blifernez-Klassen, O.; Hoekzema, Y.; Mussgnug, J.H.; Kruse, O. A novel one-stage cultivation/fermentation strategy for improved biogas production with microalgal biomass. *J. Biotechnol.* **2015**, *215*, 44–51. [CrossRef] [PubMed]
72. Xu, Y.; Huang, Y.; Wu, B.; Zhang, X.; Zhang, S. Biogas upgrading technologies: Energetic analysis and environmental impact assessment. *Chin. J. Chem. Eng.* **2015**, *23*, 247–254. [CrossRef]
73. Krischan, J.; Makaruk, A.; Harasek, M. Design and scale-up of an oxidative scrubbing process for the selective removal of hydrogen sulfide from biogas. *J. Hazard. Mater.* **2012**, *215–216*, 49–56. [CrossRef] [PubMed]
74. Wang, J.; Wan, W. Factors influencing fermentative hydrogen production: A review. *Int. J. Hydrogen Energy* **2009**, *34*, 799–811. [CrossRef]
75. Argun, H.; Kargi, F.; Kapdan, I.K.; Oztekin, R. Biohydrogen production by dark fermentation of wheat powder solution: Effects of C/N and C/P ratio on hydrogen yield and formation rate. *Int. J. Hydrogen Energy* **2008**, *33*, 1813–1819. [CrossRef]
76. Ferreira, A.F.; Marques, A.C.; Batista, A.P.; Marques, P.A.; Gouveia, L.; Silva, C. Biological hydrogen production by *Anabaena* sp.—Yield, energy and CO_2 analysis including fermentative biomass recovery. *Int. J. Hydrogen Energy* **2012**, *37*, 179–190. [CrossRef]
77. Ruiz-Marin, A.; Canedo-López, Y.; Chávez-Fuentes, P. Biohydrogen production by *Chlorella vulgaris* and *Scenedesmus obliquus* immobilized cultivated in artificial wastewater under different light quality. *AMB Express* **2020**, *10*, 1–7. [CrossRef] [PubMed]
78. Nobre, B.; Villalobos, F.; Barragán-Huerta, B.E.; Oliveira, A.; Batista, A.P.; Marques, P.; Mendes, R.; Sovova, H.; Palavra, A.; Gouveia, L. A biorefinery from *Nannochloropsis* sp. microalga—Extraction of oils and pigments. Production of biohydrogen from the leftover biomass. *Bioresour. Technol.* **2013**, *135*, 128–136. [CrossRef] [PubMed]
79. Batyrova, K.; Hallenbeck, P.C. Hydrogen production by a *Chlamydomonas reinhardtii* strain with inducible expression of photosystem II. *Int. J. Mol. Sci.* **2017**, *18*, 647. [CrossRef] [PubMed]
80. Juantorena, A.; Sebastian, P.; Santoyo, E.; Gamboa, S.; Lastres, O.; Sánchez-Escamilla, D.; Bustos, A.; Eapen, D. Hydrogen production employing *Spirulina maxima* 2342: A chemical analysis. *Int. J. Hydrogen Energy* **2007**, *32*, 3133–3136. [CrossRef]
81. Batista, A.P.; Moura, P.; Marques, P.A.; Ortigueira, J.; Alves, L.; Gouveia, L. *Scenedesmus obliquus* as feedstock for biohydrogen production by *Enterobacter aerogenes* and *Clostridium butyricum*. *Fuel* **2014**, *117*, 537–543. [CrossRef]
82. Wieczorek, N.; Kucuker, M.A.; Kuchta, K. Fermentative hydrogen and methane production from microalgal biomass (*Chlorella vulgaris*) in a two-stage combined process. *Appl. Energy* **2014**, *132*, 108–117. [CrossRef]
83. Yang, Z.; Guo, R.; Xu, X.; Fan, X.; Li, X. Enhanced hydrogen production from lipid-extracted microalgal biomass residues through pretreatment. *Int. J. Hydrogen Energy* **2010**, *35*, 9618–9623. [CrossRef]
84. Vijayaraghavan, K.; Karthik, R.; Nal, S.K. Hydrogen generation from algae: A review. *J. Plant Sci.* **2009**, *5*, 1–19. [CrossRef]
85. Rupprecht, J.; Hankamer, B.; Mussgnug, J.H.; Ananyev, G.; Dismukes, C.; Kruse, O. Perspectives and advances of biological H_2 production in microorganisms. *Appl. Microbiol. Biotechnol.* **2006**, *72*, 442–449. [CrossRef]
86. Kosourov, S.; Patrusheva, E.; Ghirardi, M.L.; Seibert, M.; Tsygankov, A. A comparison of hydrogen photoproduction by sulfur-deprived *Chlamydomonas reinhardtii* under different growth conditions. *J. Biotechnol.* **2007**, *128*, 776–787. [CrossRef]
87. Gaffron, H.; Rubin, J. Fermentative and photochemical production of hydrogen in algae. *J. Gen. Physiol.* **1942**, *26*, 219–240. [CrossRef] [PubMed]
88. Kessler, E. Hydrogenase, photoreduction, and anaerobic growth. In *Algal Physiology and Biochemistry*; University of California Press: Berkeley, CA, USA, 1974.
89. Hallenbeck, P. Fundamentals of the fermentative production of hydrogen. *Water Sci. Technol.* **2005**, *52*, 21–29. [CrossRef] [PubMed]
90. Tam, N.; Wong, Y. Effect of immobilized microalgal bead concentrations on wastewater nutrient removal. *Environ. Pollut.* **2000**, *107*, 145–151. [CrossRef]
91. Azwar, M.; Hussain, M.A.; Abdul-Wahab, A. Development of biohydrogen production by photobiological, fermentation and electrochemical processes: A review. *Renew. Sustain. Energy Rev.* **2014**, *31*, 158–173. [CrossRef]
92. Rashid, N.; Lee, K.; Han, J.-I.; Gross, M. Hydrogen production by immobilized *Chlorella vulgaris*: Optimizing pH, carbon source and light. *Bioprocess Biosyst. Eng.* **2012**, *36*, 867–872. [CrossRef]
93. Razzak, S.; Hossain, M.M.; Lucky, R.A.; Bassi, A.S.; de Lasa, H. Integrated CO_2 capture, wastewater treatment and biofuel production by microalgae culturing—A review. *Renew. Sustain. Energy Rev.* **2013**, *27*, 622–653. [CrossRef]
94. Klein, U.; Betz, A. Fermentative metabolism of hydrogen-evolving *Chlamydomonas moewusii*. *Plant Physiol.* **1978**, *61*, 953–956. [CrossRef] [PubMed]
95. Jurado-Oller, J.L.; Dubini, A.; Galván, A.; Fernández, E.; González-Ballester, D. Low oxygen levels contribute to improve photohydrogen production in mixotrophic non-stressed *Chlamydomonas* cultures. *Biotechnol. Biofuels* **2015**, *8*, 1–14. [CrossRef] [PubMed]
96. Dale, B.E. Thinking clearly about biofuels: Ending the irrelevant 'net energy' debate and developing better performance metrics for alternative fuels. *Biofuels Bioprod. Biorefin.* **2007**, *1*, 14–17. [CrossRef]

97. Demirbaş, A. Conversion of biomass using glycerin to liquid fuel for blending gasoline as alternative engine fuel. *Energy Convers. Manag.* **2000**, *41*, 1741–1748. [CrossRef]
98. Naik, S.; Goud, V.V.; Rout, P.K.; Dalai, A.K. Production of first and second generation biofuels: A comprehensive review. *Renew. Sustain. Energy Rev.* **2010**, *14*, 578–597. [CrossRef]
99. Harun, R.; Singh, M.; Forde, G.M.; Danquah, M.K. Bioprocess engineering of microalgae to produce a variety of consumer products. *Renew. Sustain. Energy Rev.* **2010**, *14*, 1037–1047. [CrossRef]
100. Radakovits, R.; Jinkerson, R.; Darzins, A.; Posewitz, M.C. Genetic engineering of algae for enhanced biofuel production. *Eukaryot. Cell* **2010**, *9*, 486–501. [CrossRef]
101. González-Fernández, C.; Ballesteros, M. Linking microalgae and cyanobacteria culture conditions and key-enzymes for carbohydrate accumulation. *Biotechnol. Adv.* **2012**, *30*, 1655–1661. [CrossRef]
102. Ueno, Y.; Kurano, N.; Miyachi, S. Ethanol production by dark fermentation in the marine green alga, *Chlorococcum littorale*. *J. Ferment. Bioeng.* **1998**, *86*, 38–43. [CrossRef]
103. Hernández, D.; Riaño, B.; Coca, M.; González, M.C.G. Saccharification of carbohydrates in microalgal biomass by physical, chemical and enzymatic pre-treatments as a previous step for bioethanol production. *Chem. Eng. J.* **2015**, *262*, 939–945. [CrossRef]
104. Magneschi, L.; Catalanotti, C.; Subramanian, V.; Dubini, A.; Yang, W.; Mus, F.; Posewitz, M.C.; Seibert, M.; Perata, P.; Grossman, A.R. A mutant in the ADH1 gene of *Chlamydomonas reinhardtii* elicits metabolic restructuring during anaerobiosis. *Plant Physiol.* **2012**, *158*, 1293–1305. [CrossRef]
105. Dexter, J.; Armshaw, P.; Sheahan, C.; Pembroke, J.T. The state of autotrophic ethanol production in cyanobacteria. *J. Appl. Microbiol.* **2015**, *119*, 11–24. [CrossRef] [PubMed]
106. Saad, M.G.; Dosoky, N.S.; Zoromba, M.S.; Shafik, H.M. Algal Biofuels: Current Status and Key Challenges. *Energies* **2019**, *12*, 1920. [CrossRef]
107. Deng, M.-D.; Coleman, J.R. Ethanol synthesis by genetic engineering in cyanobacteria. *Appl. Environ. Microbiol.* **1999**, *65*, 523–528. [CrossRef]
108. Piven, I.; Friedrich, A.; Dühring, U.; Uliczka, F.; Baier, K.; Inaba, M. *Cyanobacterium* sp. for Production of Compounds. U.S. Patent US20140178958A1, 26 June 2012.
109. Hellingwerf, K.; De Mattos, M.T. Alternative routes to biofuels: Light-driven biofuel formation from CO_2 and water based on the 'photanol' approach. *J. Biotechnol.* **2009**, *142*, 87–90. [CrossRef]
110. Gimpel, J.; Specht, E.A.; Georgianna, D.R.; Mayfield, S.P. Advances in microalgae engineering and synthetic biology applications for biofuel production. *Curr. Opin. Chem. Biol.* **2013**, *17*, 489–495. [CrossRef]
111. Sforza, E.; Gris, B.; De Farias Silva, C.E.; Morosinotto, T.; Bertucco, A. Effects of light on cultivation of scenedesmus obliquus in batch and continuous flat plate photobioreactor. *Chem. Eng. Trans.* **2014**, *38*, 211–216. [CrossRef]
112. Peccia, J.; Haznedaroglu, B.Z.; Gutierrez, J.; Zimmerman, J.B. Nitrogen supply is an important driver of sustainable microalgae biofuel production. *Trends Biotechnol.* **2013**, *31*, 134–138. [CrossRef] [PubMed]
113. Kuhad, R.C.; Gupta, R.; Khasa, Y.P.; Singh, A.; Zhang, Y.-H.P. Bioethanol production from pentose sugars: Current status and future prospects. *Renew. Sustain. Energy Rev.* **2011**, *15*, 4950–4962. [CrossRef]

Article

Seafood Processing Chitin Waste for Electricity Generation in a Microbial Fuel Cell Using Halotolerant Catalyst *Oceanisphaera arctica* YHY1

Ranjit Gurav [1], Shashi Kant Bhatia [1], Tae-Rim Choi [1], Hyun-Joong Kim [1], Hong-Ju Lee [1], Jang-Yeon Cho [1], Sion Ham [1], Min-Ju Suh [1], Sang-Hyun Kim [1], Sun-Ki Kim [2], Dong-Won Yoo [3] and Yung-Hun Yang [1,*]

[1] Department of Biological Engineering, College of Engineering, Konkuk University, Seoul 05029, Korea; rnjtgurav@gmail.com (R.G.); shashibiotechhpu@gmail.com (S.K.B.); srim1004@gmail.com (T.-R.C.); sopero2@naver.com (H.-J.K.); oneul1210@naver.com (H.-J.L.); whwkddus1123@gmail.com (J.-Y.C.); sion1403@naver.com (S.H.); semn0702@konkuk.ac.kr (M.-J.S.); tkdgus5257@nate.com (S.-H.K.)
[2] Department of Food Science and Technology, Chung-Ang University, Anseong 17546, Korea; skkim18@cau.ac.kr
[3] School of Chemical and Biological Engineering, Seoul National University, Seoul 08826, Korea; dwyoo@snu.ac.kr
* Correspondence: seokor@konkuk.ac.kr

Abstract: In this study, a newly isolated halotolerant strain *Oceanisphaera arctica* YHY1, capable of hydrolyzing seafood processing waste chitin biomass, is reported. Microbial fuel cells fed with 1% chitin and 40 g L^{-1} as the optimum salt concentration demonstrated stable electricity generation until 216 h (0.228 mA/cm^2). N-acetyl-D-glucosamine (GlcNAc) was the main by-product in the chitin degradation, reaching a maximum concentration of 192.01 mg g^{-1} chitin at 120 h, whereas lactate, acetate, propionate, and butyrate were the major metabolites detected in the chitin degradation. *O. arctica* YHY1 utilized the produced GlcNAc, lactate, acetate, and propionate as the electron donors to generate the electric current. Cyclic voltammetry (CV) investigation revealed the participation of outer membrane-bound cytochromes, with extracellular redox mediators partly involved in the electron transfer mechanism. Furthermore, the changes in structural and functional groups in chitin after degradation were analyzed using FTIR and XRD. Therefore, the ability of *O. arctica* YHY1 to utilize waste chitin biomass under high salinities can be explored to treat seafood processing brine or high salt wastewater containing chitin with concurrent electricity generation.

Keywords: chitin; electricity generation; halotolerant; microbial fuel cell; seafood processing

Citation: Gurav, R.; Bhatia, S.K.; Choi, T.-R.; Kim, H.-J.; Lee, H.-J.; Cho, J.-Y.; Ham, S.; Suh, M.-J.; Kim, S.-H.; Kim, S.-K.; et al. Seafood Processing Chitin Waste for Electricity Generation in a Microbial Fuel Cell Using Halotolerant Catalyst *Oceanisphaera arctica* YHY1. *Sustainability* **2021**, *13*, 8508. https://doi.org/10.3390/su13158508

Academic Editor: Dino Musmarra

Received: 15 June 2021
Accepted: 26 July 2021
Published: 29 July 2021

1. Introduction

The seafood processing industry produces a substantial amount of wastewater, mainly containing soluble, colloidal, and particulate matter. Among crustaceans, the total production of shrimp reached 5.03 million tons in 2020 and is estimated to increase to 7.28 million tons by 2025, with a compound annual growth rate of 6.1% from 2020 to 2025 and 67.6 billion USD turnover [1]. Asia alone contributes more than 80% of the global shrimp production, where Thailand is the major exporter of cultivated shrimp to the USA, Canada, Europe, South Korea, and Japan [2]. Depending on the market requirements, shrimp is exported or stored in frozen conditions with or without an outer shell. For shrimp processing, a huge amount of water is required, generating around 1000 L of highly polluted wastewater per ton of shrimp [1,3]. Shrimp processing produces 50–60% of solid waste comprising the head, viscera, and shell, which are discarded as the by-products generated in processing. The biochemical composition of shrimp waste mainly contains 15–46% chitin, 30–60% minerals, 10–40% protein, and 10–40% lipids [1,4]. Similarly, the salinity of the seafood processing wastewater is another important factor that mainly depends on the products or species being processed. The precooking or brine treatment for the canning of shrimp generates wastewater, with the NaCl concentration ranging between 20–30 g L^{-1} [5].

Chitin is the second most abundant natural biopolymer, consisting of monomeric units of GlcNAc linked with (1,4)-β-linkages [6]. The microbial chitinase (EC 3.2.2.14) is a glycoside hydrolytic enzyme capable of hydrolyzing chitin into GlcNAc or oligomers and utilizing it as a source of carbon for growth and development [7]. Chitin waste has been partly used as animal feed, as a component of aquaculture feed formulation, and for the recovery of bioactive molecules. However, a large amount of chitin biomass is being wasted, which increases environmental pollution [1]. However, particulate substrates like chitin can be inexpensive, easily available, and renewable feedstock in microbial fuel cells (MFCs) for electricity production [8]. The MFC is a bio-electrochemical system that can break and utilize chitin more effectively compared to normal fermentation conditions, owing to the anode working as an electron acceptor [7,8]. Several natural and synthetic substrates, including acetate, glucose, lactate, amino acids, butyrate, formate, fumarate, alcohols, or complex carbohydrates like cellulose, sucrose, molasses, starch, or industrial and domestic wastewater, were widely researched as fuels for MFCs [8,9]. However, high production costs and easy depletion have constrained the use of synthetic substrates in MFCs. Nevertheless, chitin waste can be an alternative synthetic substrate source for sustainable energy production by using bacteria as a catalyst to oxidize organic substrates directly into electrical energy. Previously, chitin has been utilized as a substrate by *Bacillus circulans*, *Arenibacter palladensis*, *Shewanella oneidensis*, *Aeromonas hydrophila*, or as sediment wastewater-based systems for electricity production [7,10].

Therefore, in the present study, a newly isolated halotolerant marine bacterium *O. arctica* YHY1 capable of hydrolyzing chitin waste was studied as a biocatalyst for electricity generation in MFCs. Furthermore, the electrochemical parameters, the structural changes in chitin before and after degradation, and the by-products or metabolites of degradation were analyzed.

2. Materials and Methods

2.1. Chitin Preparation, Isolation, and Identification of Chitin Degrading Microbes

The shrimp shell chitin was obtained from the local seafood processing unit near Seoul, South Korea. The obtained chitin biomass was washed, dried, and pre-treated according to the method reported earlier by Gurav et al. [10]. The resulting colloidal chitin was dried and stored in a refrigerator for further applications. Isolation of chitinolytic microbes was performed using marine soil collected from a beach near Geoje (34°51′16.5″ N 128°43′43.9″ E), Eastern Sea of South Korea. In brief, one gram of soil was serially diluted and smeared on the chitin agar plates containing (g L^{-1}) 0.7-KH_2PO_4, 0.3-K_2HPO_4, 5.0-NaCl, 0.5-$MgSO_4$, 0.0001-$ZnSO_4$, 0.0001-$MnSO_4$, 10-chitin, and 25-agar [6,10]. Seven morphologically different strains showing clear zones were isolated and cultured as a monoculture on chitin-agar plates. These seven strains were initially tested for electric current production in MFCs supplemented with 1% chitin. Based on its higher chitin degradation and electricity generation ability, strain YHY1 was selected for identification by using 16S rRNA gene sequencing.

2.2. MFC Setup and Electrochemical Analysis

As shown in Figure 1, a dual-chamber MFC consisting of cathodic and anodic chambers separated by a proton exchange membrane (Nafion 212, Omniscience, Yongin, Korea) was assembled. The anodic chamber was equipped with 2.25 cm^2 carbon felt and a silver reference electrode (Ag/AgCl), whereas the cathodic chamber contained 4 cm^2 platinum-coated carbon felt. Degassed growth media containing (g L^{-1}) 0.7-KH_2PO_4, 0.3-K_2HPO_4, 40-NaCl, 0.5-$MgSO_4$, 0.0001-$ZnSO_4$, 0.0001-$MnSO_4$, and 10-chitin was inoculated with 1% *v/v* inoculum with an optical density of 0.9 ± 0.05 at the anodic chamber, whereas 50 mL phosphate buffer (pH 7.0; 50 mM) containing 50 mM ferricyanide was filled at the cathode chamber. The anode and cathode were connected through a potentiostat (WizECM-8100 premium, Wizmac, Daejeon, Korea) and operated as a closed-circuit using an external resistance of 1000 Ω [7,8,11,12]. The current output density (mA/cm^2) of the system was recorded and plotted versus time (h). To investigate the electrocatalytic behavior and interaction between the redox mediator, electrode, and anodic biofilm, the CV was performed using a three-electrode system including working, counter, and reference electrodes with a 10 mV/S scan rate and a +1 V to −1 V potential range.

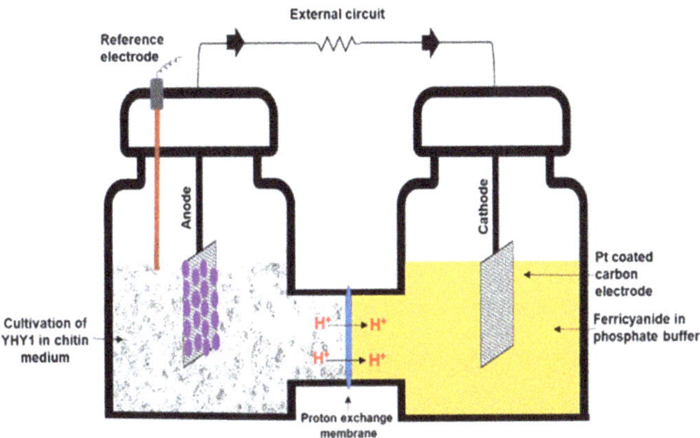

Figure 1. Setup of the dual-chamber MFC.

2.3. *Analysis of Degradation Products*

HPLC (Young Lin, YL-9100, Seoul, Korea) investigation was performed to quantify free GlcNAc in the degradation media using a C18 column (ZORBAX, SBC18) and acetonitrile: water (20:80) as the mobile phase [7,10]. Further, the metabolic profiling of the chitin degradation media was performed using HPLC (Bio-Rad, Hercules, CA, USA) equipped with Bio-Rad Aminex HPX-87H column and 5 mM H_2SO_4 as the solvent phase [13].

2.4. *SEM, XRD and FTIR Analysis*

SEM (Hitachi TM4000Plus, Tokyo, Japan) was performed to investigate the anodic biofilm [9]. The FTIR (Nicolet 6700, Thermo Fisher Scientific, Waltham, MA, USA) was executed to verify the structural changes in the chitin before and after degradation in MFCs [7,10], whereas XRD (D8 ADVANCE-DAVINCI, Bruker, Bremen, Germany) was examined to determine changes in the crystallinity of chitin. The crystalline index (CrI, %) was calculated from the XRD data using the following equations [10,14].

$$CrI_{020} = [(I_{020} - Iam)/I_{020} \times 100] \quad (1)$$

where I_{020} and I_{am} is the maximum intensity at 2θ ≅ 9° and the intensity of amorphous diffraction at 2θ ≅ 16°, respectively.

$$CrI_{110} = [(I_{110} - I_{am})/I_{110} \times 100] \quad (2)$$

where I_{110} is the maximum intensity at 2θ ≅ 20°.

3. Results

3.1. *Identification, and Salt-Tolerance in the Chitinolytic Bacterium*

Seven chitin degrading strains were newly isolated from the marine soil, where strain YHY1 showed maximum chitin hydrolysis activity with a zone diameter of 20 mm on chitin agar plates. The 16S rRNA sequencing data of strain YHY1 revealed 96% similarity with *Oceanisphaera arctica* strain V1-41; therefore, strain YHY1 was designated as *Oceanisphaera arctica* YHY1. The phylogenetic position of strain YHY1 and other allied strains is depicted in Figure 2a. The obtained gene sequence (989 bp) was submitted to Genbank under accession no. MH590704. Furthermore, the salt tolerance in strain YHY1 showed better growth on increasing the salt concentration from 5 g L^{-1} to 40 g L^{-1}, suggesting the necessity of salt for growth (Figure 2b). This strain was able to tolerate up to 80 g L^{-1} of salt concentration; however, concentrations above 80 g L^{-1} significantly inhibited the growth of the bacterium.

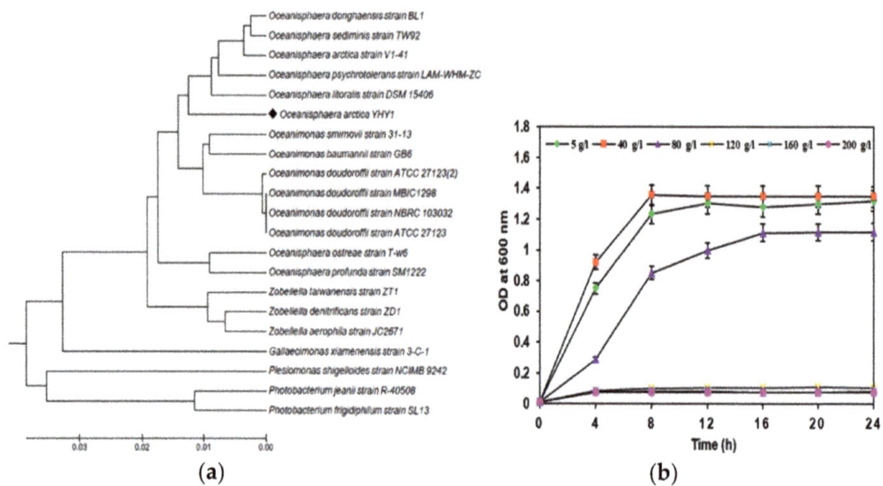

Figure 2. (**a**) Phylogenetic position of *O. arctica* YHY1; (**b**) Salt tolerance in strain YHY1.

3.2. Electrochemical Assessment of MFCs Fueled with Chitin Waste

Depleting non-renewable energy resources has enabled researchers to discover new sustainable energy assets. In MFCs, microbes act as the biocatalysts that transform the chemical energy of organic residues into electricity [10,15]. In this study, halotolerant *O. arctica* YHY1 was investigated for electricity production using seafood processing waste chitin as the carbon source. As shown in Figure 3a, the performance of MFCs fueled with 1% chitin biomass was studied, with a fixed external resistance of 1000 Ω and 40 g L^{-1} NaCl. After inoculating the MFCs with *O. arctica* YHY1, a rapid increase in electricity generation was observed in the MFCs, with a maximum current output density reaching 0.302 mA/cm^2 at 12 h. Higher current production in the initial period could be related to the free GlcNAc (12.311 mg g^{-1} chitin) that was produced during the pre-treatment of the chitin. A similar mechanism has been reported earlier in *Aeromonas hydrophila*, *Bacillus circulans*, and *Shewanella oneidensis* MFCs using chitin as the carbon source [7,11,12]. However, the current output density slightly dropped to 0.210 mA/cm^2 after 20 h; this might be due to depletion in free GlcNAc. Thereafter, the electricity generation was enhanced and remained almost constant until 216 h (0.228 mA/cm^2). Previously, chitin or pure GlcNAc has been reported as an electron donor in electricity production in MFCs using various microbial catalysts. For instance, *Aeromonas hydrophila* and *Shewanella oneidensis* fed with 0.2% chitin produced 8.77 μA/cm^2 and 4.24 μA/cm^2, whereas using pure GlcNAc produced 6.65 μA/cm^2 and 6.17 μA/cm^2 of current output density, respectively [11,12]. Similarly, *Arenibacter palladensis* and *Bacillus circulans* produced 15.15 μA/cm^2 and 26.508 μA/cm^2 current output density, respectively, in MFCs fed with 1% chitin [7,10]. However, in the present study, a stable and higher current output density of 0.228 mA/cm^2 (216 h) was observed in *O. arctica* YHY1 using 1% chitin and 40 g L^{-1} salt concentration. Salinity played an important role in the MFCs in the present study. The MFCs were supplied with 40 g L^{-1} NaCl as the optimized salt concentration for *O. arctica* YHY1. The salinity of the media has a positive impact on electricity generation as it decreases the internal resistance of the system and increases media conductivity. Furthermore, higher salinity can also prevent the acidification of the media due to the fast transfer of H^+ ions from the anode to the cathode [9,16].

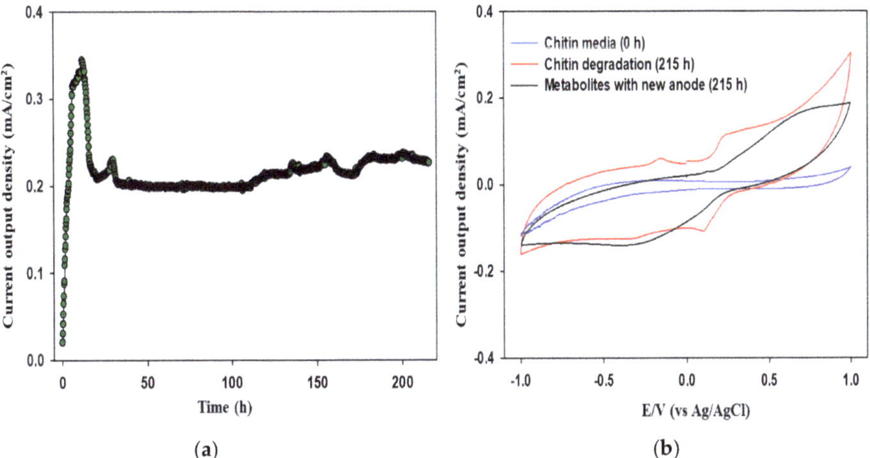

Figure 3. (**a**) Performance of strain YHY1 in MFCs fed with 1% chitin, 40 g L^{-1} NaCl, and external load of 1000 Ω; (**b**) CV of the MFCs fed with chitin. Plots represent the CV of media without inoculum (0 h), CV of media at 216 h of growth, CV of cell-free metabolites with a new anode.

CV studies were executed to reveal the electrochemical behavior of *O. arctica* YHY1 in MFCs fed with chitin. As depicted in Figure 3b, the initial CV profile of chitin media (0 h) without bacterial inoculation did not show any oxidation-reduction peaks, indicating a lack of redox mediators in the medium. However, the CV of MFCs at 216 h after inoculation showed distinct oxidation peaks at −0.15 V (vs. Ag/AgCl), and +0.20 V (vs. Ag/AgCl), reduction peaks at +0.10 V (vs. Ag/AgCl), and a broad peak at −0.40 V (vs. Ag/AgCl) (Figure 3b). During the forward scan, a higher current output density of 0.302 mA/cm^2 was observed, indicating higher oxidation reactions were recorded, as compared to the reductions reported earlier [7,17]. The CV of the cell-free degradation metabolites after the filtration and insertion of a new anode showed a broad oxidation-reduction peak at +,−0.40 V (vs. Ag/AgCl), suggesting production of a low quantity of soluble redox mediators in the chitin degraded medium. The observed formal potential of 0.200 V (vs. Ag/AgCl) can be related to the outer membrane-bound cytochrome, which can transfer the electrons directly to the electrode without any external redox mediators, as reported earlier in *Shewanella* [18]. Further, a formal potential of −0.15 V (vs. Ag/AgCl) was detected similar to *Geobacter sulfurreducens,* which generally use omcB to transfer electrons through the electrode/biofilm interface [19–22]. Similarly, the low amount of flavins might have been produced by *O. arctica* YHY1 with the formal potential of 0.40 V (vs. Ag/AgCl), as reported earlier [23]. Therefore, from the CV data, it could be predicted that *O. arctica* YHY1 mainly utilizes a direct electron transfer pathway to transfer electrons directly to the electrode surface using membrane-bound cytochromes. Furthermore, this bacterium also partly employs the indirect electron transfer pathway using extracellular redox mediators to shuttle electrons to the electrode. However, more detailed study at a molecular level is needed to find the exact mechanism used by *O. arctica* YHY1 to transfer electrons from bacteria to the electrode surface.

3.3. Chitin Degradation By-Products and Other Metabolites

Chitinolytic microbes can hydrolyze chitin biopolymer into monomeric or dimeric GlcNAc units and utilize them as the source of energy. In the present study, monomeric

GlcNAc was the main by-product of chitin hydrolysis by *O. arctica* YHY1 in MFCs. The concentration of GlcNAc at 24 h was 58.21 mg g^{-1} chitin, which was significantly increased to 192.01 mg g^{-1} at 120 h. However, after 120 h, the concentration of GlcNAc was gradually decreased to 76.22 mg g^{-1} at 216 h. From these results, it could be concluded that until 120 h, chitin was efficiently degraded, with simultaneous utilization of GlcNAc. Thereafter, depletion in the chitin concentration in media might have terminated the chitinase-producing machinery or inhibited the enzyme activity due to the generation of toxic intermediate metabolites [7,24]. However, *O. arctica* YHY1 continued to utilize the produced GlcNAc with the concurrent electricity generation until 216 h. During hydrolysis of the chitin, several metabolites were detected in the MFCs, with lactate, acetate, propionate, and butyrate as the prominent metabolites. The highest concentration of acetate, 5.901 mM (144 h), was detected, followed by butyrate, 3.572 mM (96 h), lactate, 0.932 mM (96 h), and propionate, 0.157 (120 h) (Figure 4a). Acetate, lactate, and propionate were most preferred by *O. arctica* YHY1, with 1.593 mM, 0.110 mM, and 0.099 mM concentrations, respectively, remaining unutilized at 216 h. However, butyrate was strain YHY1's less favored electron donor (2.533 mM; 216 h). Although the GlcNAc content was depleted after 120 h of growth, the current output density remained stable as the bacterium also utilized produced metabolites like lactate and acetate for growth, which correlates with the previous report on chitin as the carbon source in MFCs [7]. In *Shewanella*, lactate is first produced, followed by pyruvate, and acetate, and thus the carbon source preference can be in the following order: lactate→pyruvat→acetate [25]. The lactate concentration was higher at 96 h whereas acetate was higher at 144 h, although both concentrations decreased after reaching the maximum concentration, suggesting lactate can be oxidized to acetate [26]. During chitin hydrolysis, the production of metabolites like lactate, acetate, butyrate, succinate, format, and propionate was reported earlier in *Shewanella oneidensis*, *Bacillus circulans*, *Arenibacter palladensis*, and *Aeromonas hydrophila* [7,10–12].

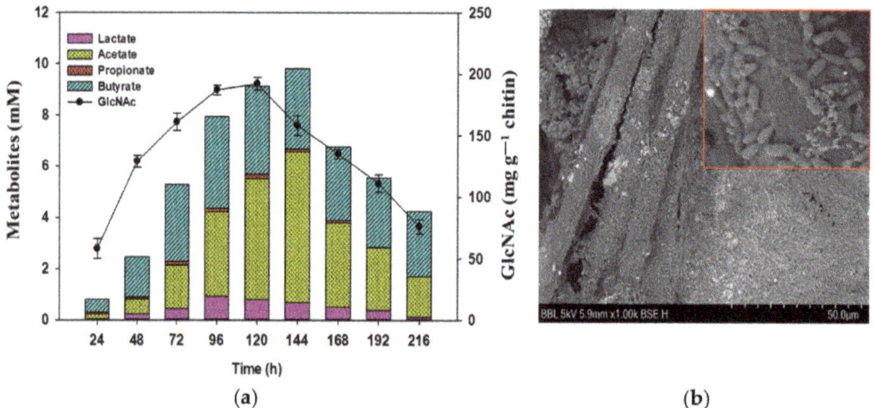

(**a**) (**b**)

Figure 4. (**a**) Production and utilization of GlcNAc and other metabolites in MFC; (**b**) SEM examination of anodic biofilm.

SEM investigation of the anodic biofilm of *O. arctica* YHY1 revealed a thick biofilm formation on the anode surface (Figure 4b). This result could also be correlated with the CV data, suggesting that membrane-bound cytochromes were principally involved in direct electron transfer to the anode.

3.4. Investigating Structural Changes in Chitin Polymer in MFCs

As shown in Figure 5a, chitin biopolymer has a typical band pattern for amide bonds at 1660 and 1630 cm^{-1} (amide I), 1558 cm^{-1} (amide II), 1318 cm^{-1} (amide III), and 693 cm^{-1} (amide V) [7,27]. Similarly, an IR peak at 1028 cm^{-1} assigned to stretching vibration for –C–O–C of the glucosamine ring, and a peak at 890 cm^{-1} related to ring stretching for β-1,4 glycosidic bonds were detected in the chitin [28]. Nevertheless, on the degradation of chitin in MFCs, the band intensity at 1660 cm^{-1} (amide I), 1378 cm^{-1} (amide III), and 1080 cm^{-1} (C–O stretching) substantially decreased, suggesting breaking of the C–O and C–H bonds [7,10]. Likewise, the XRD analysis revealed alterations in the crystallinity of the chitin before and after degradation (Figure 5b). The diffraction pattern of chitin before degradation showed distinct peaks at lattice (020), (110), (120), (101), and (130) [7]. Strong reflections were detected at 2θ 9.45°, 19.05°, and 31.46°, whereas other peaks were detected at 12.93°, 20.09°, 23.60°, and 26.51°. The crystalline index (CrI) of the chitin before and after degradation was calculated considering reflections at (020) and (110). Initial CrI(020) and CrI(110) of the chitin were 74.41% and 83.60%, respectively, which significantly decreased to 67.96% CrI(020) and 81.15%(CrI110), respectively, after the degradation of the chitin by *O. arctica* YHY1. Further, peaks at 2θ ≅ 23 related to the polysaccharide structure of chitin showed broad scattering due to the hydrolysis of the chitin, indicating a decrease in the crystallinity [7]. Likewise, a sharp peak detected at 31.46° was significantly reduced on the degradation of the chitin biomass.

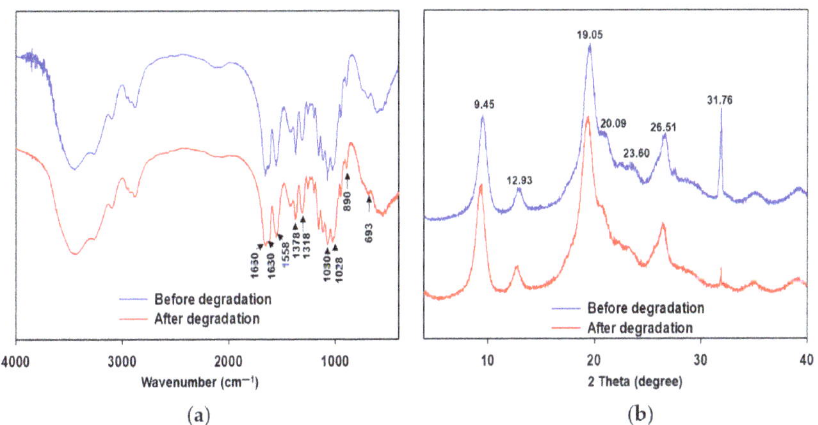

Figure 5. Structural changes in the chitin before and after degradation. (**a**) FTIR; (**b**) XRD.

4. Conclusions

The newly isolated halotolerant *O. arctica* YHY1 used in the present study generates a higher and stable electric current for a longer time by using seafood processing waste chitin as a carbon source. Monomeric GlcNAc and metabolites like acetate, lactate, and propionate produced during chitin degradation were consumed by strain YHY1 in a timely manner. The performance of the bacterium was improved by supplementing the MFCs with a salt concentration of 40 g L^{-1}. Further, *O. arctica* YHY1 can transfer electrons directly to the anode surface using membrane-bound cytochromes, with the partly involvement of extracellular redox mediators. Therefore, the findings of the present study can provide insights for utilizing chitin waste under high salt conditions, which can serve a dual purpose of recycling the seafood chitin biomass and generating electricity.

Author Contributions: Conceptualization, R.G.; Methodology, R.G.; Software, T.-R.C.; Validation, H.-J.K. and H.-J.L.; Formal analysis, J.-Y.C.; Investigation, R.G.; Resources, S.H.; Data curation, M.-J.S.; Writing—original draft preparation, R.G.; Writing—review and editing, S.K.B., S.-K.K. and D.-W.Y.; Visualization, S.-H.K.; Supervision, Y.-H.Y.; Project administration, Y.-H.Y.; Funding acquisition, Y.-H.Y. All authors have read and agreed to the published version of the manuscript.

Funding: This study was supported by the National Research Foundation of Korea (NRF) (NRF-2019R1F1A1058805 and NRF-2019M3E6A1103979) and the Research Program to Solve the Social

Issues of the NRF funded by the Ministry of Science and ICT (2017M3A9E4077234). This work was also supported by the R&D Program of MOTIE/KEIT (20016324).

Institutional Review Board Statement: Not applicable.

Informed Consent Statement: Not applicable.

Data Availability Statement: The data presented in this study are available on request from the corresponding author.

Conflicts of Interest: The authors declare no conflict of interest.

References

1. Nirmal, N.P.; Santivarangkna, C.; Rajput, M.S.; Benjakul, S. Trends in shrimp processing waste utilization: An industrial prospective. *Trends Food Sci. Technol.* **2020**, *103*, 20–35. [CrossRef]
2. Senphan, T.; Benjakul, S.; Kishimura, H. Characteristics and antioxidative activity of carotenoprotein from shells of Pacific white shrimp extracted using hepatopancreas proteases. *Food Biosci.* **2014**, *5*, 54–63. [CrossRef]
3. Djellouli, M.; López-Caballero, M.E.; Arancibia, M.Y.; Karam, N.; Martínez-Alvarez, O. Antioxidant and antimicrobial enhancement by reaction of protein hydrolysates derived from shrimp by-products with glucosamine. *Waste Biomass Valorization* **2020**, *11*, 2491–2505. [CrossRef]
4. Tan, Y.N.; Lee, P.P.; Chen, W.N. Microbial extraction of chitin from seafood waste using sugars derived from fruit waste-stream. *AMB Express* **2020**, *10*, 17. [CrossRef]
5. Ching, Y.C.; Redzwan, G. Biological treatment of fish processing saline wastewater for reuse as liquid fertilizer. *Sustainability* **2017**, *9*, 1062. [CrossRef]
6. Gurav, R.; Tang, J.; Jadhav, J. Novel chitinase producer *Bacillus pumilus* RST25 isolated from the shellfish processing industry revealed antifungal potential against phyto-pathogens. *Int. Biodeterior. Biodegrad.* **2017**, *125*, 228–234. [CrossRef]
7. Gurav, R.; Bhatia, S.K.; Choi, T.R.; Jung, H.R.; Yang, S.Y.; Song, H.S.; Park, Y.L.; Han, Y.H.; Park, J.Y.; Kim, Y.G.; et al. Chitin biomass powered microbial fuel cell for electricity production using halophilic *Bacillus circulans* BBL03 isolated from sea salt harvesting area. *Bioelectrochemistry* **2019**, *130*, 107329. [CrossRef]
8. Rezaei, F.; Richard, T.L.; Logan, B.E. Analysis of chitin particle size on maximum power generation, power longevity, and Coulombic efficiency in solid–substrate microbial fuel cells. *J. Power Sources* **2009**, *192*, 304–309. [CrossRef]
9. Gurav, R.; Bhatia, S.K.; Choi, T.R.; Kim, H.J.; Song, H.S.; Park, S.L.; Lee, S.M.; Lee, H.S.; Kim, S.H.; Yoon, J.J.; et al. Utilization of different lignocellulosic hydrolysates as carbon source for electricity generation using novel *Shewanella marisflavi* BBL25. *J. Clean. Prod.* **2020**, *277*, 124084. [CrossRef]
10. Gurav, R.; Bhatia, S.K.; Moon, Y.M.; Choi, T.R.; Jung, H.R.; Yang, S.Y.; Song, H.S.; Jeon, J.M.; Yoon, J.J.; Kim, Y.G.; et al. One-pot exploitation of chitin biomass for simultaneous production of electricity, n-acetylglucosamine and polyhydroxyalkanoates in microbial fuel cell using novel marine bacterium *Arenibacter palladensis* YHY2. *J. Clean. Prod.* **2019**, *209*, 324–332. [CrossRef]
11. Li, S.-W.; Zeng, R.J.; Sheng, G.-P. An excellent anaerobic respiration mode for chitin degradation by *Shewanella oneidensis* MR-1 in microbial fuel cells. *Biochem. Eng. J.* **2017**, *118*, 20–24. [CrossRef]
12. Li, S.W.; He, H.; Zeng, R.J.; Sheng, G.P. Chitin degradation and electricity generation by *Aeromonas hydrophila* in microbial fuel cells. *Chemosphere* **2017**, *168*, 293–299. [CrossRef]
13. Bhatia, S.K.; Gurav, R.; Choi, T.R.; Jung, H.R.; Yang, S.Y.; Moon, Y.M.; Song, H.S.; Jeon, J.M.; Choi, K.Y.; Yang, Y.H. Bioconversion of plant biomass hydrolysate into bioplastic (polyhydroxyalkanoates) using *Ralstonia eutropha* 5119. *Bioresour. Technol.* **2019**, *271*, 306–315. [CrossRef] [PubMed]
14. Focher, B.; Beltrame, P.L.; Naggi, A.; Torri, G. Alkaline *N*-deacetylation of chitin enhanced by flash treatments. Reaction kinetics and structure modifications. *Carbohydr. Polym.* **1990**, *12*, 405–418. [CrossRef]
15. Faria, A.; Gonçalves, L.; Peixoto, J.M.; Peixoto, L.; Brito, A.G.; Martins, G. Resources recovery in the dairy industry: Bioelectricity production using a continuous microbial fuel cell. *J. Clean. Prod.* **2017**, *140*, 971–976. [CrossRef]
16. Zhang, L.; Wang, J.; Fu, G.; Zhang, Z. Simultaneous electricity generation and nitrogen and carbon removal in single-chamber microbial fuel cell for high-salinity wastewater treatment. *J. Clean. Prod.* **2020**, *276*, 123203. [CrossRef]
17. Abbas, S.Z.; Rafatullah, M.; Ismail, N.; Shakoori, F.R. Electrochemistry and microbiology of microbial fuel cells treating marine sediments polluted with heavy metals. *RSC Adv.* **2018**, *8*, 18800–18813. [CrossRef]
18. Carmona-Martinez, A.A.; Harnisch, F.; Fitzgerald, L.A.; Biffinger, J.C.; Ringeisen, B.R.; Schröder, U. Cyclic voltammetric analysis of the electron transfer of *Shewanella oneidensis* MR-1 and nanofilament and cytochrome knock-out mutants. *Bioelectrochemistry* **2011**, *81*, 74–80. [CrossRef]
19. Richter, H.; Nevin, K.P.; Jia, H.; Lowy, D.A.; Lovley, D.R.; Tender, L.M. Cyclic voltammetry of biofilms of wild type and mutant *Geobacter sulfurreducens* on fuel cell anodes indicates possible roles of OmcB, OmcZ, type IV pili, and protons in extracellular electron transfer. *Energy Environ. Sci.* **2009**, *2*, 506–516. [CrossRef]
20. Qian, X.; Reguera, G.; Mester, T.; Lovley, D.R. Evidence that OmcB and OmpB of *Geobacter sulfurreducens* are outer membrane surface proteins. *FEMS Microbiol. Lett.* **2007**, *277*, 21–27. [CrossRef]

21. Stephen, C.S. Abundance of the multiheme c-type cytochrome omcB increases in outer biofilm layers of electrode-grown *Geobacter sulfurreducens*. *PLoS ONE* **2014**, *9*, 104336. [CrossRef]
22. Steidl, R.J.; Lampa-Pastirk, S.; Reguera, G. Mechanistic stratification in electroactive biofilms of *Geobacter sulfurreducens* mediated by pilus nanowires. *Nat. Commun.* **2016**, *7*, 12217. [CrossRef] [PubMed]
23. Gurav, R.; Bhatia, S.K.; Choi, T.R.; Choi, Y.K.; Kim, H.J.; Song, H.S.; Lee, S.M.; Park, S.L.; Lee, H.S.; Koh, J.; et al. Application of macroalgal biomass derived biochar and bioelectrochemical system with *Shewanella* for the adsorptive removal and biodegradation of toxic azo dye. *Chemosphere* **2021**, *264*, 128539. [CrossRef] [PubMed]
24. Lee, Y.S.; Kim, K.Y. Antagonistic potential of *Bacillus pumilus* L1 against root-knot nematode, *Meloidogyne arenaria*. *J. Phytopathol.* **2016**, *164*, 29–39. [CrossRef]
25. Mao, L.; Verwoerd, W.S. Theoretical exploration of optimal metabolic flux distributions for extracellular electron transfer by *Shewanella oneidensis* MR-1. *Biotechnol. Biofuels* **2014**, *7*, 118. [PubMed]
26. Nakagawa, G.; Kouzuma, A.; Hirose, A.; Kasai, T.; Yoshida, G.; Watanabe, K. Metabolic characteristics of a glucose-utilizing *Shewanella oneidensis* strain grown under electrode-respiring conditions. *PLoS ONE* **2015**, *10*, 0138813. [CrossRef] [PubMed]
27. Zhang, A.; Wei, G.; Mo, X.; Zhou, N.; Chen, K.; Ouyang, P. Enzymatic hydrolysis of chitin pretreated by bacterial fermentation to obtain pure *N*-acetyl-D-glucosamine. *Green Chem.* **2018**, *20*, 2320–2327. [CrossRef]
28. Kumari, S.; Rath, P.; Kumar, A.S.; Tiwari, T.N. Extraction and characterization of chitin and chitosan from fishery waste by chemical method. *Environ. Technol. Innov.* **2015**, *3*, 77–85. [CrossRef]

 *sustainability*

Review

Recent Developments in Microbial Electrolysis Cell-Based Biohydrogen Production Utilizing Wastewater as a Feedstock

Pooja Dange [1], Soumya Pandit [2,*], Dipak Jadhav [3], Poojhaa Shanmugam [1], Piyush Kumar Gupta [2], Sanjay Kumar [2], Manu Kumar [4], Yung-Hun Yang [5,6] and Shashi Kant Bhatia [5,6,*]

1 Amity Institute of Biotechnology, Amity University, Mumbai 4102016, India; poojadange99@gmail.com (P.D.); poojhaa26@gmail.com (P.S.)
2 Department of Life Sciences, School of Basic Sciences and Research, Sharda University, Greater Noida 201306, India; dr.piyushkgupta@gmail.com (P.K.G.); sanjay.syadavbiotech@gmail.com (S.K.)
3 Department of Agricultural Engineering, Maharashtra Institute of Technology, Aurangabad 431010, India; deepak.jadhav1795@gmail.com
4 Department of Life Science, Dongguk University-Seoul, 32 Dongguk-ro, Ilsandong-gu, Goyang-si 10326, Korea; manukumar007@gmail.com
5 Institute for Ubiquitous Information Technology and Application, Konkuk University, Seoul 05029, Korea; seokor@konkuk.ac.kr
6 Department of Biological Engineering, Konkuk University, Seoul 05029, Korea
* Correspondence: sounip@gmail.com (S.P.); shashibiotechhpu@gmail.com (S.K.B.)

Abstract: Carbon constraints, as well as the growing hazard of greenhouse gas emissions, have accelerated research into all possible renewable energy and fuel sources. Microbial electrolysis cells (MECs), a novel technology able to convert soluble organic matter into energy such as hydrogen gas, represent the most recent breakthrough. While research into energy recovery from wastewater using microbial electrolysis cells is fascinating and a carbon-neutral technology that is still mostly limited to lab-scale applications, much more work on improving the function of microbial electrolysis cells would be required to expand their use in many of these applications. The present limiting issues for effective scaling up of the manufacturing process include the high manufacturing costs of microbial electrolysis cells, their high internal resistance and methanogenesis, and membrane/cathode biofouling. This paper examines the evolution of microbial electrolysis cell technology in terms of hydrogen yield, operational aspects that impact total hydrogen output in optimization studies, and important information on the efficiency of the processes. Moreover, life-cycle assessment of MEC technology in comparison to other technologies has been discussed. According to the results, MEC is at technology readiness level (TRL) 5, which means that it is ready for industrial development, and, according to the techno-economics, it may be commercialized soon due to its carbon-neutral qualities.

Keywords: microbial electrolysis cells; chronological development; wastewater to hydrogen; scale-up; life-cycle assessment; MEC commercialization

Citation: Dange, P.; Pandit, S.; Jadhav, D.; Shanmugam, P.; Gupta, P.K.; Kumar, S.; Kumar, M.; Yang, Y.-H.; Bhatia, S.K. Recent Developments in Microbial Electrolysis Cell-Based Biohydrogen Production Utilizing Wastewater as a Feedstock. *Sustainability* **2021**, *13*, 8796. https://doi.org/10.3390/su13168796

Academic Editor: Tomonobu Senjyu

Received: 15 June 2021
Accepted: 29 July 2021
Published: 6 August 2021

1. Introduction

Water and energy are two inseparable commodities, which greatly influence the growth of human civilization. The worldwide population is estimated to reach 9.7 billion by 2050, while worldwide energy consumption is expected to exceed 736 quadrillion of British thermal units (BTUs) by 2040 [1]. Water is an important natural source for all life forms on Earth, and their quality of life is determined by its availability and quality [2,3]. The supply of clean water is essential for the establishment and maintenance of different human activities including households, agriculture, and industries. Freshwater is becoming one of the scarcest resources in recent years due to the increase in the world's population and industrial activities [4]. The environment-related concerns due to the emission of greenhouse gases (GHGs) from fossil reserves has resulted in a paradigm shift in industrial business strategies to design new configuration systems for wastewater treatment along with the production of green biofuels [5]. Some of the major concerns for government bodies all around the world are the increasing global population, improving their living standards, environmental change, and enhancing water demand for energy generation [6].

Wastewater is a complex mixture of diverse categories of pollutants [7]. There are several forms of water contamination such as nutrient contamination, surface water contamination, oxygen depletion, groundwater contamination, microbiological contamination, suspended matter, chemical water contamination, and oil spillage, since water originates from many sources. Wastewater generated from various effluents needs to be treated before its recycling and reuse. Wastewater is rich in nutrients and its release in the open may cause eutrophication and pose a threat to flora and fauna. Wastewater also contains many unwanted chemicals and pathogens and is able to cause gastroenteritis, skin infections, and leptospirosis-like diseases [8]. Traditionally, the main purpose of wastewater treatment was to protect downstream users from health risks using various physical (grit and flotation) and chemical (neutralization, flocculation, oxidation, etc.) methods. These methods are all expensive and result in sludge production and secondary water pollution [9]. Water reservoirs are becoming more contaminated as a result of rising levels of micropollutants such as medicines, organic polymers, and suspended particles. Powdered activated carbon (PAC) has been shown to be a viable option for water filtration with little environmental effect [10]. H_2 was formerly created using a variety of thermochemical, electrolytic, and photolytic techniques. Heat and pressure are used in thermochemical procedures to disrupt molecular bonds. Electrolysis is the process of breaking water into its parts using electricity. H_2 is extracted from microorganisms via photolytic reactions. Furthermore, thermochemical processes need fossil fuels as raw material, whereas electrolytic and photolytic processes need a lot of energy and are, thus, quite costly [11,12]. Fossil fuel burning emits greenhouse gases (CO_2, SO_2, and NOx) and toxic pollutants such as polycyclic hydrocarbons, mercury, and volatile chemicals responsible for global warming, with a negative effect on human health [13].

Biological hydrogen generation is required to solve the thermodynamic and environmental issues by utilizing wastewater for hydrogen production with simultaneous wastewater treatment. When biomass is employed as a raw material, the organic compounds dissolved in the wastewater have a high energy state, making mechanical combustion difficult. The raw ingredients such as various types of wastewaters, lignocellulosic biomass, and organic compounds, utilized in biological hydrogen generation are readily accessible, cost-effective, and waste from other sectors.

Although different kinds of water electrolysis technologies have been established, further development is needed before they can be incorporated into large-scale, cost-effective electricity networks. For example, according to a recent techno-economic study, water electrolysis utilizing solar energy is still not economically viable when compared to hydrogen generation from fossil sources. Microbial electrolysis is considered to be a valuable, novel approach within this framework. Microbial electrolysis cells were originally suggested in 2005. Microbial electrolysis cell technology provides a dual advantage of gaseous energy production and organic waste treatment in these circumstances [14].

1.1. *Sources of Wastewater*

The properties of wastewater differ depending on its source. There are various sources of wastewater such as industrial wastewater, solid waste and sewage disposal, arsenic contamination of groundwater, underground storage, tube leakages, and inadequate sanitary facilities, which directly affect our environment.

1.1.1. Industrial Wastewater

Industries generate a huge amount of garbage, which is full of harmful chemicals and contaminants including antibiotics, polyphenolic compounds, and Azo dyes. It is one of the primary causes of contamination in the water environment. Over the past century, a significant volume of agricultural wastewater has been dumped into waterways, wetlands, and marine areas. Industries such as pharmaceuticals, food, pulp and paper, textiles, tannery, pesticides, dyeing, and painting are the most significant sectors for water contamination. Industries discharge a huge amount of unprocessed industrial waste into the rivers knowingly or unknowingly and trigger pollution all over the country [15].

1.1.2. Sugar-Based Wastewater

Comparatively, a significant concentration of carbohydrates (2300–3500 mg·L^{-1}), sugars (0.65–1.18%), proteins (0.12–0.15%), and starch (65–75%) in starch processing wastewater (SPW) is discharged, which is an essential energy-rich source that can be transformed into a broad range of usable products [16]. SPW was used as a substrate to feed an energy-producing microbial consortium, achieving power generation of 0.044 mA·cm^{-2} in combination with a drop in chemical oxygen demand (COD) in 6 weeks from over 1700 mg·L^{-1} to 50 mg·L^{-1} [17]. Wastewater from breweries as a substrate in MFCs has become another favorite amongst investigators, largely due to its weak strength. While the composition of brewery wastewater differs, it is usually nearly 10-fold higher in concentration than domestic wastewater on the scale of 3000–5000 mg COD·L^{-1} [18].

1.1.3. Cellulose- and Chitin-Based Waste

Substrates such as cellulose and chitin are inexpensive biopolymeric resources that can be utilized for power generation and are easily accessible. In industrial and municipal wastewaters, these green materials also constitute a major portion of organic compounds [19]. Only a few reports on the usage of these particulate substrates in MFCs have been performed. The microorganism must be able to anaerobically hydrolyze cellulose and be electrochemically active, using an electrode as an electron acceptor, thus oxidizing cellulose hydrolysis metabolites for specific processing of cellulose to generate hydrogen in MECs.

1.1.4. Landfill Leachates

Landfill leachates are extremely contaminated landfill wastes comprising four main classes of contaminants with a diverse composition: dissolved organic and inorganic biocomponents, heavy metals, and organic xenobiotic compounds [20]. Habermann and Pommer initially documented the usage of landfill effluent in a biofuel cell for removal of COD, but no current output values were listed [21].

1.1.5. Protein-Based Wastewater

The diverse structure of fats, proteins, fibers, highly organic material, parasites, meat-processing effluents, and pharmaceuticals for veterinary purposes is known to be dangerous worldwide. Due to the vast spectrum of slaughterhouse wastewater (SWW) and pollutant levels, SWW is usually analyzed using bulk criteria. SWW comprises significant quantities of biochemical oxygen demand (BOD), chemical oxygen demand, total organic carbon (TOC), total nitrogen (TN), total phosphorus (TP), and total suspended solids (TSS) [22]. A major concern for the livestock sector is indeed the overall management of SWW to decrease its ecological consequences [23]. It includes organic matter usable for processing from microbial activity. For its purification, it is important to reduce the BOD value therein [24].

1.2. Need for Hydrogen as a Biofuel

When compared to conventional carbon-based fossil fuels, hydrogen as a biofuel is more efficient in supplying energy. Biohydrogen is what we term hydrogen that is generated using biological routes and biotechnological principles [25]. The current analysis focuses on the mechanisms that produce hydrogen, the biology that underpins them, and the use of wastewater to produce hydrogen. It is a viable option since it is easy to make from renewable resources and can be utilized in high-efficiency fuel cells. Hydrogen is now created using a variety of thermochemical, electrolytic, and photolytic techniques [13]. The traditional practices mentioned above, on the other hand, are harmful to the environment in terms of GHG emissions. Thermochemical processes use fossil fuels as a source of energy, while electrolytic and photolytic processes need a lot of energy and are, thus, quite costly [26].

The biological synthesis of hydrogen is required to solve these difficulties and reduce negative environmental repercussions. The organic elements dissolved in wastewater are in a high-energy condition when biomass is employed as a source material. As a result, they are difficult to combust mechanically. This is the point at which we must depend on biological techniques to generate H_2. The most favored technique of dark fermentative biohydrogen generation is restricted by the thermodynamic barrier. In this case, MEC technology provides a dual advantage of gaseous energy production and organic waste treatment. However, many more technological advances and a better knowledge of the role and function of microbial communities in biohydrogen generation must be accomplished before this method can be commercialized [27].

2. Biological Hydrogen Production

With the ever-rising energy demand, hydrogen can potentially be an alternative efficient and clean fuel replacement of the conventional ones [28]. An innovative approach to solve the issue of waste generation would be to utilize commercial and residential wastewater for producing electricity [29]. Biohydrogen production can come into action via four processes: (a) bio-photolysis, (b) dark fermentation, (c) photo fermentation, and (d) microbial electrolysis, as described below.

2.1. Bio-Photolysis

Bio-photolysis in a biological system generally means the dissociation of the water molecule in the presence of light (photons). In such a lytic process, photosynthetic microorganisms such as microalgae and cyanobacteria are involved, whereby photosystems (PSI, PSII) absorb light energy and the excited electrons pass through a sequence of energy carriers, eventually receiving two protons when a water molecule splits [28], as depicted in Figure 1.

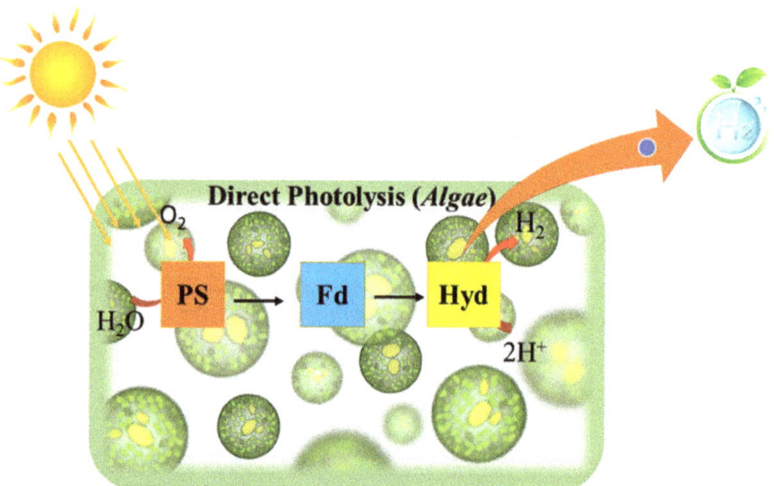

Figure 1. Schematic representation of direct photolysis for hydrogen production.

In the direct process, the system follows the following reactions:

$$2H_2O + hv \rightarrow O_2 + 4H^+ + Fd^{(4e^-)} \rightarrow Fd_{(red)}.$$

$$4e^- + 4H^+ + Fd_{(red)} \overset{Hydrogenase}{\rightarrow} Fd_{Ox} + 2H_2^{Fd\ (red)}.$$

Although photoproduction of hydrogen in the biosystem is environmentally friendly and can be undertaken by partial activation of photosystem II in *Chlamydomonas* spp., it is not economically feasible because it is inefficient for development at the industrial level. When discussing the drawbacks of this mechanism, many physiological factors are taken into consideration, such as hydrogenase O_2 sensitivity, competition with other metabolic pathways, downregulation of electron transport by non-dissipation of a proton gradient, and performance under non-saturating illumination [30].

2.2. Dark Fermentation

Dark fermentation (DF) is a process involving fermentation by dark-adapted microbes of carbohydrates under anoxic conditions to give H_2 as a resulting product along with acids, as illustrated in Figure 2 [28]. Studies have shown that *Clostridium* spp. can use a wide variety of sugars, making it a possible candidate to form H_2 from wastewater at a large scale [31]. The lowered pH due to acid output hampers the generation of H_2.

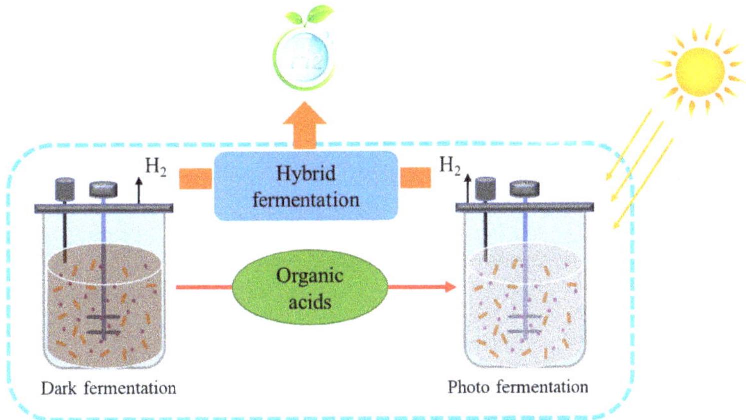

Figure 2. Diagrammatic representation of dark and photo fermentation (hybrid process).

A major economic emphasis is on biological H_2 production from wastewater using dark fermentation. The H_2 production system has many similarities with methanogenic anaerobic digestion, notably, the two gaseous compounds that can be separated from treated wastewater. The mixed microbial population present in both bioprocesses have the same properties but exhibit one major difference in biological H_2 production: bacteria such as homo-acetogens and methanogens inhibit hydrogen formation. To destroy these microbes, heat treatment is required without affecting spore-forming fermenting bacteria. Other approaches include high dilution rates or low pH for higher activity of the reactor and optimized operating environmental conditions to maximize the output of hydrogen. Unfortunately, the production of bio-H_2 depends on a comparatively limited volume of the overall equal H_2 present in wastewater. We cannot estimate recovery efficiency in high-carbohydrate wastewater to surpass 15% of the electron equivalent, under optimized conditions. Therefore, various researchers have performed two-step processes that involve the development of bio-H_2 via methanogenic anaerobic digestion to increase the overall energy output of the process. Methanogenic anaerobic digestion, as described later, is a useful and convenient process. Another promising technique is the conversion of methane to H_2 via a catalytic method. Therefore, H_2 production through the direct method is somewhat limited to the pretreatment step in large-scale energy production, while the discharge of H_2 gas through plastic enclosures and thin metal sheets represents another limitation because of the high diffuse rate of H_2 [32]. To increase the overall hydrogen yield, hybrid processes are also suggested, whereby organic acids produced in dark fermentation can be used as a feedstock in photo fermentation to produce hydrogen (Figure 2).

2.3. Photo Fermentation

The process of photo fermentation involves organic/inorganic substrate oxidation in the presence of O_2 to release electrons that end up reducing ferredoxins. These reduced ferredoxins are related directly to the production of H_2 and to the fixation of CO_2. The following reaction takes place:

$$(CH_2O_2)_2 \overset{\text{NADPH}}{\rightarrow} Fd_{(red)} + ATP \overset{\text{Nitrogenase}}{\rightarrow} H_2 + Fd_{Ox} + ADP + Pi.$$

Although this process has advantages such as enhanced potential conversion yields and the ability to absorb a wide variety of substrates, there are drawbacks such as continuously regulated area/volume ratio maintenance in photobioreactors, temperature control, and controlled agitation. For the development of bio-H_2 from wastewater, various types of reactor systems such as batch and continuous reactors are used, with each configuration providing its advantages and disadvantages [33].

3. Existing Wastewater Treatment Technologies and Their Bottlenecks

Traditional wastewater treatment usually takes place when wastewater is carried by a sewer to a centralized wastewater treatment plant, where the wastewater is then treated linearly by the end-of-pipe technology. It is time to make the switch to a closed-loop system [34], in which water is treated with simultaneous nutrient and energy recovery. Due to the superior positioning of decentralized wastewater treatment plants, centralized wastewater treatment plants can lead to a reduction in operational and capital expenditure (OPEX). Despite being in their infancy when it comes to deployment and optimization, decentralized solutions are lagging in the adoption phase when related to centralized ones. Clearly, the consequences of decentralization are visible, and developments in wastewater treatment plants are trying to make the transition from centralization to decentralization, where waste resources are reclaimed [35]. Because of a lack of effective methods for garbage disposal, management, and recycling, this problem will inevitably worsen [36]. Although centralized water systems are not relevant in many regions of the globe despite the drastic expansion in population and people favoring metropolitan areas, that trend is becoming even more prevalent as time progresses. For several reasons, from the changing demographics to building codes, rural people have to change their lifestyle, and this has placed wastewater setups under burden. Unfortunately, in several instances, this has led to efficient wastewater treatment systems not being put in place. In certain cases, it is difficult and expensive to build centralized wastewater treatment plants after the fact. With the increased use of distributed and non-networked technologies, it is more likely that we will develop a decentralized wastewater treatment infrastructure, which helps to encourage improved system robustness and lower economic and environmental costs [37].

Microbial Electrolysis Cell Mechanism for Wastewater Treatment with Simultaneous Hydrogen Production

The microbial electrolysis cell is a capable technique for removing organics while simultaneously producing hydrogen gas. When mixed with other elements, hydrogen is a plentiful element on Earth (water, hydrocarbons, etc.). To produce biohydrogen in a pure and regulated form, several industrial procedures are necessary. Among a variety of fuels (gasoline: 47.5 $MJ \cdot kg^{-1}$ higher heating value (HHV); 44.5 $MJ \cdot kg^{-1}$ lower heating value (LHV)), hydrogen has the greatest thermal efficiency (141.9 $MJ \cdot kg^{-1}$ HHV; 119.9 $MJ \cdot kg^{-1}$ LHV) and may be preserved for extended periods until being employed in fixed or mobile operations. The benefits of biohydrogen generation from waste matter, such as solid wastes and wastewater, are increased by making the process more sustainable [38]. Hydrogen is an excellent future fuel that can be used to meet the world's energy demands due to its high energy density, ecologically benign combustion profile, and ability to be used at ambient temperature and pressure [39].

MECs provide a different approach for centralized wastewater treatment systems. According to this energy conversion estimate, the chemical energy potential of organic components constituting wastewater's core is roughly 9.3-fold higher than the energy required to treat it [40]. There is a notable increase in employing biomass energy to get energy from biological anaerobic wastewater treatment because of the simplicity and resilience of this technique [41]. Anaerobic digestion (AD) is a technique that is increasingly being used to generate biogas from wastewater sludge. A process termed acetate digestion provides energy by breaking down acetate to produce CO_2 and CH_4 [42]. Alternatively, the process is quite sluggish and causes the biogas to produce a significant quantity of CO_2, which reduces the energy density. Because biogas contains a high quantity of CO_2, it must be stored, and substantial chemical treatment is performed, involving cryogenic separation, to eliminate CO_2 until it can be utilized [43]. Compared to the use of biological techniques, MECs provide an alternate method for producing both CH_4 and H_2.

In MECs, microbes are utilized as biocatalysts to reduce the activation overpotential of a certain redox process, enhancing voltage efficiency and production rate [44]. On the surface of the anode, some microbes can develop a biofilm, which can convert the chemical energy contained in organic molecules into electrical energy. At the cathode, this electrical energy is subsequently used to produce additional useful products, such as H_2 and CH_4 [45]. To withstand cellular function and growth, certain microorganisms are electrochemically active, which means that they can transfer electrons along with the electrode to keep things running [46]. The chronological development of MEC technology is illustrated in Figure 3.

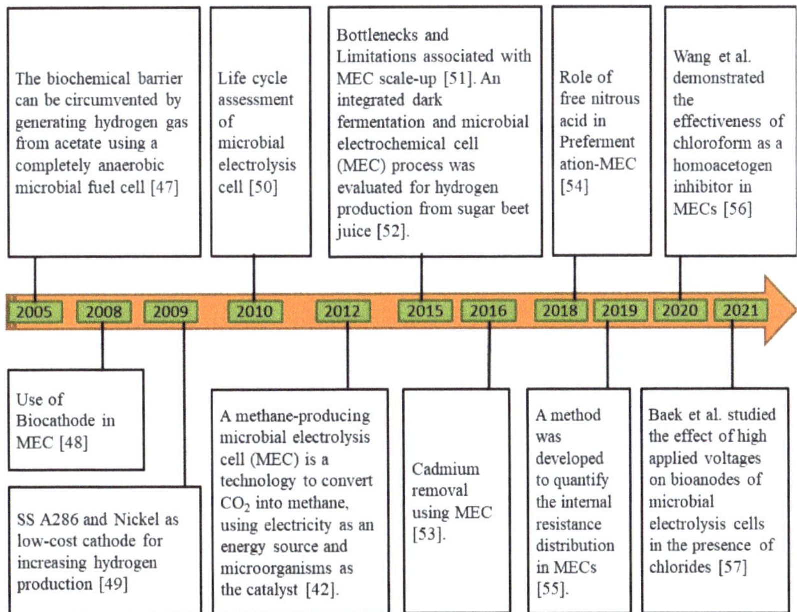

Figure 3. Chronological development of microbial electrolysis cell (MEC) technology [42,47–57].

4. Thermodynamics and Electrochemistry for Hydrogen Production Using MEC

Because of the thermodynamic limitations, major organic compounds such as volatile acids (e.g., butyrate, acetate, and propionate) and solvents (e.g., ethanol and butanol) cannot be used for fermentative H_2 production. However, additional energy is required for overcoming this limitation and for producing hydrogen [58]. In MEC, the required additional energy is provided by the voltage supplied through the power source. However, there is a necessity for higher applied potential than the equilibrium potential (E_{eq}) of the electrochemical cell (EC) for driving the MEC process. The equation is given as follows:

$$E_{eq} = E_{cat} - E_{an}, \quad (1)$$

where E_{eq} is the equilibrium potential of the EC, E_{cat} is the cathodic half-life potential, and E_{an} is the anodic half-life potential.

Therefore, to find the equilibrium voltage, determining the individual half-life potential using the Nernst equation is required.

Another well-researched electrochemical process is the hydrogen evolution reaction (HER). There are two reasons for this: (i) hydrogen evolution was originally regarded to be one of the simplest electrochemical processes, and (ii) it is a very significant process for society since hydrogen will one day replace fossil fuels as a transportation fuel, because of its simple reaction scheme.

$$2H^+ + 2e^- \rightarrow H_2 \text{ (Under acidic conditions)}.$$

$$2H_2O + 2e^- \rightarrow H_2 + 2OH^- \text{ (Under alkaline conditions)}.$$

The equilibrium potential of the hydrogen evolution process is significantly reliant on the pH at the cathode since protons (or hydroxyl ions) are involved, according to

$$E^{o}_{H2.\,pH} = -0.059 \times pH. \quad (2)$$

It has been well established that HER proceeds via two successive steps. The initial adsorption of a proton to form adsorbed hydrogen, i.e., the Volmer reaction ($H^+ + e^- + H_{ad}$), is usually considered to be fast; however, there are two possibilities for the subsequent, slower hydrogen evolution process: one is the hemolytic Tafel reaction; the other is the heterolytic Heyrovsky reaction [59].

$$H_2O + e^- + Me \rightarrow MeH_{ads} + OH^- \text{ (volmer electrochemical discharge step)}.$$

$$MeH_{ads} + H_2O + e^- \rightarrow H_2 \uparrow + OH^- + Me \text{ (Heyrovsky electrochemical disorption step)}.$$

$$MeH_{ads} \rightarrow H_2 + 2Me \text{ (Tafel catalytic recombination step)}.$$

4.1. Anodic Potential for Hydrogen Production

In the MEC method, anode respiratory bacteria such as antibiotic-resistant bacteria (ARB) or exoelectrogens residing in anodic biofilms transform organic matter into bicarbonates, electrons, and protons; the electrons are then transported to the cathode via a limited energy supply (−0.300 V vs. standard hydrogen electrode (SHE) formed by ARB), where they interact with protons to produce hydrogen gas. For example, the sodium acetate reaction at the anode is presented below.

$$CH_3COO^- + 4H_2O \rightarrow 2HCO_3^- + 9H^+ + 8e^-.$$

In terms of the conceptual anode potential (E_{an}) for oxidation of acetates under typical biological conditions (pH = 7, T = 298.15 K, $[CH_3COO^-]$ = 0.0169 mol·L^{-1} (1 g·L^{-1}), and $[HCO_3^-]$ = 0.005 mol·L^{-1}), the Nernst equation can be approximated to

$$E_{an} = E_{an}{}^{0} - \frac{RT}{8F}\ln\frac{[CH_3COO^-]}{[HCO_3^-][H^+]^9}, \quad (3)$$

$$= 0.187 - \frac{8.31 \times 298.15}{8 \times (9.65 \times 10^4)}\ln\frac{[0.0169]}{[0.005]^2[10^{-7}]^9} = -0.300\ \text{V}, \quad (4)$$

where $E_{an}{}^{0}$ is the standard electrode potential for acetate oxidation (0.187 V), R is the universal gas constant (8.31 J·mol^{-1}·K^{-1}), T is the absolute temperature (K), and F is Faraday's constant (9.65 × 10^4 C·mol^{-1}) [60].

4.2. *Performance Calculation*

4.2.1. Hydrogen Production Rate and Coulombic Efficiency

The system's efficiency was measured in terms of the rate of hydrogen production, the rate of hydrogen recovery, the Coulombic efficiency, the volumetric density, and the energy recovered [61].

On the basis of COD elimination, the total potential number of moles generated, n_{th}, is

$$n_{th} = \frac{b_{H_2/s} v_L \Delta C_S}{M_s}, \quad (5)$$

where $b_{H_{2/s}}$ = 4 mol/mol denotes the maximum amount of hydrogen that can be stoichiometrically generated from the substrate, v_L denotes the volume of liquid in the reactor, ΔS (g COD·L^{-1}) denotes the change in concentration of the substrate during one batch cycle, and Ms denotes the molecular weight of the substrate. The COD concentration (g COD·L^{-1}) was converted into moles of acetate using a conversion factor of 0.78 g COD·g^{-1} sodium acetate. According to the measured current, the moles of hydrogen recovered by n_{CE} can be calculated as follows:

$$n_{CE} = \frac{\int_{t=0}^{t} I dt}{2F}, \quad (6)$$

where I = V/R_{ex} is the current computed from the voltage across the resistor, and 2 is the conversion factor used to transform moles of electrons to hydrogen. F = 96,485 C/mol e$^-$ is Faraday's constant, and dt (s) is the interval over which data were collected. The Coulombic hydrogen recovery is expressed as

$$r_{CE} = \frac{n_{CE}}{n_{th}} = C_E, \quad (7)$$

where C_E is the Coulombic efficiency. The hydrogen recovery (in moles) at the cathode r_{Cat} is calculated as

$$r_{Cat} = \frac{n_{H_2}}{n_{CE}}, \quad (8)$$

where n_{H_2} is the number of moles of hydrogen recovered over a batch cycle. The maximum volumetric hydrogen production rate (Q) measured in m^3 H_2·m^{-3} of reactor per day (m^3 H_2·m^{-3}·day^{-1}) is calculated as

$$Q_{H_2}\left(m^3\, m^{-3}\cdot day\right) = \frac{I_v(A/m^3) r_{Cat}[1C/s)/A](0.5\, mol\cdot H_2/mol\cdot e^-)(86,400\, s/d)}{\left(F = 9.65 \times 10^4 \frac{C}{mol}\cdot e^-\right) C_g\left(mol\cdot \frac{H_2}{L}\right)\left(10^3 \frac{L}{m^3}\right)} = \frac{43.2\, I_V r_{Cat}}{F c_g(T)}. \quad (9)$$

4.2.2. Energy Recovery

The amount of energy provided by the power source to the circuit (W_E) is expressed as

$$W_E = \sum_1^n \left(IE_{ap}\Delta t - I^2R_{ex}\Delta t\right) \text{ (adjusted for losses across the resistor)}, \quad (10)$$

where E_{ap} (V) denotes the voltage applied, R_{ex} is the external resistor, and Δt (s) is the time increment for n data points measured during a batch cycle [61]. Energy balances based on combustion heats are usually used for electrolyzers and for predicting the amount of energy contained in organic matter. The amount of energy added by the substrate is expressed as

$$W_s = \Delta H_s n_s, \quad (11)$$

where ΔH_s= 870.28 kJ·mol^{-1} is the heat of combustion of the substrate, and n_S denotes the total number of consumed moles of the substrate during a batch cycle based on COD removal. The ratio of the total energy of the hydrogen produced to the input of required electrical energy is the energy efficiency relative to the electrical input (n_E).

$$n_E = \frac{n_{H_2}\Delta H_{H_2}}{W_E}, \quad (12)$$

where ΔH_{H_2} = 285.83 kJ·mol^{-1} is the energy content of hydrogen based on the heat of combustion (upper heating value), and $W_{H_2} = n_{H_2}\Delta H_{H_2}$. The efficiency relative to the added substrate (n_S) is calculated as

$$n_S = \frac{W_{H_2}}{W_S}. \quad (13)$$

The overall energy recovery based on both the electricity and the substrate inputs (η_{E+S}) is expressed as

$$\eta_{E+S} = \frac{W_{H_2}}{W_E + W_S}. \quad (14)$$

The percentages of energy contributed by the power source (e_E) and substrate (e_S) are calculated as follows:

$$e_E = \frac{W_E}{W_E + W_S}, \; e_S = \frac{W_S}{W_E + W_S}. \quad (15)$$

5. MEC Reactor Architecture

Similar to MFC, an elementary MEC architecture has two chambers that are connected employing an ion-exchange membrane. Over time, many different combinations of MECs, which are described below, have been developed for hydrogen yield improvement. The previous layouts comprised a basic H-type cell containing gas collection parts connected to a cathode chamber [62]. Eventually, various refinements were made to develop dual-chambered MECs for straightforward operation.

According to the findings of the research, based on a comparative analysis of various combinations, single-chambered MEC had higher hydrogen production rates and current densities than dual-chambered MEC. As a result, significant efforts have been made to further refine this combination for use in scale-up investigations. This surplus amount of MEC configurations indicates that the system setup for hydrogen generation in MEC is quite important. Substantial research was carried out to determine suitable MEC configurations. Several types of reactor modifications were assembled according to the results: cylindrical design, tubular reactor design, two-chamber MEC, up-flow single-chamber reactor, single-chamber membrane-less MEC, and many others. The hydrogen output and Coulombic efficiency of the MEC depend largely on the reactor configuration. Initially, researchers used dual-chambered MECs; the single-chambered MEC was introduced later for increasing the volumetric power density of the cathode and the hydrogen yield.

5.1. Single-Chambered MEC

A research group put together single-chambered MECs to synthesize hydrogen and assess the process efficiency. The main design used a glass bottle with a total capacity of 50 mL, while the secondary configuration used vials made of borosilicate glass with a total capacity of 10 mL; the cells usually used a mixed culture and pure culture of *S. oneidensis*, respectively. The anode and cathode, measuring 3.5 × 4 cm^2 and 4 × 5 cm^2, were kept 2 cm apart by plastic screws. The anode was type A carbon, and the cathode was type B carbon with platinum (0.5 $mg \cdot cm^{-2}$) as a catalyst separated by a J-cloth layer to prevent a short-circuit in both configurations [63,64]. Single-chambered MECs lack a membrane, as illustrated in Figure 4. When production rates are high, the microbial conversion of hydrogen to methane will be slow, with hydrogen being relatively insoluble in water. Energy losses of the membrane are lowered in membrane-less MECs, and the energy recovery process is high [65].

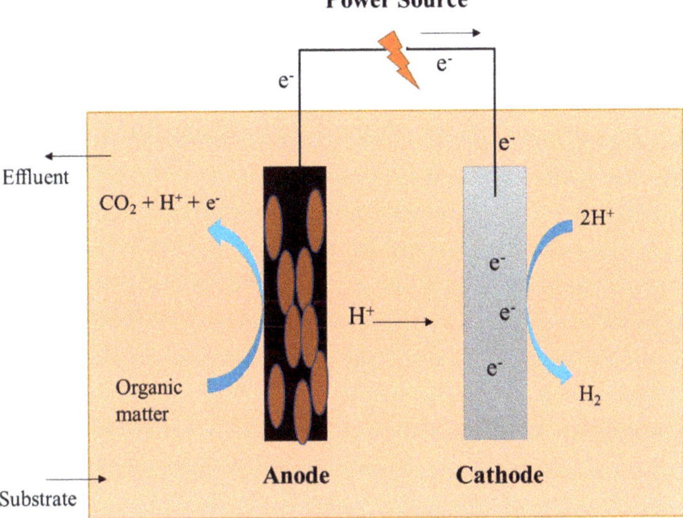

Figure 4. Schematic representation of a single-chambered microbial electrolysis cell.

5.1.1. An Up-Flow Single-Chambered MEC

Lee HS constructed an up-flow single-chamber MEC by inserting a cathode on top of the MEC and implemented out a program to monitor hydrogen and electron equivalents in batch trials to enhance the output of hydrogen gas. In a batch evaluation experiment lasting 32 h with a starting acetate concentration of 10 mM, the CE was 60% ± 1%, the H_2 yield was 59% ± 2%, and methane production was insignificant [66].

5.1.2. Smallest-Scale MEC

An MEC system with a single power source unit was designed. They used transparent glass serum vials with graphite plates functioning as anodes, which they found to be effective. It was soaked overnight, rinsed thrice in Milli-Q water, and further polished by sandpaper. Following the introduction of the unbent piece of wire through a hole drilled at the top center of the graphite plate, the bent end of the wire was inserted into a second hole and folded to form a tight connection between the wire and the plate. At 0.6 V of applied voltage, MECs having NiMo cathodes exhibited 33% better performance than NiW cathodes by accomplishing a hydrogen production rate (HPR) of 2.0 $m^3 \cdot day^{-1} \cdot m^{-3}$ at a current density of 270 $A \cdot m^{-3}$; however, this was slightly lower than MECs with a Pt catalyst which could accomplish 2.3 $m^3 \cdot day^{-1} \cdot m^{-3}$ [65].

5.1.3. A Cathode-on-Top Single-Chamber MEC

The reactor was made up of two parts: a top cover and the main chamber, both of which were constructed of glass and had a capacity of 0.4 L, as described in Figure 5 [62]. The substrate and electrolyte were pumped in via the bottom intake, and the produced gas was collected from the cathode with the use of a gasbag. HPR increased from 0.03 $L{\cdot}L^{-1}{\cdot}day^{-1}$ to 1.58 $L{\cdot}L^{-1}{\cdot}day^{-1}$ in a 24 h batch test when the applied voltage was expanded from 0.2 V to 1.0 V, and total hydrogen recoveries rose from 26.03% to 87.73% when the applied voltage was increased from 0.2 V to 1.0 V. The greatest total energy recovery was 86.78% at an applied voltage of 0.6 V [67].

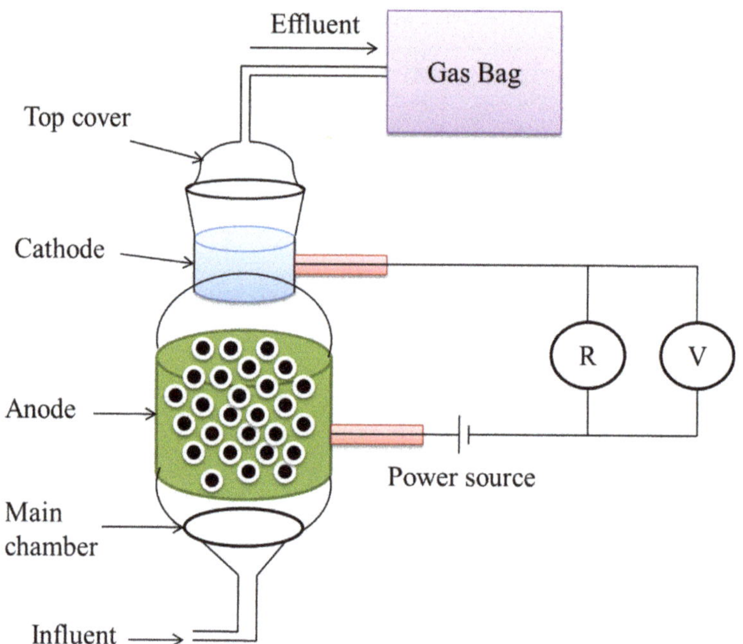

Figure 5. A cathode-on-top type of microbial dual-chambered reactor.

5.2. Dual-Chambered MEC

Dual-chambered MECs are often used in typical configurations and consist of an anodic and a cathodic chamber separated by the use of a membrane, as shown in Figure 6. The H-type MEC is a widely utilized dual-chambered MEC. Dual-chambered MECs are difficult to scale up because of their complex assemblies and large volumes with high internal resistance. The utilization of a membrane serves a dual purpose. It lowers the crossover from the anode to the cathode chamber and helps to prevent short-circuits at the cathode chamber, as well as aiding in the preservation of the purity of the product collected on the cathode side. The proton exchange membrane (PEM) is the most often used membrane since it is intended to only allow free protons to flow through while employing $-SO_3$ functional groups [68,69]. Alternative membranes, including anion-exchange membranes (AEM), such as AMI7001, bipolar membranes, and charge-mosaic membranes (CMM) [70] have also been investigated in MECs, in addition to the conventional membranes [71]. Cheng and Logan created a reactor in which an anion-exchange membrane (AEM) was used between the electrodes to achieve their desired results. The anode chamber was filled with graphite granules that were heated with ammonia gas, which increased the current densities, resulting in a reduction in the amount of time required for reactor accommodation. A carbon cloth served as the cathode, and a platinum catalyst was maintained near the membrane, which was connected to the outside circuit via a titanium wire [72].

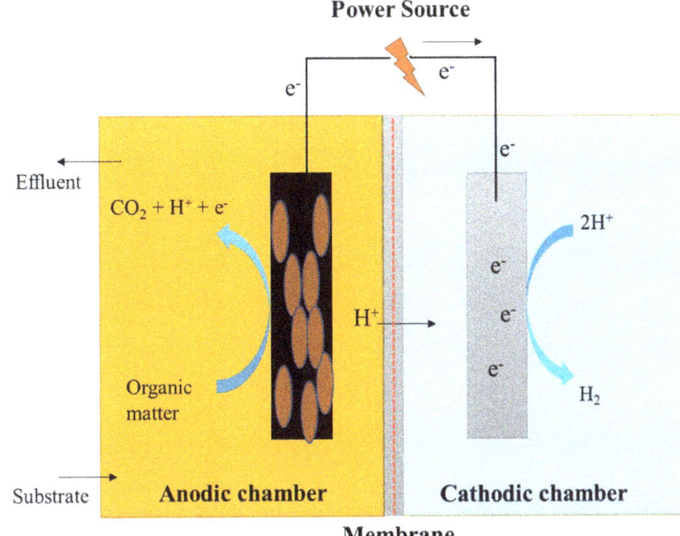

Figure 6. Schematic representation of a dual-chambered microbial electrolysis cell.

Various patents associated with MEC reactor architecture (Figure 7) are displayed in Table 1.

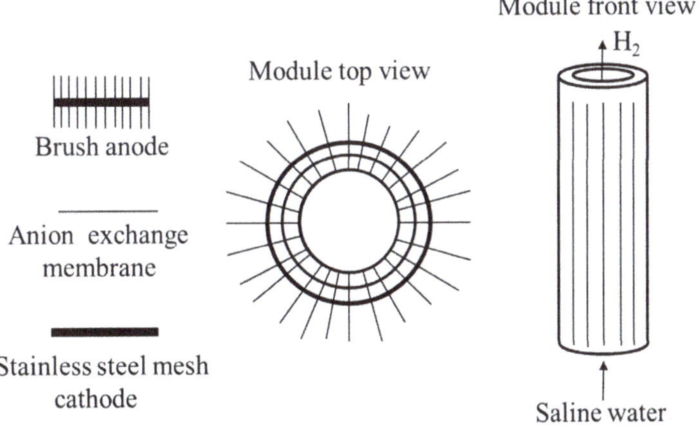

Figure 7. *Cont.*

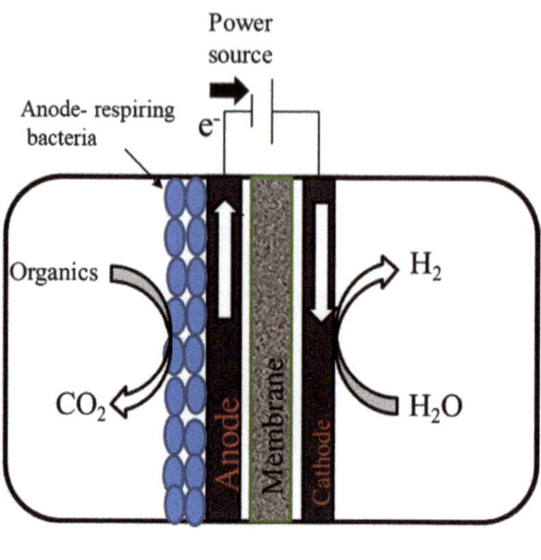

Figure 7. (**A**) Top and front view of graphite brush anode used in the microbial electrolysis cell; (**B**) microbial electrolysis cell.

Table 1. Patents associated with MEC reactor architecture.

Patent ID	Description	Reference
US8440438B2	More competition for fossil fuels and the desire to avoid carbon dioxide release through burning necessitate the development of new and sustainable technologies for energy generation and carbon capture. In accordance with aspects of the present invention, methods are given that include providing an electromethanogenic reactor containing an anode, a cathode, and a plurality of methanogenic microorganisms positioned on the cathode. Methanogens are given electrons and carbon dioxide. Even in the absence of hydrogen and/or organic carbon sources, methanogenic microbes produce methane.	[73]
US20150233001A1	Bioelectrochemical systems including a microbial fuel cell (MFC) and a microbial electrolysis cell (MEC) are provided. Both systems can ferment insoluble or soluble biomass, with the MFC capable of using a consolidated bioprocessing (CBP) organism to also hydrolyze an insoluble biomass and an electricigen to produce electricity. The MEC, on the other hand, relies on electricity input into the system, a fermentative organism, and an electricigen to produce fermentative products such as ethanol and 1,3-propanediol from a polyol biomass (e.g., containing glycerol). There are also approaches that are related.	[74]
EP2747181A1	The present invention relates to a process for inhibiting methanogenesis in single-chamber microbial electrolysis cells, which includes the initial addition of at least one methanogenesis inhibitor and is characterized by the following: dissolved hydrogen is removed from said cells after said initial addition of at least one methanogenesis inhibitor.	[75]
US7922878B2	The present invention provides a system for hydrogen gas generation that includes a hydrogen gas electrode assembly with a first anode in electrical communication with a first cathode, a microbial fuel cell electrode assembly with a second anode in electrical communication with a second cathode, a microbial fuel cell electrode assembly in electrical communication with a third cathode, and a hydrogen gas electrode assembly with a third anode in electrical communication with a third cathode. The hydrogen gas electrode assembly is at least partially contained in the interior space of a single-chamber housing.	[76]
US9216919B2	A brush anode microbial electrolysis cell is shown. At the cathode of the microbial electrolysis cell, a method for manufacturing products such as hydrogen is also given. The microbial electrolysis cell has a cylindrical shape with a concentric brush anode spirally wrapped around the outside of a cylindrical MEC, as illustrated in Figure 7. In some situations, the procedure may require sparging the anode and/or cathode with air. In rare situations, CO_2-containing gas can also be fed into a cathode chamber to lower the pH.	[77]
CN102408155A	The invention discloses a microbial electrolysis cell that combines CO_2 conversion and sewage treatment functionalities. The invention relates to the intersection of biofuel cells and the environment, as well as carbon dioxide capture and usage, and it is specifically linked to a kind of CO_2 conversion collection, with WWT in the microorganism electrolytic cell.	[78]

6. Scale-Up Reactor Designing

Researchers have been trying to successfully construct large-scale reactors, so that wastewater treatment and value-added byproduct production become easier. These products can be hydrogen gas, methane, and so on. The theoretical conclusion for perfect up-scaling of the process through works of literature and data is difficult because of the differences in substrate type and other parameters such as operation mode [79]. Most of the reactors, as concluded from reviews, are operational in a continuously fed mode. The first scale-up of the MEC system was performed by Cusick et al. [80]. Based on the previous demonstration at laboratory scale, a single-chambered 1000 L MEC reactor was designed for H_2 production from winery wastewater with high volatile fatty acid (VFA) content [81]. Along with hydrogen gas production, the production of methane was also observed at a gradually increasing level despite the MEC being functional for a short time. These observations, along with cases of cathode contamination, led to the development of a two-chamber system with a total capacity of 100 L by Heidrich [82,83]. In the tests conducted with wastewater produced domestically, methane production was inhibited successfully with the introduction of a polymeric membrane. This membrane separated the cathode from the rest of the reactor and helped in avoiding microbial crossover to the catholyte [79].

In batch mode, COD removal was high with higher H_2 production. However, when the operational mode was switched to continuous mode, the COD level decreased along with the Coulombic efficiency. The probable reason behind the performance degradation may have been inefficient substrate flows into the reaction chamber and slow transportation of produced gas to the collection chamber. Hence, the usefulness of the transportation phenomenon in a functional scale-up reactor was highlighted [79]. Cotterill explained the cassette design module with Baeza using three different cassettes [84]. They observed material decay and degradation due to the applied voltage in a MEC. The development of a hermetic system is a challenge for reactors as the up-scaled reactors have shown leaking problems that might create balance errors. The existing systems are the perfect basis for up-scaled reactors; however, more work has to be done on the reactor design to make it fit for commercial and industrial applications [85–87]. More low-cost membranes and electrode materials need to be recognized. Attaining an optimum applied voltage and operational mode can also change the future of MECs [79]. Propylene has been the material chosen for building the body of a large-scale reactor [80,83], whereas polycarbonate is used for smaller systems [88]. Due to economic reasons, stainless-steel electrodes are used. Good electrolytic activity and proper H_2 gas evolution add to its characteristics. However, nickel for the cathode and carbon materials for the anode [87] are also used.

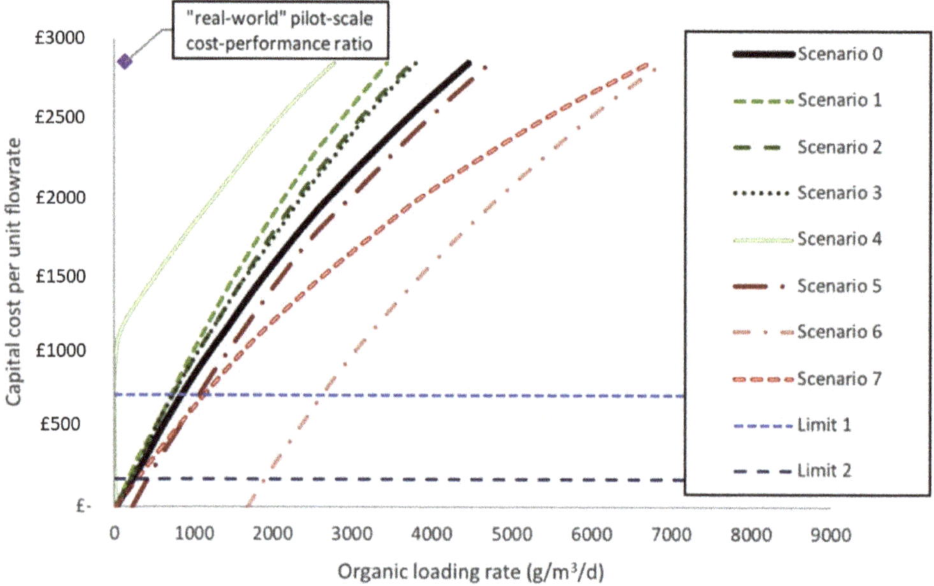

Figure 8. Cost–performance ratio curves for eight scenarios of a financially competitive MEC. Scenario 0: baseline MEC model; Scenario 1—double hydrogen yield; Scenario 2—applied voltage reduced to 0.6 V; Scenario 3—energy price changes; Scenario 4—anode and current collector value returned after 20 years; Scenario 5—membrane replaced annually; Scenario 6—membrane and cathode replaced annually; Scenario 7—additional staff member required; Limit 1—capital cost of reactor minus the anode; Limit 2—capital cost of reactor minus the anode and current collector (adapted from Aiken et al. [88]).

7. Optimizing Features Affecting the MEC System Design for Hydrogen Production and Wastewater Treatment

Rapid commercialization of such a technique can result in a significant expansion in the industry, while also helping to enhance funds and boost research for continual advancements. The factors to be considered in the analysis and comparison of MECs for wastewater treatment and energy generation include several variables. This assessment highlights the main feedstock, anode and cathode materials, reactor volume, system architecture, outputs, and expenses as the most important aspects when it comes to the system design. These factors have an impact on system performance, as well as the commercialization of the technology's economic feasibility. To be economically viable, MECs must strike a stability between overall system performance optimization and the economic proficiency of entirely basic components. MEC construction in the future should include ways to cut down on anode costs, enhance organic loading rates, and develop better knowledge about component requirements and electrode life expectancy [88]. For researchers to determine the next steps in the development of this technology, they must evaluate the parameters regarding industrial use and take into consideration the economic and manufacturing benefits, while simultaneously researching how to hasten the introduction of this method into the industry.

7.1. Feedstock

In general, wastewater strength is related to the degree of contamination of the water and determines the duration of treatment, reactor size, and quantity of energy needed and produced. This is because high-strength wastewater is rich in organic content; consequently, more energy may be recovered by treatment procedures. However, in the case of wastewater that has high strength, treating it will take a longer period, thereby increasing the amount of hydraulic retention time (HRT). An attempt should be made to enhance the HRT for a lower reactor size, thereby optimizing energy output or utilization. The loading levels and, hence, the organic loading rate (OLR) are substantially influenced by the source of wastewater.

With advancements in MEC cost, a system's sustainable organic loading rate (OLR) varies within 800–1400 mg·L^{-1} [88]. As a result of the lower concentration of COD in urban wastewater (300–500 mg·L^{-1}), it necessitates a smaller HRT (5–9 h); swine wastewater, on the other hand, requires a substantially longer HRT of 314 h due to the higher concentration of COD in the waste (18,300 mg·L^{-1}). Longer HRTs provide a substantial obstacle to industrial adoption since they reduce the volume of waste which can be processed each day and necessitate the construction of larger reactors. High-strength wastewaters, on the other hand, offer more energy and, as a consequence, are more economically viable in terms of energy production, since larger organic loads have a higher energy potential. Moreover, wastewater of low strength has a low COD (total dissolved solidity (TDS) <250 mg·L^{-1}), while the HRT is low; however, it is still difficult to economically justify treatment because of the extremely low energy output linked to its lower organic content. When wastewater strength (COD) increases even a little (by 360–400 mg·L^{-1}), treatment is possible economically, since higher energy generation offsets the higher cost of treatment.

A hydrogen-producing MEC with an OLR ranging between 1000 and 2000 mg COD·L^{-1}·day^{-1}, according to Gil-Carrera et al., is a viable alternative when compared to activated sludge in terms of treatment efficacy. For wastewater treatments with OLRs more than 2000 mg COD·L^{-1}·day^{-1}, a possible HRT for diverse waste streams has been estimated [89]. The addition of electrodes, a wider bacterial adhesion surface, and the voltage delivered all contribute to enhanced performance. Based on the 1400 mg COD·L^{-1}·day^{-1} OLR, both crude glycerol and cheese wheat, have extremely high HRT. In these circumstances, reactor design is critical to decreasing HRT while increasing solid retention to minimize reactor size. Traditional wastewater treatment requires a lot of energy, particularly activated sludge treatment, which accounts for approximately 60% of the total energy required. Compared to AD, MEC-ADs for the treatment of wastewater have led to a 1.7-fold increase in energy generation [90] and substrate removal, demonstrating the possibility for extremely effectual treatment for commercial waste streams previously employed. In addition, minimizing post-treatment needs would result in a significant reduction in the energy usage of the treatment procedure. The energy demands for full water treatment were estimated to be 0.057 kWh·m^{-3} (or 0.087 kWh·m^{-3} if ultraviolet treatment is chosen), which is about 85% less than the electrical energy consumption of a typical activated sludge process [91]. Utilizing MECs to reduce wastewater treatment energy consumption would have far-reaching global implications. There is an increase in the amount of research being done on the effectiveness of MECs in treating a variety of waste streams, which is aiding in the development of a better knowledge of how different microbial populations interact with different substrates. Because of its abundant availability and chemical composition, the waste feedstock is seen as an appealing and cost-effective substitute to pure chemicals in MECs.

7.1.1. Domestic or Residential Wastewater

The H_2 produced by MECs is less expensive than the projected commercial value of hydrogen ($6·kg^{-1} H_2), with a cost of $3.01·kg^{-1} H_2 for domestic wastewater [92]. Heidrich et al. ran a 120 L MEC on site for the treatment of domestic wastewater. According to their findings, over more than 3 months, with a Coulombic efficiency of 55%, the MEC was capable of creating pure H_2 (100% ± 6.4% purity). The reactor generated approximately 0.015 L H_2·L^{-1}·day^{-1} and retrieved around 70% of the electric power input [82,83]. According to Zhen et al. (2016), the utilization of the liquid portion of pressed municipal solid waste (LPW) for H_2 production was investigated. The maximum H_2 production (0.38 ± 0.09 m^3·m^{-3}·day^{-1} and 30.94 ± 7.03 mmol·g COD^{-1} added) was obtained at an applied voltage of 3.0 V and a pH of 5.5. Acetate, propionate, and butyrate, following their acetification, were used to achieve electrohydrogenesis, which resulted in an overall H_2 recovery of 49.5% ± 11.3% of the COD provided in the experiment [93].

7.1.2. Industrial/Food Processing Wastewater

Montpart et al. investigated the usage of glycerol, milk, and starch in varied concentrations and levels of complexity in synthetic wastewater using a single-chamber MEC. It was discovered that only milk was capable of sustaining hydrogen synthesis for a prolonged period of time. The introduction of glycerol and starch in MEC did not inhibit the total multiplication of H_2 scavengers, even under circumstances of short H_2 retention time caused by frequent nitrogen sparging [94]. Shen et al. used a continuous up-flow fixed-bed MEC to treat recalcitrant wastewater generated from hydrothermal liquefaction of cornstalks while simultaneously producing hydrogen. At 1.0 V in the cathode, a hydrogen generation rate of 3.92 $mL \cdot L^{-1} \cdot day^{-1}$ was attained, although the highest power density (305.02 $mW \cdot m^{-3}$) was achieved at 0.6 V [95]. Guo et al. studied the effect of several cathode/anode ratios in membrane-less MECs using beer wastewater. With an improved cathode/anode ratio of 4 $cm^2 \cdot cm^{-3}$ and an applied voltage of 0.9 V, methane production of 0.14 $m^3 \cdot m^{-3} \cdot day^{-1}$ was achieved [95]. Furthermore, wastewater from a soybean edible oil refinery was used to generate bioelectricity and biomethane through the utilization of MFCs and MECs [89]. In comparison to conventional anaerobic digestion, the methane yield was 45.4 ± 1.1 $L \cdot kg^{-1}$ COD, and the generation rate of MECs was 0.133 ± 0.005 $m^3 \cdot m^{-3} \cdot day^{-1}$.

7.1.3. Fermentation Effluents

The fermentation effluents comprise various byproducts such as acetate, butyrate, ethanol, lactate, and formate. These byproducts can still undergo further reactions, resulting in the production of more hydrogen. The fermentation of sugars by bacteria is a common source of H_2 production, although transformation of carbohydrates to hydrogen is insufficient. Additional hydrogen is frequently generated from the effluent of an ethanol dark fermentation reactor, according to the results of a single-chamber MEC test. At an applied voltage of E(ap) = 0.6 V, overall H_2 recovery of 83% ± 4% was achieved using a pH-controlled effluent (pH = 6.7–7.0), with a hydrogen generation rate of 1.41 ± 0.08 m^3 $H_2 \cdot m^{-3} \cdot day^{-1}$; furthermore, after combining the MEC and fermentation system, the overall H_2 recovery increased to 96%, and the system was able to produce an average of 2.11 m^3 $H_2 \cdot m^{-3} \cdot day^{-1}$, which corresponds to a voltage efficiency of 287%. At applied voltages ranging from 0.5 to 0.8 V, high cathodic hydrogen recoveries (ranging from 70% ± 5% to 94% ± 4%) were obtained [96].

When the MEC was integrated with the other fermentation system, 96% of the H_2 was recovered at a production rate of 2.11 m^3 $H_2 \cdot m^{-3} \cdot day^{-1}$, resulting in electrical energy productivity of 287%. Sosa-Hernández et al. investigated the potential of spent yeast (SY) for energy recovery in MEC. Tests were conducted on the concentrations of SY produced by bench alcoholic fermentation and ethanol, which ranged between 750 and 1500 mg $COD \cdot L^{-1}$ and between 0 and 2400 mg $COD \cdot L^{-1}$, respectively. The removal efficiency (RE), Coulombic recovery (CR), Coulombic efficiency (CE), H_2 production, and current density of the COD removal system were all measured and analyzed. The combination of 1500 mg $COD \cdot L^{-1}$ SY + 1200 mg $COD \cdot L^{-1}$ ethanol produced an appealing current density (222.0 ± 31.3 $A \cdot m^{-3}$) and H_2 generation (2.18 ± 0.66 $L_{H2} \cdot d^{-1} \cdot L_{Reactor}{}^{-1}$) [97]. Cai et al. developed a bioelectrochemically assisted anaerobic reactor and compared it to an anaerobic digestion (AD) control reactor in order to generate methane. They achieved an average methane production rate of 0.070 mL $CH_4 \cdot mL^{-1}$ reactor$\cdot day^{-1}$, which was 2.59 times greater than the AD control reactor (0.027 m^3 $CH_4 \cdot m^{-3} \cdot day^{-1}$), as well as an increase in COD removal of approximately 15% above the AD control. When the fermentation liquid is changed to sludge fermentation liquid, the rate of methane generation was raised even further, reaching 0.247 mL $CH_4 \cdot mL^{-1}$ reactor$\cdot day^{-1}$ [98].

7.1.4. Swine Wastewater

Employing a single-chamber MEC with a graphite-fiber brush anode, hydrogen gas was produced at a rate of 0.9–1.0 m^3 $H_2 \cdot m^{-2} \cdot day^{-1}$ utilizing either pure or diluted swine wastewater (which may contain complex molecules that degrade at a slower rate than hydrogen). As a result, a greater allocation of carbon should be assured in order to meet the carbon demand that cannot be fulfilled by the organisms themselves. Undiluted swine waste matter would require a carbon-to-nitrogen ratio greater than 1.7 in order to degrade completely). COD removals ranged from 8–29% in 20 h tests and 69–75% in longer tests (184 h) that used full-strength effluent to achieve the highest levels of COD removal. The gas produced contained up to 77% ± 11% H_2, with overall recoveries of up to 28% ± 6% of the COD contained within the waste matter being recovered as hydrogen [99].

7.1.5. Refinery Wastewater

Lijiao Ren was the first person to make use of refinery wastewaters [100]. The treatment of six different refinery wastewater samples was observed in mini-MECs. These waste matter samples were assessed separately and in combination with domestic waste matter as MFC feedstock to determine whether the treatability of the waste matter might be enhanced by introducing microbes and nutrients. In MEC testing, the various refinery wastewater collections differed in terms of current production and treatability. All the de-oiled wastewater samples from mixed sources performed well, with one sample providing values similar to DW. Other samples had low current densities as a result of high starting pH or a low BOD. The effectiveness of MECs was investigated using current generation throughout a number of batch-fed cycles. Refinery wastewater (RW) was treated in MECs that were previously acclimated utilizing domestic wastewater (DW) or a 50:50 mixture of RW and DW. The most effective results were obtained from de-oiled refinery wastewater collected from a single site (DOW1), which had a maximum current density of 2.1 ± 0.2 $A \cdot m^{-2}$ (maximum current density), 79% COD removal, and 82% BOD removal. The results observed were consistent with the results obtained from the domestic wastewater treatment study [100].

7.1.6. Winery Wastewater

Field site experiments were carried out at the Napa Wine Co. (NWC), which is located in Oakville, California, USA. NWC crushes and decants an annual total of 7000 tons of grapes and creates an annual total of 5.1×10^4 m^3 of wastewater. The treatment of on-site wastewater (through aerobic biological oxidation) consumes roughly 654,000 kWh of power per year. A continuous-flow MEC on a scale of 1000 L was built and analyzed for current production and COD removal using winery waste matter, with positive results. The reactor was divided into 24 modules, each containing 144 electrode pairs. The development of an exoelectrogenic biofilm took around 60 days, which is significantly lengthier than the time typically necessary for laboratory reactors. By the time the experiment was scheduled to be completed, current production attained a maximum of 7.4 $A \cdot m^{-3}$ (after 100 days). Despite the fact that the majority of the product gas was transformed to methane (86% ± 6%), the cathodic approach produced a peak of 0.19 ± 0.04 $L \cdot L^{-1} \cdot day^{-1}$. Higher techniques for isolating H_2 gas generated at the cathode will be necessary in future testing in order to increase the amount of hydrogen recovered. The current generation was adjusted by guaranteeing a sufficient volatile fatty acid content (VFA/SCOD $\geq$ 0.5) and by increasing the temperature of the waste materials (to 31 ± 1 °C) in the waste stream. In addition, the usage of MECs remains a potential technology for the integration of energy recovery and wastewater treatment [101].

Various substrates have been utilized in MEC as illustrated in Table 2.

Table 2. Performance of MEC configurations using different substrates.

Substrate Type	MEC Specification	Applied Voltage (V)	Hydrogen Production Rate ($m^3 \cdot m^{-3} \cdot day^{-1}$)	Reference
Acetate	Two-chamber MEC	0.45	0.37	[47]
Acetate	Two-chamber MEC	0.5	0.02	[102]
Domestic wastewater	Bioelectrochemically assisted microbial reactor	0.5	0.01	[103]
Acetate	Modified MFC	0.6	1.1	[104]
Acetate	Single-chamber membrane-free MEC	0.6	0.69	[64]
Acetate	Single-chamber membrane-free MEC	0.8	3.12	[81]
Acetate	Single-chamber MEC with anion-exchange membrane	1.0	0.3	[71]
Acetate	Two-chamber MEC	1	50	[105]
Glycerol	Two-chamber MEC	0.9	3.9	
Municipal solid waste	Single-chamber MEC	3.0	0.38 ± 0.09	[106]
Glycerol, milk, starch	Single-chamber MEC	0.8	0.94	[94]
Palm oil mill effluent	Single-chamber MEC	0.1–0.8	205 mL $H_2 \cdot g$ COD^{-1}	[107]

7.2. Inoculation

Inoculating microorganisms into a MEC in which they would survive and grow, producing a microbial biofilm on the electrodes, is known as system inoculation. Because microorganisms are inoculated into the system, inoculum selection is essential for organic matter degradation [108]. Because the response of an MEC system is governed by the microorganisms involved, inoculation directly controls the start-up time of the system. The start-up period refers to the time it takes for a system to begin producing H_2 or biogas. Long start-up durations diminish system proficiency because it takes longer for the system to begin generating energy; consequently, choosing the correct inoculum source is critical to maximizing system efficiency. Methods for reducing the time it takes for microbial electrochemical systems (MESs) to start up have been investigated. Selection of inoculum could be helpful to reduce start-up time while simultaneously improving efficiency. Biogas production was significantly increased by 18.5% when a mixture of inoculum sources containing a 1:4 ratio of activated sludge and municipal sewage was used as compared to a mono-inoculated treatment, implying that diversification is required for nutrient enrichment, which is essential for digestion [108]. In other words, the development of the methanogenic consortium in the system was also more rapid, which indicates that this mixed bacterial consortium helps enhance the hydrogenotrophic methanogen population. The amount of inoculum at the start of the experiment is equally critical. In the study conducted by Escapa et al., the authors discovered that the reactor failed to start when residential wastewater of low strength (230 mg $COD \cdot L^{-1}$) was used as the inoculum [51]. It is said that, with the addition of gas production, the start-up time for a large pilot system might be anywhere between 50 and 90 days. A simple technique to shorten the time it takes to set up a large-scale system is through "pre-acclimatization" of electrodes by applying a beneficial microbe to the electrodes before putting them into the system. Carbon electrodes that were inoculated with hydrogenotrophic bacteria extracted from natural bog sediment and pre-charged with hydrogen-rich water started up faster and produced more methane than those that were not [109].

A significant start-up time can also provide problems if repayment is not completed within a certain time frame. Pre-inoculation of electrodes can give a low-cost option for lowering start-up time, but this method comes with the added benefit of avoiding expensive modifications. Additional research into the best beginning circumstances will increase the technology's appeal for industrial use, providing the basis for MECs to be commercialized for worldwide wastewater treatment.

7.3. Electrode Material

7.3.1. Anode Material

The anode's performance is critical for bio-electrochemical systems that rely on bio-electrochemical reactions, which take place at the anode, and which can be replaced by MFCs, MECs, and MEC-ADs. It was established that anode activity is a restrictive component in the overall performance of the system [110]. As an electro-active bacterium (EAB) attaches itself to the bioanode surface, producing a biofilm, it provides the bioanode with energy. As a result of this oxidation by EAB, organic molecules are converted to CO_2. Because of their remarkable ability to adhere to EAB, their huge surface area, and their abundance, carbon-based materials have become the most extensively used electrode material [111]. It has been shown that carbon compounds help to increase interfacial microbial colonization and, thus, the production of biofilms. Electrically conductive current collectors of metals are employed as electron acceptors, to overwhelm poor conductivity. Titanium wire is often utilized because of its corrosion resistance [40]. Additionally, the capability to simulate interfacial microbial colonies allows for improved current density by developing a beneficial microenvironment for electron transport that compensates for the decreased conductivity [112]. Graphite is affordable, plentiful, and conductive, and, because of this, it has become one of the most extensively used electrode materials [113]. Graphite electrodes have been implemented in many ways, including brush, granular, rods, felts, and foams [112]. Nevertheless, graphite's molecular structure and morphology are both planar in comparison to other carbon materials due to its low surface porosity required for bacterial adhesion. Surface area-increasing porous 3D carbon materials, such as carbon brushes, felts, meshes, and foams, have been the focus of recent research [114]. Carbon fiber (CF) electrodes have previously proven to be effective and are currently being used to achieve good outcomes [115]. Carbon nanotubes (CNTs) have incredible electrical, mechanical, biological, and thermal properties, making them ideal for real-time applications. Despite extensive study and application possibilities of carbon nanotubes, many issues such as biodegradability, biotoxicity, and biosafety remain difficult to address and should be addressed with caution prior to design and manufacturing [116]. Anodes of a mesh-like design (i.e., porous, woven, or multilobed) tend to generate greater current densities (more current flows) than flat or plate-shaped anodes because of improved mass transfer, surface area, and biofilm growth. Carbon fiber brushes give excellent test results; however, because of their very expensive cost, they are seldom employed in large-scale BES. Based on an independent study, which used recycled carbon fiber anodes and found that, in comparison to graphite felt anodes, the use of recycled carbon fiber electrodes produced better results while also being cheaper, Carlotta-Janes et al. found that it was possible to improve performance while cutting costs if recycled carbon fiber anodes were used. As there is a considerable portion of the anode in the current model without a biofilm, increasing the surface area of the anode is more likely to result in a greater increase in biofilm density and adherence on the anode. Reduction in anode size will be advantageous for commercial viability since the anode material constitutes about 70% of the whole system, which will need a 90% drop in cost to make it profitable [88]. Additionally, molybdenum anodes showed excellent overall durability, neither corroding nor lowering in current production for over 350 days. Another important consideration is the endurance of the electrode materials. Unfortunately, there are no data on electrode materials' long-term durability, and most experiments last about 1 year. Material dissipation in electrodes is often underestimated, which may lead to significant issues when determining which materials to utilize commercially. Stainless steel is also good since it has several characteristics which can be utilized [117]. In terms of conductivity and scale-up potential, stainless steel outperforms carbon anodes owing to lower capital expenditures, despite a relatively flat surface, which reduces its biocompatibility. Stainless steel has a high nickel concentration and may efficiently catalyze the HER. Stainless-steel brush cathodes, for example, produced hydrogen at a rate of 1.7 $m^3 \cdot m^{-3} \cdot day^{-1}$ and had a cathodic efficiency of 84%, comparable to Pt cathodes in single-chamber MECs. The high Ni content (8–11%) and the large specific surface area were also implicated for the rapid hydrogen generation (810 $m^2 \cdot m^{-3}$). Flame spray oxidation improves the biocompatibility of stainless steel by producing an iron oxide coating on the surface, which facilitates the adherence of iron-reducing bacteria and increases surface roughness without sacrificing corrosion resistance [118]. Because stainless steel has yet to be tested on pilot systems, more research into its durability is needed. Cotterill et al. compared a 30 L tank to a 175 L tank to examine how tank capacity influences H_2 production. H_2 generation was fourfold greater in the small MEC in comparison to the bigger MEC, when the anode surface area was reduced from 1 m^2 to 0.06 m^2. The larger MEC had a lesser performance, demonstrating a negative relationship between scale and gas output, implying that efficiency decreases as size increases [119]. As part of the commercialization process, a cost–benefit analysis of anode materials must be completed, which considers the material's availability, corrosion resistance, and capacity to scale up.

7.3.2. Cathode Material

The necessity to create either CH_4 or H_2 dictates reactions at the cathode, and this is dependent on the need and potential to manufacture and utilize hydrogen or methane on-site. The rate at which H_2 is consumed is determined by the amount of methanogenic activity present. The hydrogen will very probably be consumed throughout the reaction if the device is operated as a single chamber without a membrane, and the biogas generated will be in form of CH_4 [80]. Temperature has a direct influence on methanogen activity, with temperatures exceeding 35 °C considerably boosting methanogenic activity [120]. Hydrogenotrophic methanogens are more prevalent in MECs where CH_4 is produced, according to the study. Hydrogenotrophic methanogens produce CH_4 through the intermediate synthesis of H_2. The CH_4 synthesis route reveals that the cathode material's hydrogen evolution capacity is a critical design factor. As a result, the cathode serves as both a biocatalyst and an electrocatalyst, enhancing hydrogen evolution reactions (HER) by increasing electrode–microbe electron transfer [121]. The presence of hydrogen-scavenging bacteria in waste streams necessitates the use of membranes if pure hydrogen is required. Multiple investigations have shown that membrane systems can achieve hydrogen purity above 98% [122]. Corrosion resistance, good conductivity, high specific surface area, biocompatibility, and outstanding mechanical qualities are all required of successful cathode materials [123]. Furthermore, cathodic materials must minimize significant hydrogen evolution overpotentials. Cathode fabrication for industrial application must be low-cost, utilizing easily accessible materials and conventional production procedures, for the large-scale deployment to be practicable. Metals have been investigated because they conduct electricity more efficiently than carbon-based materials [124], and they have greater biocompatibility, as well as cathode potential, which prevents corrosion. Platinum has the strongest HER activity, which leads to improved H_2 evolution [125]. Platinum, on the other hand, has some disadvantages, including being expensive and having substantial mining environmental effects; as a result, the invention of new metallic electrode materials is required [126]. Stainless steel and nickel have performed well as nonprecious metals [127].

Stainless steel is a typical material for electrode construction because it is a relatively inexpensive metal. When it comes to hydrogen production, stainless steel with a large specific surface area can be as effective as a platinum catalytic electrode containing carbon. Because of its high conductivity as a transient metal, stainless-steel mesh is thought to have outstanding ohmic resistance and electron transport resistance. Meshes and brushes made of stainless steel have a low cost and excellent performance, making them an ideal cathode made of a non-precious metal for further evaluation and scale-up operations. The findings are consistent with the use of meshes and wool in pilot-scale systems to produce a low-cost, high-surface-area cathode with a low cost and large surface area. Nickel, like other non-platinum metals, has high corrosion resistance, as well as high hermetic electron transfer activity. Nickel is also more corrosion-resistant than stainless steel, which is important for an electrode because it must be long-lasting in order to be commercially viable. Hydrogenotrophic methanogens play a role in the enhanced performance, implying that nickel's high HER activity relative to other materials helps it perform better [128]. HER activity must be a key focus of study to maximize the efficiency of cathodes. Stainless steel is now commonly used. Due to its availability, a pilot study has shown it to be the ideal cathode material for scale-up. Cost and machinability are two factors to consider. On a pilot scale, a comparison was made between nickel and stainless steel. It would be wise to experiment with nickel cathodes to determine if the improved performance justifies the additional expense. Various cathode materials are described in Table 3.

7.4. Effect of Electrolyte pH

Because the HER at the cathode depends on electrolyte pH and has the most crucial impact on overall performance of MECs. High overpotentials can occur owing to a difference in redox potentials between anode and cathode chambers; it was observed that more cation instead of proton percolates through the cation-exchange membrane. Consequently, the cathode becomes alkaline while the anode becomes acidic. Theoretically, 59 mV of voltage loss is incurred due to a difference in pH level of 1 between the anode and cathode. Microbial activities are pH-dependent; microbes are highly sensitive to surrounding pH, and its variability may cause modifications in microbial respiration and, consequently, extracellular electron transfer. In fact, because microbes are mostly active at neutral pH, most MEC studies are conducted at pH 7. Moreover, many other parameters (ion transfer, conductivity, substrate oxidation, etc.) are directly or indirectly associated with pH. Researchers have reported that low cathode and high anode pH improved hydrogen production [129–131]. Protons accumulate under high pH, thereby increasing the electrogen proliferation due to conducive environment. Research suggests that periodic polarity reversal can be used to stabilize pH in two-chambered MECs [132]. An electrolyte, including a weak acid, operates as an electric charge at high pH, increasing MEC characteristics, and the deprotonation process may increase the conductivity of the electrolyte while lowering the impedance between the anode and cathode. However, we must evaluate the possible impacts of weak acid catalysis and solution resistance for a lower pH electrolyte to determine whether the reactor can function more effectively. However, certain experimental findings revealed that the presence of phosphate species and some weak electrolyte acids, as a charge carrier for improving conductivity, had a beneficial impact on a stainless-steel brush cathode and also reduced the Pt/C cathode's overpotential. Merrill et al. [133] found that lowering the pH improves MEC performance by lowering solution resistance and cathode overpotential. Munoz et al. [134] found that using phosphate as an electrolyte may increase the rate of hydrogen generation and current density in MECs. Yossan et al. [135] investigated five kinds of catholytes in MECs, namely, deionized water, tap water, NaCl solution, acidified water, and a phosphate buffer. Due to its greater buffer capacity, a 100 mM phosphate catholyte in a MEC exhibited the best rate of hydrogen generation. As a result, phosphate is the most often utilized electrolyte in MECs.

7.5. Temperature

Temperature is a key element in MECs that affects their function because it improves exoelectrogen selectivity and production. Most microbes prefer an optimum temperature range of 35–40 °C for growth, enzyme activity, and the development of a durable biofilm, which increases substrate degradation, mass transfer, and power generation. According to the COD removal efficiency and microbe loading at the anode, Omidi et al. reported that 31 °C is the most efficient operating temperature for MECs [136]. As a result, the test temperature of an MEC is typically kept at about 30 °C. Lu et al., on the other hand, demonstrated that utilizing a single-chambered MEC produces hydrogen at low temperatures such as 4 °C, while simultaneously reducing the generation of methane [137]. Additionally, the anode biofilm and hydrodynamic force both impact hydrogen generation in MECs [138], with the hydrodynamic force having a larger effect on hydrogen production than the anode biofilm [139]. Furthermore, by placing two anodes on each side of the cathode, additional hydrogen may be generated.

7.6. Applied Potentials

As explained in the above sections, a minimum of 0.2 V is required to break the thermodynamic barrier for feasibly producing hydrogen at the MEC cathode. Large cathodic overpotentials reduce the efficiency of the overall process. Even though hydrogen evolution increases with increasing applied potential [140], optimum potential ranging from 0.2–0.8 V must be applied for achieving process scalability [141]. Researchers have reported that varying applied potential can decrease cell metabolism and increase cell lysis [142].

8. Bottlenecks in Commercialization of MECs for Biogas Production during Wastewater Treatment

It is essential to understand the functioning of MECs on a large scale for wastewater treatment. This also poses a barrier in adopting MECs owing to the little understanding of these systems. More modest frameworks produce high energy yield per volume when standardized. For metropolitan wastewater, the benchtop investigation produced standardized net energy of 25.96 $kWh{\cdot}m^{-3}{\cdot}day^{-1}$ [88]. On the other hand, a pilot-scale framework produced standardized net energy of 0.11 $kWh{\cdot}m^{-3}{\cdot}day^{-1}$. This demonstrates that energy creation cannot be accurately scaled, and that proficiency decreases as the size of the system increases. The net energy of a 1000 L pilot-scale framework with a cathodic surface area of 18.1 $m^2{\cdot}m^{-3}$ was determined to be 2.11 $kWh{\cdot}m^{-3}{\cdot}day^{-1}$ for vineyard wastewater [80]. One obstruction is the trouble of precisely contrasting the various arrangements in the examination. The reactor size is just a single boundary and does not give a genuine portrayal of the framework's versatility.

Table 3. Different cathode materials along with catalysts used in MECs.

Cathode Material	Catalyst	I_d	Q ($m^3{\cdot}m^{-3}{\cdot}day^{-1}$)	E_{app} (v)	References
Activated carbon	Nitrogen	NA	0.0060 ($m^3{\cdot}m^{-2}$)	0.8 $kWh{\cdot}m^{-3}$	[143]
Stainless steel	Nickel oxide	NA	0.76	0.6 $kWh{\cdot}m^{-3}$	[49]
	Molybdenum disulfide	10.7 Am^{-2}	NA	NA	[144]
Carbon cloth	Nickel/molybdenum	2.1 Am^{-2}	1.25	0.6 $kWh{\cdot}m^{-3}$	[145]
	Molybdenum disulfide/carbon nanotubes	NA	0.01	0.8 $kWh{\cdot}m^{-3}$	[146]
	Nickel powder	NA	1.2	0.6 $kWh{\cdot}m^{-3}$	[69]
Carbon paper	Nickel-tungsten	200 Am^{-2}	1.5		[66]
	Nickel powder	NA	2.6 ($L{\cdot}m^{-3}{\cdot}day$)	1.0 $kWh{\cdot}m^{-3}$	[147]
	Nano-Mg $(OH)_2$/graphene	18.3 Am^{-2}	0.63	0.7 $kWh{\cdot}m^{-3}$	[148]
	Palladium nanoparticles	NA	2.6	0.6 $kWh{\cdot}m^{-3}$	[149]
		22.8 Am^{-2}	50	1.0 $kWh{\cdot}m^{-3}$	[105]
Nickel foam	Nickel/phosphorous	NA	2.29	0.9 $kWh{\cdot}m^{-3}$	[150]
	Nickel/molybdenum	NA	0.13	0.6 $kWh{\cdot}m^{-3}$	[151]
	Nickel/tungsten		0.14		[152]
	Graphene	NA	1.31	0.8 $kWh{\cdot}m^{-3}$	[153]
	Nickel/iron layered double hydroxide	197 Am^{-2}	2.12	0.8 $kWh{\cdot}m^{-3}$	[154]
Gas diffusion electrode	Nickel powder	4.6 Am^{-2}	5.4	1.0 $kWh{\cdot}m^{-3}$	[155]

8.1. Economic and Cost Analysis

The confirmation that innovation is financially viable for a large scope and has productivity that is virtually identical to or better than other arrangements accessible is critical to the passage of innovation from the lab to the field. Performing techno-economic and life-cycle assessments is important for progressing innovation used in the financial analysis of energy advancements to identify and evaluate operating and capital expenditures (OPEX and CAPEX) over their entire life cycle. The yield considers financial aid evaluators and levelized cost of energy objectives, such as investment return and net present value. To analyze the natural influence, the LCCA appraisal must be conducted similarly to the life-cycle assessment, with the same assumptions. MECs will benefit from using these projections since they will be better able to comprehend future costs, execution procedures, and risk factors [156]. When we say CAPEX, we are referring to the expenditure of performance; however, when we say OPEX, we are referring to the operational expense throughout the performance. Even though simple arrangements have a low CAPEX, they may be wasteful and require additional support, resulting in a higher OPEX than more complex arrangements. As a result, while evaluating the complete LCCA of innovations, it is necessary to consider both the CAPEX and the OPEX costs. Furthermore, when considering the profit from the venture from the energy produced, a high CAPEX might be acceptable in innovation such as MEC-AD for wastewater treatment, where energy is given [157].

In the ebb and flow UK market, 1 kWh of electrical expense is equivalent to £0.144, and, as a result, the price of treating 1 m^3 of metropolitan wastewater is equivalent to £0.072 per kWh of electricity. As a result, for every 1 kg of COD neutralized, activated sludge treatment produces 0.4 kg of sludge, necessitating extra processing, which is often done using AD in order to recover a fraction of the energy [157]. MECs, on the other hand, can both reduce COD and generate energy. MECs currently require a significant amount of CAPEX to be implemented and are several times more costly (248-fold) than activated sludge frameworks [88].

According to the results of an analysis, the overall cost of the MEC framework was around £2344·m^{-3}. In aspects of toxicity, MEC was evaluated by comparing to activated sludge. Although the MEC CAPEX was twice that of the active sludge, the consumption of energy was 10-fold lower [88]. Because of the potential for MECs to generate bioenergy, they can be either energy-neutral or energy-positive, whereas activated sludge is the most energy-concentrated of the currently available wastewater treatment techniques, according to the EPA [82,83]. Generally speaking, the expenses of MECs need to decrease by 84% to £375·m^{-3} [158].

The positive cash flows generated at 2020 pricing (£15,000·$year^{-1}$) (Figure 8) had a small influence on the net present value (NPV) under scenario 1 (a doubling of current yields to 30 L·m^{-3}) due to the cost of materials (assuming current performances). MEC capital expenses are projected to exceed their revenue from hydrogen production, as a result. For the baseline scenario (0) to be economically viable, either hydrogen prices must be raised to £5.09·kg^{-1} (higher than the EU's target of £3.55 by 2020 and £2.66 by 2030) or yields must be increased from 15 L·m^{-3} to 21.5 L·m^{-3} by 2020 and 28.7 L·m^{-3} by 2030 (increases of 43% and 91%, respectively). Due to the relatively high cathodic efficiencies, it will be necessary to raise Coulombic efficiencies and the organic loading rate (OLR) in order to improve yields. Although hydrogen must be sold at a loss in order to be competitive with other sources of energy, power costs in MECs are negligible when compared to those of activated sludge (AS) [88].

8.2. Scale-Up Strategies

An exploration was directed in which a pilot-scale MEC took care of winery wastewater utilizing different stacked modules and discovered that there was equivalent execution across every one of the 24 modules that were practically identical to the limited lab-scale test. It was tracked down that weak electric association among modules brought about considerable ohmic misfortunes, bringing about various flow densities among modules, featuring the need to utilize great arrangements among modules [80]. Although less expensive than utilizing large tanks, this methodology can be significantly more expensive than doing so because as many materials are needed to transport the numerous modules. This increases the CAPEX of the framework and serves as an obstacle to the commercialization of the product. Another examination distinguished that, while considering a piled strategy for the increase of hydrogen-generating MECs, the partition film between the anode and cathode poses a huge hindrance. The investigation tracked down that high internal resistance inside the cell is brought about by the sluggish development of electrons over the membrane layer. Moreover, because membranes are costly, using a large number of films could increase the cost of the framework, creating a significant barrier to commercialization. This shows that frameworks devoid of membranes might be effectively versatile.

The technique for arranging modest cells in series is equivalently less reported; numerous little vessels masterminded in series against one bigger tank implies that the framework can exploit improved blending, strong maintenance, diminished ohmic misfortunes, and enhanced microbial consortium separation over the treatment cycle. Large, permanent treatment vessels are frequently used in conventional wastewater treatment frameworks, and their treatment capacity is frequently restricted. Organizing reactor vessels in succession allows sediments to settle throughout the framework, reducing strong maintenance duration and developing microenvironments inside the framework based on the number of organics in each vessel [159]. A higher concentration of larger, undigested organics will typically be found in tanks toward the beginning of a series rather than tanks at the end of a series. Microbial colonies form in response to these specific conditions, just as they do in response to fluctuations in distributed oxygen, pH, volatile substances, and unsaturated fatty acids [160].

Pilot-Scale Limitations

MEC technology is at the technological readiness level (TRL) 5, which indicates that the system is being evaluated in a particular context. Instead of focusing solely on illustrating the result, future research should seek to standardize it. According to the literature study of the various pilot projects, there are two major areas where MECs must improve in order to become a viable wastewater treatment technology. It would be critical to achieve the target volumetric treatment rates at the beginning of this process in order to determine the feasibility of this strategy. Existing research suggests that increasing the rate at which organic materials are loaded into the system can improve this volumetric technique. This can be performed by either increasing the intensity of the wastewater or increasing the flow rate of the wastewater. Any of these modifications would increase the rate at which organics are distributed into the reactor within the wastewater, which in turn would increase the rate at which organics are removed from the wastewater. By fully appreciating and optimizing this mass transfer, it is able to design the optimal reactor length for the given flow rate while also accurately predicting the costs. Furthermore, the issue of scale is a significant concern [161]. The highest MEC recorded so far was 1000 L, achieved with a hydraulic retention time of 1 day. When it comes to wastewater treatment activities, it is far from the scale that is required. Therefore, several BESs have been built with compact electrodes, numerous units of which can be mounted in any established tank. This is crucial because it allows for treatment in urban environments with a small land footprint to be achieved with these current tanks because of their depth. The design of an electrode that can stretch to a depth of 3 m would have to take into account the influence of hydrostatic pressure on the development of the biofilm, the output, and the structural integrity of the electrode itself. It would, therefore, be necessary to cope with the advancements in the kinetic and thermodynamic features of biological and electrochemical mechanisms that have occurred throughout time. More research is required to investigate the most recent reactor designs that keep the same land footprint as existing properties, which is currently lacking [162].

8.3. Investigation of Methane and Hydrogen Generation and Its Implications for Industrial Application

Purified hydrogen can be utilized as an important asset for fuel and assembling different synthetic substances. Because hydrogen has a greater energy content than methane, it has the potential to improve the efficiency of hydrogen generation in comparison to methane. MECs for the evolution of hydrogen consist of a double-chamber reactor with a film placed at the cathode to prevent hydrogenotrophic methanogenesis from occurring, thereby improving the framework's cost, and complications will only increase their expense and complexity [161]. In typical wastewaters, a rich source of microorganisms is present, rendering the cathode susceptible to biofouling, which could require replacement of the cathode, thus increasing the costs [88]. Hydrogen likewise requires modern storage spaces and handling hardware. On account of the on-site production of energy, the worth of methane and hydrogen will decrease to the levelized cost of energy (LCOE) and levelized cost of heat (LCOH) produced and used. There are broad foundations and equipment to preserve and change methane into electrical and nuclear power. A pilot-scale framework was developed to provide methane while also generating a significant amount of net energy [80]. In comparison, the standardized net energy announced for a hydrogen-delivery MEC was 76.2 $kWh{\cdot}m^{-3}{\cdot}day^{-1}$ at 98% purity, which is higher than the previous estimate [127].

The gas storage foundation additionally assumes a critical part in the business reasonability of utilizing MECs for energy production. It has been recommended that MECs can be combined with inexhaustible resources [162]. Renewable energy provides the MEC with the ability to generate gas, which can be used as a kind of energy storage. Compared to hydrogen, methane has a greater energy thickness when it comes to barometrical pressing factors. In addition to its increased energy thickness and greater atomic size, methane's increased practicability makes it the most feasible candidate for non-compressed energy storage. Biogas generated by AD is of low quality and difficult to store, with storage energy accounting for only 10% of the total energy produced [163]. MEC-ADs generate biogas of good quality with CH_4 content coming to 86% $\pm$ 6% [80], featuring the upside of MEC-ADs over AD frameworks with respect to energy storage capacity. MEC-ADs designed to generate methane can contend with AD innovation; however, this is generally neglected on a wide scale owing to expenses. In contrast to the generation of power from hydrogen, there are various innovations that can use biogas straightforwardly with responding motors, microturbines, energy units, gas turbines, steam turbines, and consolidated cycle frameworks. Moreover, biogas can be converted into biomethane, which can then be infused into the matrix [163]. Lastly, when compared to the development of methane storage and conversion innovation, the development of hydrogen storage and conversion innovation is less advanced.

Hydrogen Production Technology and Inventory

The production of hydrogen is primarily accomplished through two methods: steam methane reforming (SMR) of natural gas and electrolysis of water. Recently, there has been increased interest in hydrogen production from biomass resources. This is due to the large amount of biomass waste generated by various industrial and agricultural activities, which has the potential to be converted into useful energy in a very short period of time. A further interesting alternative for hydrogen production from renewable resources is high-temperature electrolysis (HTE) by solid oxide electrolysis cells (SOECs). These are the hydrogen production techniques that were investigated in this study, together with their inventories, which are summarized and provided in Table 4. In the literature, you can find a thorough overview of each of the different H_2 technologies [164]. Although many key input parameters have been changed in this study, several of the underlying assumptions used in the analysis are still based on the default assumptions of the sub-models used for inventory data (the hydrogen production analysis models—H_2A and the GREET model) and the Eco-invent. The emphasis of this study was on the unit-operational level, and, for the sake of simplicity, only the major unit (i.e., electrolyzer, industrial reformer, furnace) for each unit source was reconstructed utilizing generic data from the Eco-invent database for the manufacturing infrastructure. Because of the lack of data availability, several secondary systems were ruled out. It is widely acknowledged that such stages have minor consequences if they are spread out over the course of their operating lifespan [165]. Several different system configurations were simulated in order to test the sensitivity of the findings.

Table 4. Resources required to produce 1 kg of H_2 from different production technologies and pathways [166].

Type	Conversion Pathway	Electricity ($kWh \cdot kg^{-1}$ H_2)	Water ($kg \cdot kg^{-1}$ H_2)	Ammonia ($kg \cdot kg^{-1}$ H_2)	Glucose ($kg \cdot kg^{-1}$ H_2)	Corn Liquor ($kg \cdot kg^{-1}$ H_2)
Thermo-chemical	Steam methane reforming (SMR)	1.11	21.869	-	-	-
	Biomass gasification (BMG)	0.98	305.5	-	-	-
	Coal gasification (CG)	1.72	2.91	-	-	-
Biological	Dark fermentation + microbial electrolysis cell (MEC), without ER	21.6	104.225	0.102	0.335	0.008
	Dark fermentation + microbial electrolysis cell (MEC), with ER	6.03				
	Dark fermentation + microbial electrolysis cell (MEC), with H_2 recovery	21.6				

9. Integrated MEC Systems

MECs have been shown to be viable options for dealing with the issue of wastewater treatment while also producing hydrogen. However, there have been certain challenges in commercializing these MEC technologies. Thus, combining new technologies with traditional ones may be used to overcome the thermodynamic limits, as well as material prices, methanogens, substrate concentration, and other issues that MES faces on its own. The performance of several integrated systems is highlighted in Table 5. Various patents based on integrated MEC technology are described in Table 6.

Table 5. Current integrated MEC configurations.

Type of Integration	Applied Voltage	Type of Reactor	Substrate	Hydrogen Production	Ref.
MEC-AD	−0.2 V	Dual-chamber	AD effluent	1.7 ± 0.2 L $H_2 \cdot L^{-1} \cdot day^{-1}$	[167]
	1.2V	Single-chamber	Fermentation sludge	0.16 m^3 $H_2 \cdot m^{-3} \cdot day^{-1}$	[168]
	−0.2 V	Single-chamber	Food waste	3.48 L $H_2 \cdot L^{-1} \cdot day^{-1}$	[169]
MEC/pyrolysis	Batch mode: 0.8 V	Dual-chamber	Pyrolyzed biomass effluent	2.5 L $H_2 \cdot L^{-1} \cdot day^{-1}$	[170]
	Continuous mode: 0.96 V	Dual-chamber	Aqueous phase of bio-oil	4.3 L $H_2 \cdot L^{-1} \cdot day^{-1}$	
MEC/dark fermentation	0.8 V	Single-chamber	DF effluent	100 mL $H_2 \cdot g$ COD^{-1} 700 mL $H_2 \cdot g$ COD^{-1}	[171]
	200 mV	Dual-chamber	Fruit juice wastewater	1609 mL $H_2 \cdot g$ COD^{-1}	[172]
	550 mV	Single chamber	DF effluent	81 mL $H_2 \cdot L^{-1} \cdot day^{-1}$	[173]
	0.6 V	Single-chamber	Cassava starch processing wastewater	465 mL $H_2 \cdot g$ COD^{-1}	[174]
MEC/MFC/dark fermentation	0.2 V	Single-chamber	Cellulose	0.24 m^3 $H_2 \cdot m^{-3} \cdot day^{-1}$	[175]

Table 6. Various patents associated with integrated MEC technology.

Patent ID.	Description	Reference
KR101575790B1	Using a bioelectrochemical device, the present invention pertains to an apparatus and a method for boosting the efficiency of an anaerobic digestion tank used for treating waste water or slurry-type waste with a high content of organic materials. In the anaerobic digestion tank, an electrode device is installed that consists of electrode modules that contain an oxidation electrode and a reduction electrode on which microorganisms with electric activity cling and develop and are integrally joined to have a separation membrane therebetween.	[176]
CN201416000Y	Electrolysis is used to enhance the fermentation for hydrogen production in experimental installation. The utility model belongs to the biomass ferment hydrogen production device.	[177]
CN211339214U	For landfill leachate treatment, a microbial electrolysis cell/membrane bioreactor combination device is used. The utility model is for a microbial electrolysis cell/membrane bioreactor combination treatment device for treating landfill leachate in the field of wastewater treatment.	[178]
US20140285007A1	A power management unit (PMU) is used in various versions of the invention to control the production of hydrogen and electricity for external usage in an MFC/MEC coupled system. A PWM controller and low-voltage electronic switches using MOSFETs are included in one embodiment of the PMU. The PWM controller generates the necessary timing waveform to drive the switches. In other instances, the switches can be replaced with any switching regulator capable of producing high efficiency at low operating voltages and currents. A wastewater treatment plant might use such a technology.	[179]
CN104141147B	Microbiological fuel cell with self-driven microorganism electrolysis cell for hydrogen production and storage. The current invention relates to a method for reclaiming elemental mercury by using a microbiological fuel cell made from industrial effluent containing Hg^{2+}. The method for hydrogen storage is driven microbial electrolysis cell hydrogen synthesis.	[180]

10. Comparison between MEC Technology and Water Electrolysis

When compared to water electrolysis, the primary advantage of MECs is that they do not produce oxygen [181–184], which is regarded to be a part of the process safety; this process safety asset is essential when producing pressurized hydrogen, such as during water electrolysis, because, when producing pressurized hydrogen at high pressure, there is an increased risk of hydrogen crossover at the anode compartment [163]. A further advantage of MECs is that the anode produces chlorine at a very low potential, eliminating the need for the electrolyte to be dissociated from chlorine, which is required for water electrolysis. This is another advantage of MECs [185,186]. MECs have another operational parameter, which is pH. The pH of the electrolyte is neutral, although many microbial biofilms do not tolerate extremes in pH [183]. Furthermore, the neutral pH offers the additional process safety benefit of allowing electrolytes to be handled without the need for severe safeguards. When disposing of and replacing a phosphate buffer with neutral pH, it is much simpler than when disposing of and replacing pH 14 potassium hydroxide, which is used in water electrolysis [184–186].

Comparison in Terms of Theoretical Energy Yields

MECs are also more efficient than water electrolysis in terms of energy generation, and two different sources of energy must be considered, with electricity being the most prominent of the two. The reaction in MECs that produces hydrogen consumes a significantly lower amount of energy than water electrolysis, resulting in lower power consumption. It is an endothermic reaction that requires heat in order for the reaction to occur at equilibrium cell voltage. The energy yields from multichannel electrolysis cells (MECs) are up to 10–11.3 γ_G greater based on MEC configurations.

Chemical energy is the other source of energy, and it is obtained when the oxidation of various compounds present at the anode is completed. During the side production of methane, the energy yield for G, including hydrogen generation, decreased below 10; nevertheless, the thermal energy yield for both increased to 12 Y_H. This yield demonstrates that hydrogen produced by MECs can store up to 10 times the electrical energy in the form of Gibbs free energy [187]. Table 7 lists the characteristics of some of these pilot studies including electrical and thermal energy yields.

Table 7. Characteristics of MEC pilot studies (volume, ionic conductivity, methane, and hydrogen production rates) with final electrical (γ_G) and thermal (γ_H) energy yields.

Anolyte	Volume (L)	Current Density ($A{\cdot}m^{-2}$)	CH_4 ($L{\cdot}day^{-1}{\cdot}m^{-2}$)	H_2 ($L{\cdot}day^{-1}{\cdot}m^{-2}$)	γ_G	γ_H	Ref.
Wastewater acetate	120	0.3	0.06	3.6	0.69	-	[82,83]
Sewage sludge	33	0.01	0.28	0	-	6.45	[188]
Synthetic wastewater	30	0.78	-	-	-	-	[189]
Synthetic, acetate, saline	4	42.5	Very less	102	0.37	-	[85]
Sucrose, wastewater	18	0.1	0.82	0.45	0.69	5.55	[190]
Wastewater effluent	16	0.72	-	-	-	-	[186]
Acetate, pig slurry	16	1.75	<2	17.8	1.25	1.5	[189]
Wastewater	130	0.3	<5	2.52	1.03	1.24	[84]

11. Conclusions

To summarize, the development of MEC technology has shown positive results, primarily by lowering the total cost of wastewater treatment and energy generation. Simultaneously, it delivers a significant advantage via the creation of value-added fuels such as hydrogen. Furthermore, MEC technology is still in its early stage, since it faces several obstacles such as mass transfer restrictions, energy loss, and other issues that must be thoroughly investigated on a pilot and industrial scale utilizing real-world wastewaters. However, when it comes to the actual applications of MEC technology, it should be noted that establishing unique configurations in both anode and cathode structural design, as well as membrane structural design, should be given top priority to maximize the technology's cost-effectiveness. However, in recent years, reports have emerged that the technology's prospects are promising, as seen by the successful construction of many pilot-scale MEC reactors, indicating that the first commercial encounter with the technology is on the way. There is a significant gap between the literature results and the reality of scalable research, and the real challenge today is now to study the credibility of pilot-scale studies. Moreover, there is no comparative analysis of the cost estimation with respect to conventional technologies and MEC, which could help in the commercialization. The obstacles to bringing this technology forward from TRL 5 are not only technological but are also connected to innovation policy incentives. The stringent regulatory efficiency restrictions and the profitability of water utilities do not support developments in systems that are unlikely to satisfy these standards immediately.

Biofuels and bioenergy are being developed using this approach, which has been proposed as a means of transforming electrical energy from renewable energy sources such as solar and wind into biofuels and bioenergy. As a result, the development of integrated MECs with hydrolysis has the potential to increase the rate of breakdown of nonbiodegradable complex organics and, as a result, improve the overall efficiency of production.

Author Contributions: Conceptualization, P.D. and P.S.; software, P.D.; validation, P.S., S.K., M.K., S.P., Y.-H.Y. and S.K.B.; writing—original draft preparation, P.D. and P.S.; writing—review and editing, P.S., D.J., and P.K.G.; supervision, P.S., S.K., S.P. and S.K.B. All authors read and agreed to the published version of the manuscript.

Funding: This study was supported by the Research Program to solve social issues of the National Research Foundation of Korea (NRF)s funded by the Ministry of Science and ICT (grant number 2021R1F1A1050325). This study was also performed with the support of the R&D Program of MOTIE/KEIT [grant number 20014350 and 20009508].

Institutional Review Board Statement: Not Applicable.

Informed Consent Statement: Not Applicable.

Acknowledgments: The authors would like to acknowledge the KU Research Professor Program of Konkuk University, Seoul, South Korea. The authors would like to acknowledge the Sharda University Seed grant for research, India.

Conflicts of Interest: The authors declare no conflict of interest.

References

1. Baruah, J.; Nath, B.K.; Sharma, R.; Kumar, S.; Deka, R.C.; Baruah, D.C.; Kalita, E. Recent Trends in the Pretreatment of Lignocellulosic Biomass for Value-Added Products. *Front. Energy Res* **2018**, *6*, 141. [CrossRef]
2. Savla, N.; Pandit, S.; Khanna, N.; Mathuriya, A.S.; Jung, S.P. Microbially Powered Electrochemical Systems Coupled with Membrane-Based Technology for Sustainable Desalination and Efficient Wastewater Treatment. *J. Korean Soc. Environ. Eng.* **2020**, *42*, 360–380. [CrossRef]
3. Pandit, S.; Savla, N.; Jung, S.P. 16—Recent advancements in scaling up microbial fuel cells. In *Integrated Microbial Fuel Cells for Wastewater Treatment*; Abbassi, R., Yadav, A.K., Khan, F., Garaniya, V., Eds.; Butterworth-Heinemann: Oxford, UK, 2020; pp. 349–368, ISBN 978-0-12-817493-7.
4. Singh Asiwal, R.; Kumar Sar, S.; Singh, S.; Sahu, M. Wastewater Treatment by Effluent Treatment Plants. *IJCE* **2016**, *3*, 19–24. [CrossRef]
5. Dey, P.; Pal, P.; Kevin, J.D.; Das, D.B. Lignocellulosic Bioethanol Production: Prospects of Emerging Membrane Technologies to Improve the Process—A Critical Review. *Rev. Chem. Eng.* **2020**, *36*, 333–367. [CrossRef]
6. Gude, V.G. Energy and Water Autarky of Wastewater Treatment and Power Generation Systems. *Renew. Sustain. Energy Rev.* **2015**, *45*, 52–68. [CrossRef]
7. Bhatia, S.K.; Mehariya, S.; Bhatia, R.K.; Kumar, M.; Pugazhendhi, A.; Awasthi, M.K.; Atabani, A.E.; Kumar, G.; Kim, W.; Seo, S.-O.; et al. Wastewater Based Microalgal Biorefinery for Bioenergy Production: Progress and Challenges. *Sci. Total Environ.* **2021**, *751*, 141599. [CrossRef]
8. Tiwari, R.R. Occupational Health Hazards in Sewage and Sanitary Workers. *Indian J. Occup. Environ. Med.* **2008**, *12*, 112–115. [CrossRef]

9. Guimarães, N.R.; Filho, S.S.F.; Hespanhol, B.P.; Piveli, R.P. Evaluation of Chemical Sludge Production in Wastewater Treatment Processes. *Desalin. Water Treat.* **2016**, *57*, 16346–16352. [CrossRef]
10. Sher, F.; Hanif, K.; Rafey, A.; Khalid, U.; Zafar, A.; Ameen, M.; Lima, E.C. Removal of Micropollutants from Municipal Wastewater Using Different Types of Activated Carbons. *J. Environ. Manag.* **2021**, *278*, 111302. [CrossRef]
11. Agasam, T.; Savla, N.; Kalburge, S.J.; T.k., S.; Pandit, S.; Jadhav, D.A. Chapter 7—Microbial electrosynthesis: Carbon dioxide sequestration via bioelectrochemical system. In *The Future of Effluent Treatment Plants*; Shah, M., Rodriguez-Couto, S., Mehta, K., Eds.; Elsevier: Amsterdam, The Netherlands, 2021; pp. 113–132, ISBN 978-0-12-822956-9.
12. Bhagchandanii, D.D.; Babu, R.P.; Sonawane, J.M.; Khanna, N.; Pandit, S.; Jadhav, D.A.; Khilari, S.; Prasad, R. A Comprehensive Understanding of Electro. *Fermenta* **2020**, *6*, 92. [CrossRef]
13. Bhatia, S.K.; Jagtap, S.S.; Bedekar, A.A.; Bhatia, R.K.; Rajendran, K.; Pugazhendhi, A.; Rao, C.V.; Atabani, A.E.; Kumar, G.; Yang, Y.-H. Renewable Biohydrogen Production from Lignocellulosic Biomass Using Fermentation and Integration of Systems with Other Energy Generation Technologies. *Sci. Total Environ.* **2021**, *765*, 144429. [CrossRef]
14. Rousseau, R.; Etcheverry, L.; Roubaud, E.; Basséguy, R.; Délia, M.-L.; Bergel, A. Microbial Electrolysis Cell (MEC): Strengths, Weaknesses and Research Needs from Electrochemical Engineering Standpoint. *Appl. Energy* **2020**, *257*, 113938. [CrossRef]
15. Gaur, V.K.; Sharma, P.; Sirohi, R.; Awasthi, M.K.; Dussap, C.-G.; Pandey, A. Assessing the Impact of Industrial Waste on Environment and Mitigation Strategies: A Comprehensive Review. *J. Hazard. Mater.* **2020**, *398*, 123019. [CrossRef]
16. Jin, B.; van Leeuwen, H.J.; Patel, B.; Yu, Q. Utilisation of Starch Processing Wastewater for Production of Microbial Biomass Protein and Fungal α-Amylase by Aspergillus Oryzae. *Bioresour. Technol.* **1998**, *66*, 201–206. [CrossRef]
17. Kim, B.H.; Park, H.S.; Kim, H.J.; Kim, G.T.; Chang, I.S.; Lee, J.; Phung, N.T. Enrichment of Microbial Community Generating Electricity Using a Fuel-Cell-Type Electrochemical Cell. *Appl. Microb. Biotechnol.* **2004**, *63*, 672–681. [CrossRef]
18. Vijayaraghavan, K.; Ahmad, D.; Lesa, R. Electrolytic Treatment of Beer Brewery Wastewater. *Ind. Eng. Chem. Res.* **2006**, *45*, 6854–6859. [CrossRef]
19. Rezaei, F.; Richard, T.L.; Logan, B.E. Analysis of Chitin Particle Size on Maximum Power Generation, Power Longevity, and Coulombic Efficiency in Solid–Substrate Microbial Fuel Cells. *J. Power Sources* **2009**, *192*, 304–309. [CrossRef]
20. Kjeldsen, P.; Barlaz, M.A.; Rooker, A.P.; Baun, A.; Ledin, A.; Christensen, T.H. Present and Long-Term Composition of MSW Landfill Leachate: A Review. *Crit. Rev. Environ. Sci. Technol.* **2002**, *32*, 297–336. [CrossRef]
21. Habermann, W.; Pommer, E.H. Biological Fuel Cells with Sulphide Storage Capacity. *Appl. Microbiol. Biotechnol.* **1991**, *35*, 128–133. [CrossRef]
22. Bustillo-Lecompte, C.F.; Mehrvar, M. Slaughterhouse Wastewater Characteristics, Treatment, and Management in the Meat Processing Industry: A Review on Trends and Advances. *J. Environ. Manag.* **2015**, *161*, 287–302. [CrossRef]
23. Fornero, J.J.; Rosenbaum, M.; Angenent, L.T. Electric Power Generation from Municipal, Food, and Animal Wastewaters Using Microbial Fuel Cells. *Electroanalysis* **2010**, *22*, 832–843. [CrossRef]
24. Venkata Mohan, S.; Saravanan, R.; Raghavulu, S.V.; Mohanakrishna, G.; Sarma, P.N. Bioelectricity Production from Wastewater Treatment in Dual Chambered Microbial Fuel Cell (MFC) Using Selectively Enriched Mixed Microflora: Effect of Catholyte. *Bioresour. Technol.* **2008**, *99*, 596–603. [CrossRef] [PubMed]
25. Kumar, M.D.; Kavitha, S.; Tyagi, V.K.; Rajkumar, M.; Bhatia, S.K.; Kumar, G.; Banu, J.R. Macroalgae-Derived Biohydrogen Production: Biorefinery and Circular Bioeconomy. *Biomass Conv. Bioref.* **2021**. [CrossRef]
26. Staffell, I.; Scamman, D.; Abad, A.V.; Balcombe, P.; Dodds, P.E.; Ekins, P.; Shah, N.; Ward, K.R. The Role of Hydrogen and Fuel Cells in the Global Energy System. *Energy Environ. Sci.* **2019**, *12*, 463–491. [CrossRef]
27. Cardeña, R.; Cercado, B.; Buitrón, G. Chapter 7—Microbial Electrolysis Cell for Biohydrogen Production. In *Biohydrogen*, 2nd ed.; Biomass, Biofuels, Biochemicals; Pandey, A., Mohan, S.V., Chang, J.-S., Hallenbeck, P.C., Larroche, C., Eds.; Elsevier: Amsterdam, The Netherlands, 2019; pp. 159–185, ISBN 978-0-444-64203-5.
28. Nayak, B.K.; Pandit, S.; Das, D. Biohydrogen. In *Air Pollution Prevention and Control*; Kennes, C., Veiga, M.C., Eds.; John Wiley & Sons, Ltd: Chichester, UK, 2013; pp. 345–381, ISBN 978-1-118-52336-0.
29. Rajesh Banu, J.; Ginni, G.; Kavitha, S.; Yukesh Kannah, R.; Adish Kumar, S.; Bhatia, S.K.; Kumar, G. Integrated Biorefinery Routes of Biohydrogen: Possible Utilization of Acidogenic Fermentative Effluent. *Bioresour. Technol.* **2021**, *319*, 124241. [CrossRef]
30. Ghirardi, M.L.; King, P.W.; Mulder, D.W.; Eckert, C.; Dubini, A.; Maness, P.-C.; Yu, J. Hydrogen Production by Water Biophotolysis. In *Microbial BioEnergy: Hydrogen Production*; Advances in Photosynthesis and Respiration; Zannoni, D., De Philippis, R., Eds.; Springer: Dordrecht, The Netherlands, 2014; Volume 38, pp. 101–135, ISBN 978-94-017-8553-2.
31. Savla, N.; Shinde, A.; Sonawane, K.; Mekuto, L.; Chowdhary, P.; Pandit, S. Microbial Hydrogen Production: Fundamentals to Application. In *Microorganisms for Sustainable Environment and Health*; Chowdhary, P., Raj, A., Verma, D., Akhter, Y., Eds.; Elsevier: Amsterdam, The Netherlands, 2020; pp. 343–365, ISBN 978-0-12-819001-2.
32. Tang, J.H. A Review of Suitable Substrates for Hydrogen Production in Microbial Electrolysis Cells. *IOP Conf. Ser. Earth Environ. Sci.* **2021**, *621*, 012145. [CrossRef]
33. Lin, C.-Y.; Lay, C.-H.; Sen, B.; Chu, C.-Y.; Kumar, G.; Chen, C.-C.; Chang, J.-S. Fermentative Hydrogen Production from Wastewaters: A Review and Prognosis. *Int. J. Hydrog. Energy* **2012**, *37*, 15632–15642. [CrossRef]
34. Werner, C.M.; Katuri, K.P.; Hari, A.R.; Chen, W.; Lai, Z.; Logan, B.E.; Amy, G.L.; Saikaly, P.E. Graphene-Coated Hollow Fiber Membrane as the Cathode in Anaerobic Electrochemical Membrane Bioreactors—Effect of Configuration and Applied Voltage on Performance and Membrane Fouling. *Environ. Sci. Technol.* **2016**, *50*, 4439–4447. [CrossRef]

35. Garrido-Baserba, M.; Vinardell, S.; Molinos-Senante, M.; Rosso, D.; Poch, M. The Economics of Wastewater Treatment Decentralization: A Techno-Economic Evaluation. *Environ. Sci. Technol.* **2018**, *52*, 8965–8976. [CrossRef]
36. Rasheed, T.; Shafi, S.; Bilal, M.; Hussain, T.; Sher, F.; Rizwan, K. Surfactants-Based Remediation as an Effective Approach for Removal of Environmental Pollutants—A Review. *J. Mol. Liq.* **2020**, *318*, 113960. [CrossRef]
37. Starkl, M.; Brunner, N.; Feil, M.; Hauser, A. Ensuring Sustainability of Non-Networked Sanitation Technologies: An Approach to Standardization. *Environ. Sci. Technol.* **2015**, *49*, 6411–6418. [CrossRef]
38. Nikolaidis, P.; Poullikkas, A. A Comparative Overview of Hydrogen Production Processes. *Renew. Sustain. Energy Rev.* **2017**, *67*, 597–611. [CrossRef]
39. Parkhey, P.; Gupta, P. Improvisations in Structural Features of Microbial Electrolytic Cell and Process Parameters of Electrohydrogenesis for Efficient Biohydrogen Production: A Review. *Renew. Sustain. Energy Rev.* **2017**, *69*, 1085–1099. [CrossRef]
40. Kim, K.N.; Lee, S.H.; Kim, H.; Park, Y.H.; In, S.-I. Improved Microbial Electrolysis Cell Hydrogen Production by Hybridization with a TiO2 Nanotube Array Photoanode. *Energies* **2018**, *11*, 3184. [CrossRef]
41. Escapa, A.; Mateos, R.; Martínez, E.J.; Blanes, J. Microbial Electrolysis Cells: An Emerging Technology for Wastewater Treatment and Energy Recovery. From Laboratory to Pilot Plant and Beyond. *Renew. Sustain. Energy Rev.* **2016**, *55*, 942–956. [CrossRef]
42. Eerten-Jansen, M.C.A.A.V.; Heijne, A.T.; Buisman, C.J.N.; Hamelers, H.V.M. Microbial Electrolysis Cells for Production of Methane from CO_2: Long-Term Performance and Perspectives. *Int. J. Energy Res.* **2012**, *36*, 809–819. [CrossRef]
43. Awe, O.W.; Zhao, Y.; Nzihou, A.; Minh, D.P.; Lyczko, N. A Review of Biogas Utilisation, Purification and Upgrading Technologies. *Waste Biomass Valor* **2017**, *8*, 267–283. [CrossRef]
44. Kitching, M.; Butler, R.; Marsili, E. Microbial Bioelectrosynthesis of Hydrogen: Current Challenges and Scale-Up. *Enzyme Microb. Technol.* **2017**, *96*, 1–13. [CrossRef]
45. Pant, D.; Singh, A.; Bogaert, G.V.; Olsen, S.I.; Nigam, P.S.; Diels, L.; Vanbroekhoven, K. Bioelectrochemical Systems (BES) for Sustainable Energy Production and Product Recovery from Organic Wastes and Industrial Wastewaters. *RSC Adv.* **2012**, *2*, 1248–1263. [CrossRef]
46. Sleutels, T.H.J.A.; Ter Heijne, A.; Buisman, C.J.N.; Hamelers, H.V.M. Bioelectrochemical Systems: An Outlook for Practical Applications. *ChemSusChem* **2012**, *5*, 1012–1019. [CrossRef]
47. Liu, H.; Grot, S.; Logan, B.E. Electrochemically Assisted Microbial Production of Hydrogen from Acetate. *Environ. Sci. Technol.* **2005**, *39*, 4317–4320. [CrossRef] [PubMed]
48. Rozendal, R.A.; Jeremiasse, A.W.; Hamelers, H.V.M.; Buisman, C.J.N. Hydrogen Production with a Microbial Biocathode. *Environ. Sci. Technol.* **2008**, *42*, 629–634. [CrossRef]
49. Selembo, P.; Merrill, M.; Logan, B. The Use of Stainless Steel and Nickel Alloys as Low-Cost Cathodes in Microbial Electrolysis Cells. *J. Power Sources* **2009**, *190*, 271–278. [CrossRef]
50. Foley, J.; Rozendal, R.; Hertle, C.; Lant, P.; Rabaey, K. Life Cycle Assessment of High-Rate Anaerobic Treatment, Microbial Fuel Cells, and Microbial Electrolysis Cells. *Environ. Sci. Technol.* **2010**, *44*, 3629–3637. [CrossRef]
51. Escapa, A.; San-Martín, M.I.; Mateos, R.; Morán, A. Scaling-up of Membraneless Microbial Electrolysis Cells (MECs) for Domestic Wastewater Treatment: Bottlenecks and Limitations. *Bioresour. Technol.* **2015**, *180*, 72–78. [CrossRef]
52. Dhar, B.R.; Elbeshbishy, E.; Hafez, H.; Lee, H.-S. Hydrogen Production from Sugar Beet Juice Using an Integrated Biohydrogen Process of Dark Fermentation and Microbial Electrolysis Cell. *Bioresour. Technol.* **2015**, *198*, 223–230. [CrossRef]
53. Colantonio, N.; Kim, Y. Cadmium (II) Removal Mechanisms in Microbial Electrolysis Cells. *J. Hazard. Mater.* **2016**, *311*. [CrossRef]
54. Liu, Z.; Zhou, A.; Zhang, J.; Wang, S.; Luan, Y.; Liu, W.; Wang, A.; Yue, X. Hydrogen Recovery from Waste Activated Sludge: Role of Free Nitrous Acid in a Prefermentation–Microbial Electrolysis Cells System. *ACS Sustain. Chem. Eng.* **2018**, *6*, 3870–3878. [CrossRef]
55. Miller, A.; Singh, L.; Wang, L.; Liu, H. Linking Internal Resistance with Design and Operation Decisions in Microbial Electrolysis Cells. *Environ. Int.* **2019**, *126*, 611–618. [CrossRef]
56. Wang, L.; Chen, Y.; Long, F.; Singh, L.; Trujillo, S.; Xiao, X.; Liu, H. Breaking the Loop: Tackling Homoacetogenesis by Chloroform to Halt Hydrogen Production-Consumption Loop in Single Chamber Microbial Electrolysis Cells. *Chem. Eng. J.* **2020**, *389*, 124436. [CrossRef]
57. Baek, G.; Shi, L.; Rossi, R.; Logan, B.E. The Effect of High Applied Voltages on Bioanodes of Microbial Electrolysis Cells in the Presence of Chlorides. *Chem. Eng. J.* **2021**, *405*, 126742. [CrossRef]
58. Guwy, A.J.; Dinsdale, R.M.; Kim, J.R.; Massanet-Nicolau, J.; Premier, G. Fermentative Biohydrogen Production Systems Integration. *Bioresour. Technol.* **2011**, *102*, 8534–8542. [CrossRef] [PubMed]
59. Morales-Guio, C.G.; Stern, L.-A.; Hu, X. Nanostructured Hydrotreating Catalysts for Electrochemical Hydrogen Evolution. *Chem. Soc. Rev.* **2014**, *43*, 6555–6569. [CrossRef]
60. Yasri, N.; Roberts, E.P.L.; Gunasekaran, S. The Electrochemical Perspective of Bioelectrocatalytic Activities in Microbial Electrolysis and Microbial Fuel Cells. *Energy Rep.* **2019**, *5*, 1116–1136. [CrossRef]
61. Fetyan, A.; El-Nagar, G.A.; Lauermann, I.; Schnucklake, M.; Schneider, J.; Roth, C. Detrimental Role of Hydrogen Evolution and Its Temperature-Dependent Impact on the Performance of Vanadium Redox Flow Batteries. *J. Energy Chem.* **2019**, *32*, 57–62. [CrossRef]

62. Kadier, A.; Simayi, Y.; Abdeshahian, P.; Azman, N.F.; Chandrasekhar, K.; Kalil, M.S. A Comprehensive Review of Microbial Electrolysis Cells (MEC) Reactor Designs and Configurations for Sustainable Hydrogen Gas Production. *Alex. Eng. J.* **2016**, *55*, 427–443. [CrossRef]
63. Guo, K.; Tang, X.; Du, Z.; Li, H. Hydrogen Production from Acetate in a Cathode-on-Top Single-Chamber Microbial Electrolysis Cell with a Mipor Cathode. *Biochem. Eng. J.* **2010**, *51*, 48–52. [CrossRef]
64. Hu, H.; Fan, Y.; Liu, H. Hydrogen Production Using Single-Chamber Membrane-Free Microbial Electrolysis Cells. *Water Res.* **2008**, *42*, 4172–4178. [CrossRef]
65. Call, D.F.; Logan, B.E. A Method for High Throughput Bioelectrochemical Research Based on Small Scale Microbial Electrolysis Cells. *Biosens. Bioelectron.* **2011**, *26*, 4526–4531. [CrossRef] [PubMed]
66. Hu, H.; Fan, Y.; Liu, H. Hydrogen Production in Single-Chamber Tubular Microbial Electrolysis Cells Using Non-Precious-Metal Catalysts. *Int. J. Hydrog. Energy* **2009**, *34*, 8535–8542. [CrossRef]
67. Guo, X.; Liu, J.; Xiao, B. Bioelectrochemical Enhancement of Hydrogen and Methane Production from the Anaerobic Digestion of Sewage Sludge in Single-Chamber Membrane-Free Microbial Electrolysis Cells. *Int. J. Hydrog. Energy* **2013**, *38*, 1342–1347. [CrossRef]
68. Rozendal, R.A.; Hamelers, H.V.M.; Buisman, C.J.N. Effects of Membrane Cation Transport on PH and Microbial Fuel Cell Performance. *Environ. Sci. Technol.* **2006**, *40*, 5206–5211. [CrossRef]
69. Selembo, P.A.; Merrill, M.D.; Logan, B.E. Hydrogen Production with Nickel Powder Cathode Catalysts in Microbial Electrolysis Cells. *Int. J. Hydrog. Energy* **2010**, *35*, 428–437. [CrossRef]
70. Rozendal, R.A.; Sleutels, T.H.J.A.; Hamelers, H.V.M.; Buisman, C.J.N. Effect of the Type of Ion Exchange Membrane on Performance, Ion Transport, and PH in Biocatalyzed Electrolysis of Wastewater. *Water Sci. Technol.* **2008**, *57*, 1757–1762. [CrossRef] [PubMed]
71. Rozendal, R.A.; Hamelers, H.V.M.; Molenkamp, R.J.; Buisman, C.J.N. Performance of Single Chamber Biocatalyzed Electrolysis with Different Types of Ion Exchange Membranes. *Water Res.* **2007**, *41*, 1984–1994. [CrossRef]
72. Kim, J.R.; Cheng, S.; Oh, S.-E.; Logan, B.E. Power Generation Using Different Cation, Anion, and Ultrafiltration Membranes in Microbial Fuel Cells. *Environ. Sci. Technol.* **2007**, *41*, 1004–1009. [CrossRef]
73. Cheng, S.; Logan, B. Electromethanogenic Reactor and Processes for Methane Production 2013. Available online: https://patents.google.com/patent/US8440438/en?oq=microbial+electrolysis+cell (accessed on 28 June 2021).
74. Reguera, G.; Speers, A.M.; Young, J.M.; Awate, B. Microbial Electrochemical Cells and Methods for Producing Electricity and Bioproducts Therein 2015. Available online: https://patents.google.com/patent/US20150233001/en?oq=microbial+electrolysis+cell (accessed on 28 June 2021).
75. Baiget, S.G.; Fernández, M.L.V.; Cirucci, J.; Planell, N.M.; Sancho, F.J.L.; Labat, J.A.B.; Canudas, A.G. Process for the Methanogenesis Inhibition in Single Chamber Microbial Electrolysis Cells 2014. Available online: https://patents.google.com/patent/EP2747181A1/en?oq=EP2747181A1 (accessed on 29 June 2021).
76. Logan, B. Electrohydrogenic Reactor for Hydrogen Gas Production 2011. Available online: https://patents.google.com/patent/US7922878B2/en?oq=US7922878B2 (accessed on 28 June 2021).
77. Popat, S.; Parameswaran, P.; Torres, C.; Rittmann, B. Microbial Electrolysis Cells and Methods for the Production of Chemical Products 2015. Available online: https://patents.google.com/patent/US9216919B2/en?oq=US9216919B2 (accessed on 28 June 2021).
78. 78. He, Y.; Wang, Y.; Xi, H.; Yao, S.; Wang, B. A Microbial Electrolytic Cell Integrating CO_2 Conversion and Sewage Treatment 2012. Available online: https://patents.google.com/patent/CN102408155A/en?oq=CN102408155A (accessed on 28 June 2021).
79. Rivera, I.; Schröder, U.; Patil, S.A. Chapter 5.8—Microbial Electrolysis for Biohydrogen Production: Technical Aspects and Scale-Up Experiences. In *Microbial Electrochemical Technology*; Biomass, Biofuels and Biochemicals; Mohan, S.V., Varjani, S., Pandey, A., Eds.; Elsevier: Amsterdam, The Netherlands, 2019; pp. 871–898, ISBN 978-0-444-64052-9.
80. Cusick, R.D.; Bryan, B.; Parker, D.S.; Merrill, M.D.; Mehanna, M.; Kiely, P.D.; Liu, G.; Logan, B.E. Performance of a Pilot-Scale Continuous Flow Microbial Electrolysis Cell Fed Winery Wastewater. *Appl. Microbiol. Biotechnol.* **2011**, *89*, 2053–2063. [CrossRef] [PubMed]
81. Call, D.; Logan, B.E. Hydrogen Production in a Single Chamber Microbial Electrolysis Cell Lacking a Membrane. *Environ. Sci. Technol.* **2008**, *42*, 3401–3406. [CrossRef] [PubMed]
82. Heidrich, E.S.; Edwards, S.R.; Dolfing, J.; Cotterill, S.E.; Curtis, T.P. Performance of a Pilot Scale Microbial Electrolysis Cell Fed on Domestic Wastewater at Ambient Temperatures for a 12month Period. *Bioresour. Technol.* **2014**, *173*, 87–95. [CrossRef]
83. Heidrich, E.S.; Dolfing, J.; Scott, K.; Edwards, S.R.; Jones, C.; Curtis, T.P. Production of Hydrogen from Domestic Wastewater in a Pilot-Scale Microbial Electrolysis Cell. *Appl. Microbiol. Biotechnol.* **2013**, *97*, 6979–6989. [CrossRef]
84. Baeza, J.A.; Martínez-Miró, À.; Guerrero, J.; Ruiz, Y.; Guisasola, A. Bioelectrochemical Hydrogen Production from Urban Wastewater on a Pilot Scale. *J. Power Sources* **2017**, *356*, 500–509. [CrossRef]
85. Carmona-Martínez, A.A.; Trably, E.; Milferstedt, K.; Lacroix, R.; Etcheverry, L.; Bernet, N. Long-Term Continuous Production of H2 in a Microbial Electrolysis Cell (MEC) Treating Saline Wastewater. *Water Res.* **2015**, *81*, 149–156. [CrossRef] [PubMed]
86. Savla, N.; Khilari, S.; Pandit, S.; Jung, S.P. Effective Cathode Catalysts for Oxygen Reduction Reactions in Microbial Fuel Cell. In *Bioelectrochemical Systems: Vol.1 Principles and Processes*; Kumar, P., Kuppam, C., Eds.; Springer: Singapore, 2020; pp. 189–210, ISBN 9789811568725.

87. Baudler, A.; Schmidt, I.; Langner, M.; Greiner, A.; Schröder, U. Does It Have to Be Carbon? Metal Anodes in Microbial Fuel Cells and Related Bioelectrochemical Systems. *Energy Environ. Sci.* **2015**, *8*, 2048–2055. [CrossRef]
88. Aiken, D.C.; Curtis, T.P.; Heidrich, E.S. Avenues to the Financial Viability of Microbial Electrolysis Cells [MEC] for Domestic Wastewater Treatment and Hydrogen Production. *Int. J. Hydrog. Energ.* **2019**, *44*, 2426–2434. [CrossRef]
89. Yu, Z.; Leng, X.; Zhao, S.; Ji, J.; Zhou, T.; Khan, A.; Kakde, A.; Liu, P.; Li, X. A Review on the Applications of Microbial Electrolysis Cells in Anaerobic Digestion. *Bioresour. Technol.* **2018**, *255*, 340–348. [CrossRef]
90. Bo, T.; Zhu, X.; Zhang, L.; Tao, Y.; He, X.; Li, D.; Yan, Z. A New Upgraded Biogas Production Process: Coupling Microbial Electrolysis Cell and Anaerobic Digestion in Single-Chamber, Barrel-Shape Stainless Steel Reactor. *Electrochem. Commun.* **2014**, *45*, 67–70. [CrossRef]
91. Gikas, P. Towards Energy Positive Wastewater Treatment Plants. *J. Environ. Manag.* **2017**, *203*, 621–629. [CrossRef]
92. Cusick, R.D.; Kiely, P.D.; Logan, B.E. A Monetary Comparison of Energy Recovered from Microbial Fuel Cells and Microbial Electrolysis Cells Fed Winery or Domestic Wastewaters. *Int. J. Hydrog. Energy* **2010**, *35*, 8855–8861. [CrossRef]
93. Zhen, G.; Kobayashi, T.; Lu, X.; Kumar, G.; Hu, Y.; Bakonyi, P.; Rózsenberszki, T.; Koók, L.; Nemestóthy, N.; Bélafi-Bakó, K.; et al. Recovery of Biohydrogen in a Single-Chamber Microbial Electrohydrogenesis Cell Using Liquid Fraction of Pressed Municipal Solid Waste (LPW) as Substrate. *Int. J. Hydrog. Energy* **2016**, *41*, 17896–17906. [CrossRef]
94. Montpart, N.; Rago, L.; Baeza, J.A.; Guisasola, A. Hydrogen Production in Single Chamber Microbial Electrolysis Cells with Different Complex Substrates. *Water Res.* **2015**, *68*, 601–615. [CrossRef] [PubMed]
95. Shen, R.; Liu, Z.; He, Y.; Zhang, Y.; Jianwen, L.; Zhu, Z.; Si, B.; Zhang, C.; Xing, X.-H. Microbial Electrolysis Cell to Treat Hydrothermal Liquefied Wastewater from Cornstalk and Recover Hydrogen: Degradation of Organic Compounds and Characterization of Microbial Community. *Int. J. Hydrog. Energy* **2016**, *41*, 4132–4142. [CrossRef]
96. Bhatia, S.C. 23—Biohydrogen. In *Advanced Renewable Energy Systems*; Bhatia, S.C., Ed.; Woodhead Publishing India: New Delhi, India, 2014; pp. 627–644, ISBN 978-1-78242-269-3.
97. Sosa-Hernández, O.; Popat, S.C.; Parameswaran, P.; Alemán-Nava, G.S.; Torres, C.I.; Buitrón, G.; Parra-Saldívar, R. Application of Microbial Electrolysis Cells to Treat Spent Yeast from an Alcoholic Fermentation. *Bioresour. Technol.* **2016**, *200*, 342–349. [CrossRef] [PubMed]
98. Cai, W.; Han, T.; Guo, Z.; Varrone, C.; Wang, A.; Liu, W. Methane Production Enhancement by an Independent Cathode in Integrated Anaerobic Reactor with Microbial Electrolysis. *Bioresour. Technol.* **2016**, *208*, 13–18. [CrossRef] [PubMed]
99. Lu, L.; Xing, D.; Xie, T.; Ren, N.-Q.; Logan, B. Hydrogen Production from Proteins via Electrohydrogenesis in Microbial Electrolysis Cells. *Biosens. Bioelectron.* **2010**, *25*, 2690–2695. [CrossRef]
100. Ren, L.; Siegert, M.; Ivanov, I.; Pisciotta, J.M.; Logan, B.E. Treatability Studies on Different Refinery Wastewater Samples Using High-Throughput Microbial Electrolysis Cells (MECs). *Bioresour. Technol.* **2013**, *136*, 322–328. [CrossRef] [PubMed]
101. Katuri, K.P.; Ali, M.; Saikaly, P.E. The Role of Microbial Electrolysis Cell in Urban Wastewater Treatment: Integration Options, Challenges, and Prospects. *Curr. Opin. Biotechnol.* **2019**, *57*, 101–110. [CrossRef]
102. Rozendal, R.A.; Hamelers, H.V.M.; Euverink, G.J.W.; Metz, S.J.; Buisman, C.J.N. Principle and Perspectives of Hydrogen Production through Biocatalyzed Electrolysis. *Int. J. Hydrog. Energy* **2006**, *31*, 1632–1640. [CrossRef]
103. Ditzig, J.; Liu, H.; Logan, B.E. Production of Hydrogen from Domestic Wastewater Using a Bioelectrochemically Assisted Microbial Reactor (BEAMR). *Int. J. Hydrog. Energy* **2007**, *32*, 2296–2304. [CrossRef]
104. Cheng, S.; Logan, B.E. Ammonia Treatment of Carbon Cloth Anodes to Enhance Power Generation of Microbial Fuel Cells. *Electrochem. Commun.* **2007**, *9*, 492–496. [CrossRef]
105. Jeremiasse, A.W.; Hamelers, H.V.M.; Saakes, M.; Buisman, C.J.N. Ni Foam Cathode Enables High Volumetric H2 Production in a Microbial Electrolysis Cell. *Int. J. Hydrog. Energy* **2010**, *35*, 12716–12723. [CrossRef]
106. Zhen, G.; Lu, X.; Kobayashi, T.; Kumar, G.; Xu, K. Promoted Electromethanosynthesis in a Two-Chamber Microbial Electrolysis Cells (MECs) Containing a Hybrid Biocathode Covered with Graphite Felt (GF). *Chem. Eng. J.* **2016**, *284*, 1146–1155. [CrossRef]
107. Krishnan, S.; Md Din, M.F.; Taib, S.M.; Nasrullah, M.; Sakinah, M.; Wahid, Z.A.; Kamyab, H.; Chelliapan, S.; Rezania, S.; Singh, L. Accelerated Two-Stage Bioprocess for Hydrogen and Methane Production from Palm Oil Mill Effluent Using Continuous Stirred Tank Reactor and Microbial Electrolysis Cell. *J. Clean. Prod.* **2019**, *229*, 84–93. [CrossRef]
108. Achinas, S.; Euverink, G.J.W. Effect of Combined Inoculation on Biogas Production from Hardly Degradable Material. *Energies* **2019**, *12*, 217. [CrossRef]
109. Xu, S.; Zhang, Y.; Luo, L.; Liu, H. Startup Performance of Microbial Electrolysis Cell Assisted Anaerobic Digester (MEC-AD) with Pre-Acclimated Activated Carbon. *Bioresour. Technol. Rep.* **2019**, *5*, 91–98. [CrossRef] [PubMed]
110. Lim, S.S.; Yu, E.H.; Daud, W.R.W.; Kim, B.H.; Scott, K. Bioanode as a Limiting Factor to Biocathode Performance in Microbial Electrolysis Cells. *Bioresour. Technol.* **2017**, *238*, 313–324. [CrossRef]
111. Savla, N.; Anand, R.; Pandit, S.; Prasad, R. Utilization of Nanomaterials as Anode Modifiers for Improving Microbial FuelCells Performance. *J. Renew. Mater.* **2020**, *8*, 1581–1605. [CrossRef]
112. Li, S.; Cheng, C.; Thomas, A. Carbon-Based Microbial-Fuel-Cell Electrodes: From Conductive Supports to Active Catalysts. *Adv. Mater.* **2017**, *29*, 1602547. [CrossRef] [PubMed]
113. Liu, D.; Roca-Puigros, M.; Geppert, F.; Caizán-Juanarena, L.; Na Ayudthaya, S.P.; Buisman, C.; ter Heijne, A. Granular Carbon-Based Electrodes as Cathodes in Methane-Producing Bioelectrochemical Systems. *Front. Bioeng. Biotechnol.* **2018**, *6*. [CrossRef]

114. Guo, K.; Prévoteau, A.; Patil, S.A.; Rabaey, K. Engineering Electrodes for Microbial Electrocatalysis. *Curr. Opin. Biotechnol.* **2015**, *33*, 149–156. [CrossRef]
115. Carlotta-Jones, D.I.; Purdy, K.; Kirwan, K.; Stratford, J.; Coles, S.R. Improved Hydrogen Gas Production in Microbial Electrolysis Cells Using Inexpensive Recycled Carbon Fibre Fabrics. *Bioresour. Technol.* **2020**, *304*, 122983. [CrossRef]
116. Rasheed, T.; Hassan, A.A.; Kausar, F.; Sher, F.; Bilal, M.; Iqbal, H.M.N. Carbon Nanotubes Assisted Analytical Detection—Sensing/Delivery Cues for Environmental and Biomedical Monitoring. *Trends Anal. Chem.* **2020**, *132*, 116066. [CrossRef]
117. Pocaznoi, D.; Calmet, A.; Etcheverry, L.; Erable, B.; Bergel, A. Stainless Steel Is a Promising Electrode Material for Anodes of Microbial Fuel Cells. *Energy Environ. Sci.* **2012**, *5*, 9645–9652. [CrossRef]
118. Guo, K.; Donose, B.C.; Soeriyadi, A.H.; Prévoteau, A.; Patil, S.A.; Freguia, S.; Gooding, J.J.; Rabaey, K. Flame Oxidation of Stainless Steel Felt Enhances Anodic Biofilm Formation and Current Output in Bioelectrochemical Systems. *Environ. Sci. Technol.* **2014**, *48*, 7151–7156. [CrossRef] [PubMed]
119. Cotterill, S.E.; Dolfing, J.; Jones, C.; Curtis, T.P.; Heidrich, E.S. Low Temperature Domestic Wastewater Treatment in a Microbial Electrolysis Cell with 1 M2 Anodes: Towards System Scale-Up. *Fuel Cells* **2017**, *17*, 584–592. [CrossRef]
120. Ahn, Y.; Im, S.; Chung, J.-W. Optimizing the Operating Temperature for Microbial Electrolysis Cell Treating Sewage Sludge. *Int. J. Hydrog. Energy* **2017**, *42*, 27784–27791. [CrossRef]
121. Zhang, T.; Nie, H.; Bain, T.S.; Lu, H.; Cui, M.; Snoeyenbos-West, O.L.; Franks, A.E.; Nevin, K.P.; Russell, T.P.; Lovley, D.R. Improved Cathode Materials for Microbial Electrosynthesis. *Energy Environ. Sci.* **2012**, *6*, 217–224. [CrossRef]
122. Mohammed, A.J.; Ismail, Z.Z. Slaughterhouse Wastewater Biotreatment Associated with Bioelectricity Generation and Nitrogen Recovery in Hybrid System of Microbial Fuel Cell with Aerobic and Anoxic Bioreactors. *Ecol. Eng.* **2018**, *125*, 119–130. [CrossRef]
123. Wang, L.; He, Z.; Guo, Z.; Sangeetha, T.; Yang, C.; Gao, L.; Wang, A.; Liu, W. Microbial Community Development on Different Cathode Metals in a Bioelectrolysis Enhanced Methane Production System. *J. Power Sources* **2019**, *444*, 227306. [CrossRef]
124. Santoro, C.; Arbizzani, C.; Erable, B.; Ieropoulos, I. Microbial Fuel Cells: From Fundamentals to Applications. A Review. *J. Power Sources* **2017**, *356*, 225–244. [CrossRef]
125. Sangeetha, T.; Guo, Z.; Liu, W.; Cui, M.; Yang, C.; Wang, L.; Wang, A. Cathode Material as an Influencing Factor on Beer Wastewater Treatment and Methane Production in a Novel Integrated Upflow Microbial Electrolysis Cell (Upflow-MEC). *Int. J. Hydrog. Energy* **2016**, *41*, 2189–2196. [CrossRef]
126. Siegert, M.; Yates, M.D.; Spormann, A.M.; Logan, B.E. Methanobacterium Dominates Biocathodic Archaeal Communities in Methanogenic Microbial Electrolysis Cells. *ACS Sustain. Chem. Eng.* **2015**, *3*, 1668–1676. [CrossRef]
127. San-Martín, M.I.; Sotres, A.; Alonso, R.M.; Díaz-Marcos, J.; Moran, A.; Escapa, A. Assessing Anodic Microbial Populations and Membrane Ageing in a Pilot Microbial Electrolysis Cell. *Int. J. Hydrog. Energy* **2019**, *44*, 17304–17315. [CrossRef]
128. Siegert, M.; Yates, M.D.; Call, D.F.; Zhu, X.; Spormann, A.; Logan, B.E. Comparison of Nonprecious Metal Cathode Materials for Methane Production by Electromethanogenesis. *ACS Sustain. Chem. Eng.* **2014**, *2*, 910–917. [CrossRef] [PubMed]
129. Ki, D.; Parameswaran, P.; Popat, S.C.; Rittmann, B.E.; Torres, C.I. Maximizing Coulombic Recovery and Solids Reduction from Primary Sludge by Controlling Retention Time and PH in a Flat-Plate Microbial Electrolysis Cell. *Environ. Sci. Water Res. Technol.* **2017**, *3*, 333–339. [CrossRef]
130. Liu, W.; Huang, S.; Zhou, A.; Zhou, G.; Ren, N.; Wang, A.; Zhuang, G. Hydrogen Generation in Microbial Electrolysis Cell Feeding with Fermentation Liquid of Waste Activated Sludge. *Int. J. Hydrog. Energy* **2012**, *37*, 13859–13864. [CrossRef]
131. Zhang, Y.; Angelidaki, I. Microbial Electrolysis Cells Turning to Be Versatile Technology: Recent Advances and Future Challenges. *Water Res.* **2014**, *56*, 11–25. [CrossRef]
132. Borole, A.P.; Mielenz, J.R. Estimating Hydrogen Production Potential in Biorefineries Using Microbial Electrolysis Cell Technology. *Int. J. Hydrog. Energy* **2011**, *36*, 14787–14795. [CrossRef]
133. Merrill, M.D.; Logan, B.E. Electrolyte Effects on Hydrogen Evolution and Solution Resistance in Microbial Electrolysis Cells. *J. Power Sources* **2009**, *191*, 203–208. [CrossRef]
134. Munoz, L.D.; Erable, B.; Etcheverry, L.; Riess, J.; Basséguy, R.; Bergel, A. Combining Phosphate Species and Stainless Steel Cathode to Enhance Hydrogen Evolution in Microbial Electrolysis Cell (MEC). *Electrochem. Commun.* **2010**, *12*, 183–186. [CrossRef]
135. Yossan, S.; Xiao, L.; Prasertsan, P.; He, Z. Hydrogen Production in Microbial Electrolysis Cells: Choice of Catholyte. *Int. J. Hydrog. Energy* **2013**, *38*, 9619–9624. [CrossRef]
136. Omidi, H.; Sathasivan, A. Optimal Temperature for Microbes in an Acetate Fed Microbial Electrolysis Cell (MEC). *Int. Biodeterior. Biodegrad.* **2013**, *85*, 688–692. [CrossRef]
137. Lu, L.; Ren, N.; Zhao, X.; Wang, H.; Wu, D.; Xing, D. Hydrogen Production, Methanogen Inhibition and Microbial Community Structures in Psychrophilic Single-Chamber Microbial Electrolysis Cells. *Energy Environ. Sci.* **2011**, *4*, 1329–1336. [CrossRef]
138. Ajayi, F.F.; Kim, K.-Y.; Chae, K.-J.; Choi, M.-J.; Kim, I.S. Effect of Hydrodymamic Force and Prolonged Oxygen Exposure on the Performance of Anodic Biofilm in Microbial Electrolysis Cells. *Int. J. Hydrog. Energy* **2010**, *35*, 3206–3213. [CrossRef]
139. Wang, A.; Liu, W.; Ren, N.; Cheng, H.; Lee, D.-J. Reduced Internal Resistance of Microbial Electrolysis Cell (MEC) as Factors of Configuration and Stuffing with Granular Activated Carbon. *Int. J. Hydrog. Energy* **2010**, *35*, 13488–13492. [CrossRef]
140. Sun, M.; Sheng, G.-P.; Zhang, L.; Xia, C.-R.; Mu, Z.-X.; Liu, X.-W.; Wang, H.-L.; Yu, H.-Q.; Qi, R.; Yu, T.; et al. An MEC-MFC-Coupled System for Biohydrogen Production from Acetate. *Environ. Sci. Technol.* **2008**, *42*, 8095–8100. [CrossRef] [PubMed]
141. Nam, J.-Y.; Tokash, J.C.; Logan, B.E. Comparison of Microbial Electrolysis Cells Operated with Added Voltage or by Setting the Anode Potential. *Int. J. Hydrog. Energy* **2011**, *36*, 10550–10556. [CrossRef]

142. Chae, K.-J.; Choi, M.-J.; Kim, K.-Y.; Ajayi, F.F.; Chang, I.-S.; Kim, I.S. A Solar-Powered Microbial Electrolysis Cell with a Platinum Catalyst-Free Cathode To Produce Hydrogen. *Environ. Sci. Technol.* **2009**, *43*, 9525–9530. [CrossRef] [PubMed]
143. Zhang, B.; Wen, Z.; Ci, S.; Chen, J.; He, Z. Nitrogen-Doped Activated Carbon as a Metal Free Catalyst for Hydrogen Production in Microbial Electrolysis Cells. *RSC Adv.* **2014**, *4*, 49161–49164. [CrossRef]
144. Tokash, J.C.; Logan, B.E. Electrochemical Evaluation of Molybdenum Disulfide as a Catalyst for Hydrogen Evolution in Microbial Electrolysis Cells. *Int. J. Hydrog. Energy* **2011**, *36*, 9439–9445. [CrossRef]
145. Hu, H.; Fan, Y.; Liu, H. Optimization of NiMo Catalyst for Hydrogen Production in Microbial Electrolysis Cells. *Int. J. Hydrog. Energy* **2010**, *35*, 3227–3233. [CrossRef]
146. Yuan, H.; Li, J.; Yuan, C.; He, Z. Facile Synthesis of MoS2@CNT as an Effective Catalyst for Hydrogen Production in Microbial Electrolysis Cells. *ChemElectroChem* **2014**, *1*, 1828–1833. [CrossRef]
147. Manuel, M.-F.; Neburchilov, V.; Wang, H.; Guiot, S.R.; Tartakovsky, B. Hydrogen Production in a Microbial Electrolysis Cell with Nickel-Based Gas Diffusion Cathodes. *J. Power Sources* **2010**, *195*, 5514–5519. [CrossRef]
148. Dai, H.; Yang, H.; Liu, X.; Jian, X.; Liang, Z. Electrochemical Evaluation of Nano-Mg(OH)2/Graphene as a Catalyst for Hydrogen Evolution in Microbial Electrolysis Cell. *Fuel* **2016**, *174*, 251–256. [CrossRef]
149. Huang, Y.-X.; Liu, X.-W.; Sun, X.-F.; Sheng, G.-P.; Zhang, Y.-Y.; Yan, G.-M.; Wang, S.-G.; Xu, A.-W.; Yu, H.-Q. A New Cathodic Electrode Deposit with Palladium Nanoparticles for Cost-Effective Hydrogen Production in a Microbial Electrolysis Cell. *Int. J. Hydrog. Energy* **2011**, *36*, 2773–2776. [CrossRef]
150. Li, F.; Liu, W.; Sun, Y.; Ding, W.; Cheng, S. Enhancing Hydrogen Production with Ni–P Coated Nickel Foam as Cathode Catalyst in Single Chamber Microbial Electrolysis Cells. *Int. J. Hydrog. Energy* **2017**, *42*, 3641–3646. [CrossRef]
151. Mitov, M.; Chorbadzhiyska, E.; Nalbandian, L.; Hubenova, Y. Nickel-Based Electrodeposits as Potential Cathode Catalysts for Hydrogen Production by Microbial Electrolysis. *J. Power Sources* **2017**, *356*, 467–472. [CrossRef]
152. Wang, L.; Liu, W.; He, Z.; Guo, Z.; Zhou, A.; Wang, A. Cathodic Hydrogen Recovery and Methane Conversion Using Pt Coating 3D Nickel Foam Instead of Pt-Carbon Cloth in Microbial Electrolysis Cells. *Int. J. Hydrog. Energy* **2017**, *42*, 19604–19610. [CrossRef]
153. Cai, W.; Liu, W.; Han, J.; Wang, A. Enhanced Hydrogen Production in Microbial Electrolysis Cell with 3D Self-Assembly Nickel Foam-Graphene Cathode. *Biosens. Bioelectron.* **2016**, *80*, 118–122. [CrossRef]
154. Lu, L.; Hou, D.; Fang, Y.; Huang, Y.; Ren, Z.J. Nickel Based Catalysts for Highly Efficient H2 Evolution from Wastewater in Microbial Electrolysis Cells. *Electrochim. Acta* **2016**, *206*, 381–387. [CrossRef]
155. Hrapovic, S.; Manuel, M.-F.; Luong, J.H.T.; Guiot, S.R.; Tartakovsky, B. Electrodeposition of Nickel Particles on a Gas Diffusion Cathode for Hydrogen Production in a Microbial Electrolysis Cell. *Int. J. Hydrog. Energy* **2010**, *35*, 7313–7320. [CrossRef]
156. Savla, N.; Suman; Pandit, S.; Verma, J.P.; Awasthi, A.K.; Sana, S.S.; Prasad, R. Techno-Economical Evaluation and Life Cycle Assessment of Microbial Electrochemical Systems: A Review. *Curr. Res. Green Sustain. Chem.* **2021**, *4*, 100111. [CrossRef]
157. Rabaey, K.; Verstraete, W. Bacterial Potential for Electricity Generation. *Trends Biotechnol.* **2005**, *6*, 291–298. [CrossRef]
158. Escapa, A.; Gómez, X.; Tartakovsky, B.; Morán, A. Estimating Microbial Electrolysis Cell (MEC) Investment Costs in Wastewater Treatment Plants: Case Study. *Int. J. Hydrog. Energy* **2012**, *37*, 18641–18653. [CrossRef]
159. Guo, H.; Kim, Y. Stacked Multi-Electrode Design of Microbial Electrolysis Cells for Rapid and Low-Sludge Treatment of Municipal Wastewater. *Biotechnol. Biofuels* **2019**, *12*, 23. [CrossRef]
160. Wang, T.; Li, C.; Zhu, G. Performance, Process Kinetics and Functional Microbial Community of Biocatalyzed Electrolysis-Assisted Anaerobic Baffled Reactor Treating Carbohydrate-Containing Wastewater. *RSC Adv.* **2018**, *8*, 41150–41162. [CrossRef]
161. Muoio, R.; Palli, L.; Ducci, I.; Coppini, E.; Bettazzi, E.; Daddi, D.; Fibbi, D.; Gori, R. Optimization of a Large Industrial Wastewater Treatment Plant Using a Modeling Approach: A Case Study. *J. Environ. Manag.* **2019**, *249*, 109436. [CrossRef]
162. Wan, L.-L.; Li, X.-J.; Zang, G.-L.; Wang, X.; Zhang, Y.-Y.; Zhou, Q.-X. A Solar Assisted Microbial Electrolysis Cell for Hydrogen Production Driven by a Microbial Fuel Cell. *RSC Adv.* **2015**, *5*, 82276–82281. [CrossRef]
163. Bastidas-Oyanedel, J.-R.; Schmidt, J.E. Increasing Profits in Food Waste Biorefinery—A Techno-Economic Analysis. *Energies* **2018**, *11*, 1551. [CrossRef]
164. Lampert, D.J.; Cai, H.; Wang, Z.; Keisman, J.; Wu, M.; Han, J.; Dunn, J.; Sullivan, J.L.; Elgowainy, A.; Wang, M.; et al. *Development of a Life Cycle Inventory of Water Consumption Associated with the Production of Transportation Fuels*; Argonne National Lab. (ANL): Argonne, IL, USA, 2015.
165. Bargigli, S.; Raugei, M.; Ulgiati, S. Comparison of Thermodynamic and Environmental Indexes of Natural Gas, Syngas and Hydrogen Production Processes. *Energy* **2004**, *29*, 2145–2159. [CrossRef]
166. Mehmeti, A.; Angelis-Dimakis, A.; Arampatzis, G.; McPhail, S.J.; Ulgiati, S. Life Cycle Assessment and Water Footprint of Hydrogen Production Methods: From Conventional to Emerging Technologies. *Environments* **2018**, *5*, 24. [CrossRef]
167. Beegle, J.R.; Borole, A.P. An Integrated Microbial Electrolysis-Anaerobic Digestion Process Combined with Pretreatment of Wastewater Solids to Improve Hydrogen Production. *Environ. Sci. Water Res. Technol.* **2017**, *3*, 1073–1085. [CrossRef]
168. Hassanein, A.; Witarsa, F.; Lansing, S.; Qiu, L.; Liang, Y. Bio-Electrochemical Enhancement of Hydrogen and Methane Production in a Combined Anaerobic Digester (AD) and Microbial Electrolysis Cell (MEC) from Dairy Manure. *Sustainability* **2020**, *12*, 8491. [CrossRef]
169. Huang, J.; Feng, H.; Huang, L.; Ying, X.; Shen, D.; Chen, T.; Shen, X.; Zhou, Y.; Xu, Y. Continuous Hydrogen Production from Food Waste by Anaerobic Digestion (AD) Coupled Single-Chamber Microbial Electrolysis Cell (MEC) under Negative Pressure. *Waste Manag.* **2020**, *103*, 61–66. [CrossRef]

170. Lewis, A.J.; Ren, S.; Ye, X.; Kim, P.; Labbe, N.; Borole, A.P. Hydrogen Production from Switchgrass via an Integrated Pyrolysis–Microbial Electrolysis Process. *Bioresour. Technol.* **2015**, *195*, 231–241. [CrossRef]
171. Li, X.-H.; Liang, D.-W.; Bai, Y.-X.; Fan, Y.-T.; Hou, H.-W. Enhanced H2 Production from Corn Stalk by Integrating Dark Fermentation and Single Chamber Microbial Electrolysis Cells with Double Anode Arrangement. *Int. J. Hydrog. Energy* **2014**, *39*, 8977–8982. [CrossRef]
172. Marone, A.; Ayala-Campos, O.R.; Trably, E.; Carmona-Martínez, A.A.; Moscoviz, R.; Latrille, E.; Steyer, J.-P.; Alcaraz-Gonzalez, V.; Bernet, N. Coupling Dark Fermentation and Microbial Electrolysis to Enhance Bio-Hydrogen Production from Agro-Industrial Wastewaters and by-Products in a Bio-Refinery Framework. *Int. J. Hydrog. Energy* **2017**, *42*, 1609–1621. [CrossRef]
173. Rivera, I.; Buitrón, G.; Bakonyi, P.; Nemestóthy, N.; Bélafi-Bakó, K. Hydrogen Production in a Microbial Electrolysis Cell Fed with a Dark Fermentation Effluent. *J. Appl. Electrochem.* **2015**, *45*, 1223–1229. [CrossRef]
174. Khongkliang, P.; Kongjan, P.; Utarapichat, B.; Reungsang, A.; O-Thong, S. Continuous Hydrogen Production from Cassava Starch Processing Wastewater by Two-Stage Thermophilic Dark Fermentation and Microbial Electrolysis. *Int. J. Hydrog. Energy* **2017**, *42*, 27584–27592. [CrossRef]
175. Wang, A.; Sun, D.; Cao, G.; Wang, H.; Ren, N.; Wu, W.-M.; Logan, B.E. Integrated Hydrogen Production Process from Cellulose by Combining Dark Fermentation, Microbial Fuel Cells, and a Microbial Electrolysis Cell. *Bioresour. Technol.* **2011**, *102*, 4137–4143. [CrossRef]
176. Song, Y.-C.; Woo, W.-H. Anaerobic Digestion Apparatus Equipped with Landscape Bioelectrochemical Apparatus and Anaerobic Digestion Method of Organic Waste Using the Same 2015. Available online: https://patents.google.com/patent/KR101575790B1/en?oq=KR101575790B1 (accessed on 29 June 2021).
177. Li, J.; Zhang, W. Experimental Device for Hydrogen Production by Electrolysis Assisted Fermentation 2010. Available online: https://patents.google.com/patent/CN201416000Y/en?oq=CN201416000Y (accessed on 29 June 2021).
178. Jisheng, L.; Chao, S.; Li, X.; Haifeng, Z.; Bei, L.; Yu, G.; Shi, H.; Yichen, H.; Yang, L.; Haocheng, Y. Microbial Electrolytic Cell-Membrane Bioreactor Combined Treatment Device for Landfill Leachate 2020. Available online: https://patents.google.com/patent/CN211339214U/en?oq=CN211339214U (accessed on 29 June 2021).
179. Krishnan, A. Power Management System for a Microbial Fuel Cell and Microbial Electrolysis Cell Coupled System 2014. Available online: https://patents.google.com/patent/US20140285007A1/en?oq=US20140285007A1 (accessed on 29 June 2021).
180. Liang, Z.; Dai, H.; Zhao, Y.; Yang, H.; Liu, X. Microbial Fuel Cell Self-Driving Microbial Electrolytic Cell Hydrogen Production and Storage Method 2016. Available online: https://patents.google.com/patent/CN104141147B/en?q=MFC+%2b+MEC&oq=MFC+%2b+MEC (accessed on 29 June 2021).
181. Chen, L.; Dong, X.; Wang, Y.; Xia, Y. Separating Hydrogen and Oxygen Evolution in Alkaline Water Electrolysis Using Nickel Hydroxide. *Nat. Commun.* **2016**, *7*, 11741. [CrossRef] [PubMed]
182. Symes, M.D.; Cronin, L. Decoupling Hydrogen and Oxygen Evolution during Electrolytic Water Splitting Using an Electron-Coupled-Proton Buffer. *Nat. Chem.* **2013**, *5*, 403–409. [CrossRef]
183. Dopson, M.; Ni, G.; Sleutels, T.H. Possibilities for Extremophilic Microorganisms in Microbial Electrochemical Systems. *FEMS Microbiol. Rev.* **2016**, *40*, 164–181. [CrossRef]
184. Patil, S.A.; Harnisch, F.; Koch, C.; Hübschmann, T.; Fetzer, I.; Carmona-Martínez, A.A.; Müller, S.; Schröder, U. Electroactive Mixed Culture Derived Biofilms in Microbial Bioelectrochemical Systems: The Role of PH on Biofilm Formation, Performance and Composition. *Bioresour. Technol.* **2011**, *102*, 9683–9690. [CrossRef] [PubMed]
185. Popat, S.C.; Torres, C.I. Critical Transport Rates That Limit the Performance of Microbial Electrochemistry Technologies. *Bioresour. Technol.* **2016**, *215*, 265–273. [CrossRef]
186. Torres, C.I.; Marcus, A.K.; Rittmann, B.E. Proton Transport inside the Biofilm Limits Electrical Current Generation by Anode-Respiring Bacteria. *Biotechnol. Bioeng.* **2008**, *100*, 872–881. [CrossRef] [PubMed]
187. Moscoviz, R.; Toledo-Alarcón, J.; Trably, E.; Bernet, N. Electro-Fermentation: How To Drive Fermentation Using Electrochemical Systems. *Trends Biotechnol.* **2016**, *34*, 856–865. [CrossRef]
188. Sugnaux, M.; Happe, M.; Cachelin, C.P.; Gasperini, A.; Blatter, M.; Fischer, F. Cathode Deposits Favor Methane Generation in Microbial Electrolysis Cell. *Chem. Eng. J.* **2017**, *324*, 228–236. [CrossRef]
189. Brown, R.K.; Harnisch, F.; Wirth, S.; Wahlandt, H.; Dockhorn, T.; Dichtl, N.; Schröder, U. Evaluating the Effects of Scaling up on the Performance of Bioelectrochemical Systems Using a Technical Scale Microbial Electrolysis Cell. *Bioresour. Technol.* **2014**, *163*, 206–213. [CrossRef]
190. Luo, S.; Jain, A.; Aguilera, A.; He, Z. Effective Control of Biohythane Composition through Operational Strategies in an Innovative Microbial Electrolysis Cell. *Appl. Energy* **2017**, *206*, 879–886. [CrossRef]

Article

Bio-Electrochemical Enhancement of Hydrogen and Methane Production in a Combined Anaerobic Digester (AD) and Microbial Electrolysis Cell (MEC) from Dairy Manure

Amro Hassanein [1,2,*], Freddy Witarsa [3], Stephanie Lansing [1,*], Ling Qiu [2] and Yong Liang [2,4]

1 Department of Environmental Science and Technology, University of Maryland, College Park, MD 20742, USA
2 College of Mechanical and Electrical Engineering, Northwest Agriculture and Forestry University, Yangling 712100, China; qling@nwafu.edu.cn (L.Q.); yongliang@glut.edu.cn (Y.L.)
3 Department of Physical and Environmental Sciences, Colorado Mesa University, Grand Junction, CO 81501, USA; fwitarsa@coloradomesa.edu
4 Guangxi Scientific Experiment Center of Mining, Metallurgy and Environment, Guilin University of Technology, Guilin 541004, China
* Correspondence: ahassane@umd.edu (A.H.); slansing@umd.edu (S.L.); Tel.: +1-301-405-0138 (A.H.); +1-301-405-1197 (S.L.)

Received: 15 September 2020; Accepted: 12 October 2020; Published: 14 October 2020

Abstract: Anaerobic digestion (AD) is a biological-based technology that generates methane-enriched biogas. A microbial electrolysis cell (MEC) uses electricity to initiate bacterial oxidization of organic matter to produce hydrogen. This study determined the effect of energy production and waste treatment when using dairy manure in a combined AD and MEC (AD-MEC) system compared to AD without MEC (AD-only). In the AD-MEC system, a single chamber MEC (150 mL) was placed inside a 10 L digester on day 20 of the digestion process and run for 272 h (11 days) to determine residual treatment and energy capacity with an MEC included. Cumulative H_2 and CH_4 production in the AD-MEC (2.43 L H_2 and 23.6 L CH_4) was higher than AD-only (0.00 L H_2 and 10.9 L CH_4). Hydrogen concentration during the first 24 h of MEC introduction constituted 20% of the produced biogas, after which time the H_2 decreased as the CH_4 concentration increased from 50% to 63%. The efficiency of electrical energy recovery (ηE) in the MEC was 73% (ηE min.) to 324% (ηE max.), with an average increase of 170% in total energy compared to AD-only. Chemical oxygen demand (COD) removal was higher in the AD-MEC (7.09 kJ/g COD removed) system compared to AD-only (6.19 kJ/g COD removed). This study showed that adding an MEC during the digestion process could increase overall energy production and organic removal from dairy manure.

Keywords: biogas; MEC; bio-hydrogen; manure; digestion

---

1. Introduction

Microbial electrolysis cell (MEC) is a bioelectrochemical technology that uses concepts from microbial fuel cell (MFC) research. While MFCs use microbial decomposition of organic compounds to produce an electric current, in an MEC, an electric current is applied to reverse the reaction to convert organic material to hydrogen (H_2) and/or methane (CH_4). Recently, MECs have been explored as a clean energy source and a promising innovative technology for H_2 production using bioelectrochemical properties. Hydrogen gas is formed in an MEC from two sources of energy: (1) bacterial oxidization of organic matter and (2) electric input [1–4]. Converting organic matter into H_2 in an MEC requires: (1) exo-electrogenic anodal microbes to release electrons and protons from organic material (oxidation reaction), and (2) an external electricity input (voltage > 0.114 V) to push the reaction to be favorable, as the reaction may not be thermodynamically favorable without the electrical input [5].

Methane-enriched biogas is produced during anaerobic digestion (AD) by anaerobic bacteria decomposing organic material, such as manure, sewage, municipal waste, and/or food waste [6]. During AD, the waste is converted into biogas, a renewable energy source that consists of 55–75% CH_4, 45–25% CO_2, and small amounts of hydrogen sulfide (H_2S), hydrogen (H_2), and other gases [7,8]. The microbial process in the AD process is divided into two main phases: acidogenic and methanogenic [6,9].

In order to enrich H_2, methanogenic bacterial growth should be decreased, as methanogens can use H_2 as a pathway for CH_4 production. While methanogenic production can be limited by low pH conditions, slightly acidified conditions (pH 5–6) were not effective in controlling methanogenesis in an MEC-only treatment in which H_2 production was the desired product [10,11]. In most MEC studies, methanogens have been found to use the produced H_2 for CH_4 generation [11–15], with CH_4 production continuing when the voltage was no longer applied. In an MEC study that washed methanogens from the MEC reactor using a low hydraulic retention time (5.3 h), CH_4 production was still detected [15]. A recent experiment confirmed that methanogens are a preventive factor for H_2 production from wastewater treatment plants using MEC [16], as CH_4 production reduces H_2 concentration and purity. Few MEC studies have been conducted using actual waste material [17,18]. Furthermore, MEC treatment volumes are often small (<0.3 L), with a high electrode surface area in the MEC, which would make scaling expensive. Liu et al. (2012) studied the effect of feeding an MEC with waste activated sludge fermentation liquid using bi-frequency ultrasonic and alkaline addition as a pre-treatment step to suppress methanogenic activity and increase H_2 production, which eliminated CH_4 production in favor of H_2, but greatly increased the process complexity [18].

In our previous work [4], we evaluated incorporating an MEC with AD in the same reactor to digest food waste to first increase H_2 production, and then use the H_2 substrate to further increase the CH_4 concentration. This work increased the energy production output, with >90% CH_4, without an increase in the process complexity. In this previous work, three treatments were tested for 23 days: (1) a merged AD and MEC system with the MEC operating for the duration of the experiment; (2) a merged AD and MEC system with the MEC operating for only the first five days, followed by the AD for the remaining 18 days; and (3) an AD-only system operating for the entire 23-day experiment. Our previous results showed that, incorporating our unique MEC design within the AD reactor enhanced the biogas quantity and quality, total energy production, and food waste treatment. At this point, there has not been a study that incorporated an MEC with AD treatment at the end of dairy manure digestion to determine the effect of MEC inclusion to increase residual energy potential, nor has there been an MEC study focused on dairy manure, which is a readily available substrate used in AD systems worldwide.

In this research, an AD and MEC were combined to demonstrate the performance of using MEC during the last 11 days of digestion when biogas production had decreased but solids and chemical oxygen demand (COD) still persisted and could be further reduced given enough time. The current study aim was to determine the effect of a combined AD-MEC to enhance the treatment performance (i.e., COD and solids removal) and energy production (CH_4 + H_2) during the last 11 days of digestion of dairy manure. The MEC was designed to comprise a small volume compared to digester volume (0.0191 m^3/m^3). This study demonstrated a new application of MEC to increase both organic waste treatment and energy production performance.

2. Materials and Methods

2.1. Substrate and Inoculum

The dairy manure substrate was obtained from the Northwest Agriculture and Forestry University Dairy Farm, Yangling, Shaanxi, China. Prior to use, the dairy manure was stored for seven days at −4 °C. The total solids (TS) and volatile solids (VS) content of the dairy manure were 13.3% and 10.9% on a wet weight basis, respectively. An inoculum to substrate (ISR) VS ratio of 1:2 was used for the experiment. The inoculum was collected from the effluent of a previous AD-MEC experiment, with 5.6% TS and 3.3% VS, on a wet weight basis. After 20 days of digestion (before starting up the MEC), the contents of both treatment reactors were mixed together and distributed back to the digesters. The COD, TS, and VS concentrations of dairy manure and inoculum mixture after 20 days of digestion (before starting up the MEC) were 46.7 g/L, 6.89%, and 4.70%, respectively. The pH, ORP, and conductivity were 7.88, −374 mV, and 11.12 ms/cm, respectively. The total active volume of each digester (AD-MEC and AD-only) was 8 L.

2.2. Reactor Design

Two reactors were used in the experiment: (1) AD-MEC: a digester that contained an MEC connected to an electricity source, and (2) AD-only: a digester with an MEC included to provide equal surface area and reactor volume, but the MEC was not connected to an electricity source (substrate was treated using only the digestion process) (Figure 1). Both reactors were made of stainless steel, with each reactor having an inner diameter of 18 cm and length of 46 cm, resulting in a total volume of 11 L (8 L active volume, and 3 L head space) (Auzone BMR-A10L, Shanghai Auzone Bio-engineering Equipment CO., LTD., Shanghai, China). Each digester was digitally programmed for temperature and mixing speed. Heating of the digesters was conducted using water that was heated and pumped into water jackets surrounding the reactors. The temperature within the digester was maintained at 16 °C for the first 10 days of digestion, then increased to 35 °C. At day 20, the MEC was connected to the power supply in the AD-MEC treatment for 11 days (from days 20–31), while the AD-only treatment operated for 11 additional days (days 20–31) without MEC as an AD-only treatment. A pH probe was inserted into each reactor (Mettler Toledo 52003679, Beijing, China), as well as an Ag/AgCl reference electrode. There was an outlet (1 cm diameter) located 14 cm from the bottom of the reactor for collecting liquid samples.

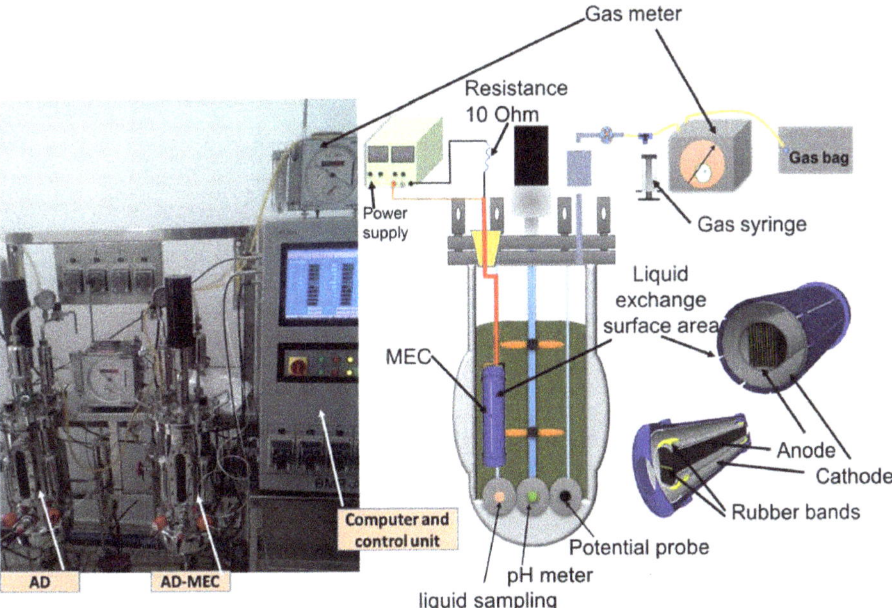

Figure 1. Photograph and schematic diagram of combined AD and MEC (AD-MEC) system compared to AD without MEC (AD-only) treatment.

2.3. Microbial Electrolysis Cell Design

The MEC was fabricated according to our previous research [4] and included a PVC pipe closed with PVC caps as the MEC sleeve, with a diameter of 3.49 cm and length of 15.50 cm, resulting in a total MEC volume of 0.153 L. To guarantee the flow of liquid between the inside and the outside of the MEC, ten vertical holes of approximately 5.5 cm in length and 0.35 cm in width were created on the sides of the PVC sleeve [4]. An additional hole with a diameter of 1.6 cm was created in the center of each PVC cap to allow gas produced within the MEC to escape (Figure 1).

Each anode was made of eleven graphite plates composed of ≥99.95% pure graphite with high electric resistivity (≤1000 μΩ·cm), with dimensions of 15.0 cm (length) × 1.50 cm (width) × 0.1 cm (thickness) per plate. Rubber bands were used to separate each graphite plate to create a space between the plates, which formed the first part of the anode [4]. The second part of the anode was made of a stainless-steel cylinder (grade 201) with 15.0 cm length, 1.85 cm diameter, and 0.05 cm thickness and low electric resistivity (≤68.5 μΩ·cm) [4]. The graphite plates (first part of the anode) were inserted into a stainless-steel cylinder (second part of the anode). The integration of a stainless-steel cylinder and graphite plates was used to increase the electric conductivity between the graphite plates and to reduce the ohmic resistance compared to a graphite-only anode [4,19,20]. Furthermore, the utilization of grade 201 stainless steel in the anode helped to decrease the distance between the electrode plates and increase the conductivity. Moreover, the stainless steel grade 201 contains manganese (5.5–7.5%), which has been shown to increase the electrical energy produced by a factor of 10 [21,22].

The anode was positioned inside the cathode (stainless-steel cylinder with a 3.13 cm diameter, 15.0 cm length, and 0.05 cm thickness) and rubber bands were used to isolate the anode and cathode to avoid short circuiting (Figure 1) [4]. Insulated wires were used to connect the anode and cathode to the circuit.

2.4. MEC Voltage

A programmed direct current power-supply (YH-305D, Yi Hua Inc., Shanghai, China) was used to apply the voltage (1.2 V) across the anode and the cathode. The anode with a 10 Ω resistor was connected serially to the power supply positive lead, while the cathode was connected to the negative lead. The voltage drops across the 10 Ω resistor were measured using a multi-meter (model ATW9205L; ATTEN Instruments Inc., Shanghai, China). Measurements of voltage drop were conducted approximately 7 times per day.

Ohm's law was used to calculate the current (I = V/R, where V is the voltage drop measured across the resistor (R)). The energy recovery efficiency was based on electricity (η_e) (%) input based on the measured current compared to the energy difference in the usable gas production between the AD-MEC and AD-only reactors. The volume of CH_4 and H_2 produced daily were normalized to the reactor active volume (m^3 gas/m^3/d). The volumetric current density (IV, A/m^3) was normalized by the MEC liquid volume (0.150 L), and the total volume of manure + inoculum (8 L). Following Call and Logan (2008), the energy dissipation in the 10 Ω resistor (W_E) was accounted for in order to determine the actual energy supplied [23]. Gibbs' free energy equation was used to calculate the energy production from CH_4 (ΔG_{CH4} = 890.4 kJ/mol) and H_2 (ΔG_{H2} = 237.1 kJ/mol) [4,24,25].

2.5. Analytical Methods

The dairy manure and inoculum TS and VS concentrations were measured using Standard Methods [26]. The COD was measured using colorimetry, using Method 410.4 [27]. Wet gas meters (W-NK-0.5, Shinagawa Co., Tokyo, Japan) were used to measure biogas production from both treatments (AD-MEC and AD). A gas bag was attached to the gas meter outlet to ensure air could not enter the gas meter while collecting the gas sample. A gas chromatograph (GC) (GC2014C, Shimadzu Co, Chiyoda-ku, Japan) was used to analyze the gas samples, which were collected from the reactor using a gas-tight syringe.

Scanning electron microscopy (SEM) (TM3000 Tabletop, HITACHI, Fukuoka, Japan) was used for scanning the electrodes, the biofilm, and the liquid between the electrodes. Prior to SEM, the samples were air dried for approximately 20 min inside a closed incubator, and then coated with gold for 20 s.

3. Results and Discussion

3.1. Organic Matter Reduction: COD, TS, and VS Removal

The COD concentration after 20 days of digestion and before MEC use was 46.7 g/L. The AD-MEC had 38.3% further reduction in COD (28.8 g/L) after the 272 h of the combined MEC and AD digestion (days 20–31) compared to 19.1% additional COD reduction in the AD-only treatment (37.8 g/L). The greater increase in COD reduction (50.1% higher) in the AD-MEC reactor compared to AD-only was higher than previous studies with AD-MECs, which showed increased COD reductions of 5% to 15% compared to AD-only [4,28,29]. Asztalos and Kim (2015) included three bioanodes and a stainless-steel mesh cathode to treat waste activated sludge at ambient temperature and found the VS and COD removal were only 5–10% higher than AD-only [30]. In our work, not only was the overall COD reduction higher, but the COD conversion efficiency to renewable energy was higher in the AD-MEC (7.09 kJ/g COD removed) compared to AD-only (6.19 kJ/g COD removed). The TS and VS concentrations in the reactors before MEC use were 6.89% and 4.70%, respectively. The additional TS reduction in the last 10 days of digestion was also higher in the AD-MEC treatment (35.2%) compared to AD-only (13.4%), and the additional VS reduction was 41.9% for AD-MEC treatment compared to 19.0% for AD-only. The results showed that the AD-MEC treatment increased the reduction of COD, TS, and VS concentrations during digestion compared to AD-only, with the additional organic matter removal likely occurring through oxidization of the organic material using exoelectrogenic bacteria attached to the anode and H_2 production at the cathode [30].

3.2. Anode Bacterial Attachment

The scanning electron microscopy (SEM) images on the anode surface showed bacterial cell colonization and growth. These cells were likely exoelectrogens, electricigens, and anodophilic bacteria that promoted substrate breakdown and electricity production (Figure 2) [31,32]. The anode (graphite) surface was completely covered with attached biofilm after the 11 days of the MEC inclusion, as confirmed by the SEM analysis. Coccoid and rod-shaped bacteria dominated the biofilm. It appears that 11 days were sufficient for the bacteria to colonize and acclimate to the anode graphite plates to form the biofilm needed to degrade the organic material (dairy manure). The SEM analysis also showed a denser microbial adherence on the graphite compared to stainless steel due to the porous structure of graphite materials [33].

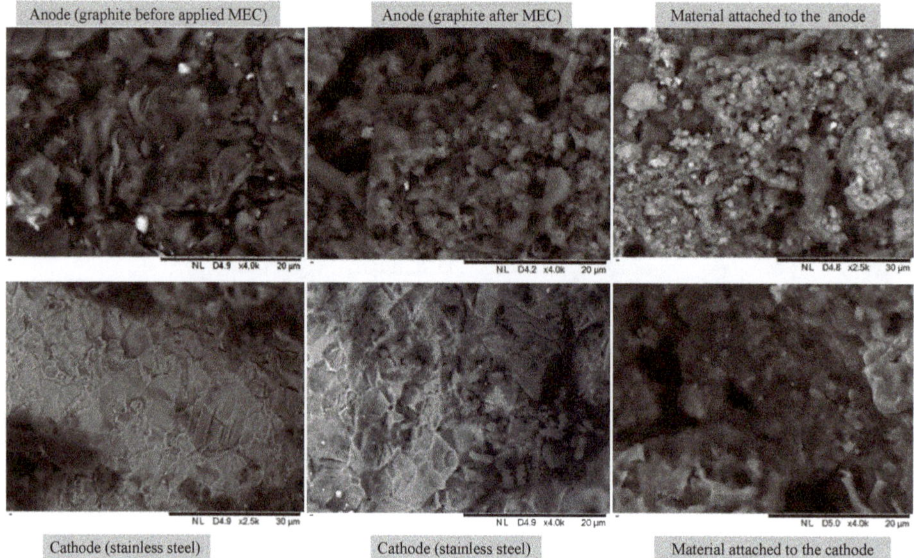

Figure 2. Morphological characteristics of the anode and cathode as visualized by the scanning electron microscope (SEM).

3.3. Biogas Production

Cumulative biogas production from the AD-MEC treatment was 93.0% higher than the AD-only treatment (Figure 3) during the MEC-inclusion period (days 20–31). The AD-MEC treatment produced 26.0 L of useful gases (CH_4 + H_2) over the additional 11-day digestion period, with 116% more CH_4 (23.6 L) and 51,804% more H_2 (2.43 L) than the AD-only treatment, which had 10.9 L CH_4 and 0.04 L H_2 (Figures 3 and 4). These results are similar to previous studies [34,35] that have shown that MEC inclusion into AD increased biogas and CH_4 production by 80–100%. The AD-MEC also reduced the CO_2 concentration (37.9% of the total gas) compared to AD-only (40.2% of the total gas), illustrating how MEC incorporation can both decrease CO_2 concentration and increase the generation of useful gases (CH_4 and H_2) in the biogas [4]. Our MEC novel design confined the cathode and the anode inside a single chamber, which reduced the cathode and anode distance, and thus, potential ohmic losses, but still allowed liquid to flow easily between the MEC and AD chambers. Recent studies [36–38] have reported that limiting the distance between the electrodes can increase the gas production, as large distances between the electrodes can inhibit electron flow between the anode and the cathode, increasing ohmic losses. Additionally, the manganese used in the stainless grade 201 could have increased the gas production, as Mg has been shown to enhance electrical energy produced in MFCs by a factor of ten [21,22].

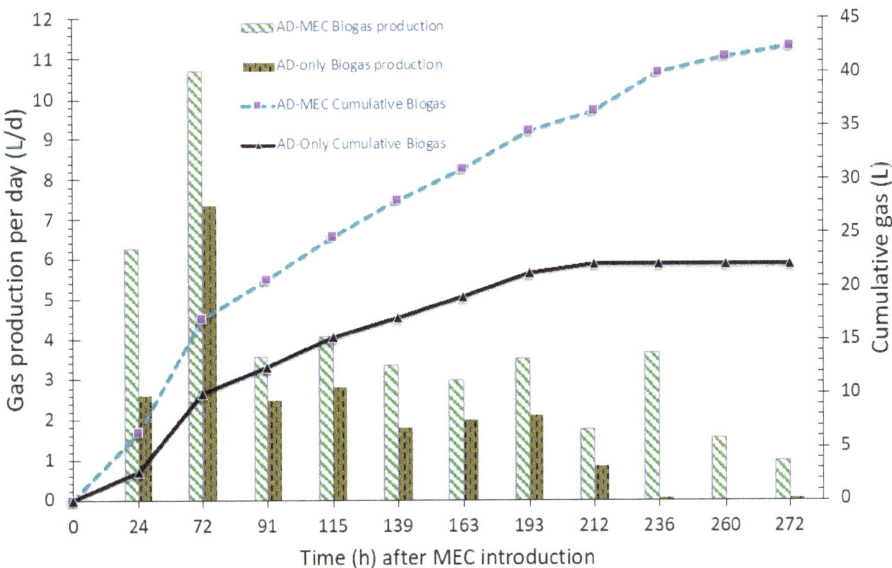

Figure 3. Cumulative and daily biogas production for AD-MEC and AD-only for days 20–31 (272 cumulative h) after MEC introduction (days 0–20 not shown, as treatments acted as duplicate treatments).

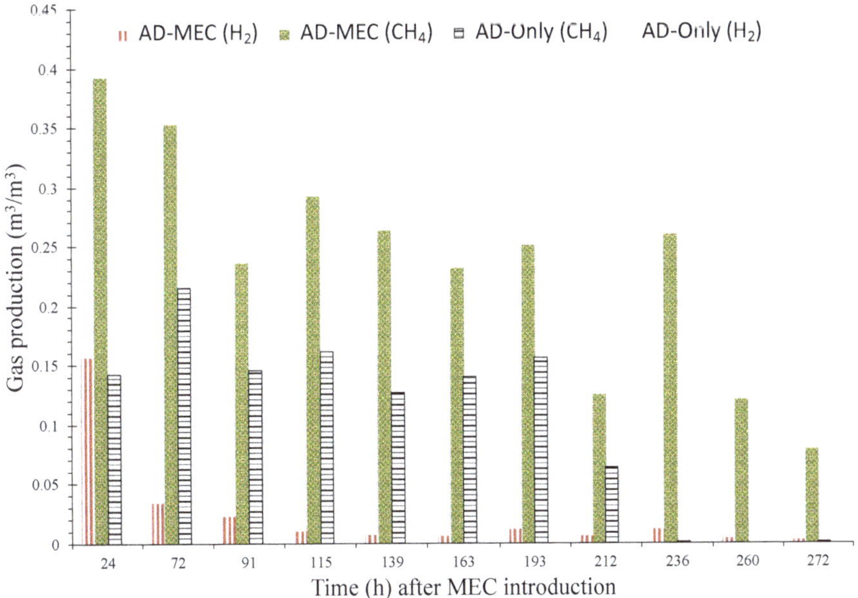

Figure 4. Methane (CH_4) and hydrogen (H_2) production for AD-MEC and AD-only for days 20–31 (272 cumulative h) after MEC introduction (days 0–20 not shown, as reactors acted as duplicate reactors).

Overall, the AD-MEC produced 149% more daily useful gases (H_2 and CH_4) (0.26 $m^3/m^3/d$) compared to AD-only (0.11 $m^3/m^3/d$) over the 11 additional days of digestion. The daily production rate of H_2 and CH_4 in the AD-MEC treatment reached a maximum of 0.16 $m^3H_2/m^3/d$ and 0.39 $m^3CH_4/m^3/d$

within the first 24 h. During the 11-day period, a cumulative $H_2 + CH_4$ production rate of 2.87 m^3/m^3 was observed for the AD-MEC treatment, compared to 1.15 m^3/m^3 observed for the AD-only treatment.

Hydrogen concentration recorded in the first 24 h of the MEC introduction reached 20% of biogas volume, but was then reduced to 2% H_2, as CH_4 increased from 50% to 56.9% after 5 days and to 63% CH_4 after 11 days. Most MEC studies observed high H_2 generation during experimental startup, which would gradually decrease while CH_4 concentration increased [13,14,24,29,39]. Recent studies have shown that hydrogenotrophic methanogens can survive in harsh environmental conditions, with continued utilization of the produced H_2 with CO_2 to form CH_4 [40–42]. It has also been shown that hydrogenotrophic methanogens can directly receive electrons from MEC electrodes [43] or H_2 [14] to enhance CH_4 production [44].

3.4. Energy Recovery

A relatively constant current, averaging at 111 mA, was recorded during the duration of the experiment, with a maximum of 383 mA recorded in the first 24 h. With an AD-MEC liquid volume of 8 L and a cathode surface area of 294 cm^2, the volumetric current and current density were calculated to be 13 A/m^3, and 7.4 A/m^2, respectively. These results were 70.9% higher than the current density (4.33 A/m^2) in a previous study [34] that resulted in 112% additional biogas with MEC inclusion. Biofilm on the electrodes also plays an important role (Figure 2) in increasing the current density [45,46].

Higher energy production was observed for the AD-MEC treatment ($W_{AD\text{-}MEC}$ = 964 kJ) compared to the AD-only treatment ($W_{AD\text{-}only}$ = 434 kJ). The total energy needed (electricity) to run the MEC ($W_{e(AD\text{-}MEC)}$ = 107 kJ) was only 20.2% of the extra energy produced by the AD-MEC treatment, compared to the AD-only treatment ($W_{AD\text{-}MEC} - W_{AD\text{-}only}$), and 11.1% of the AD-MEC's total energy production ($W_{AD\text{-}MEC}$) (Figure 5).

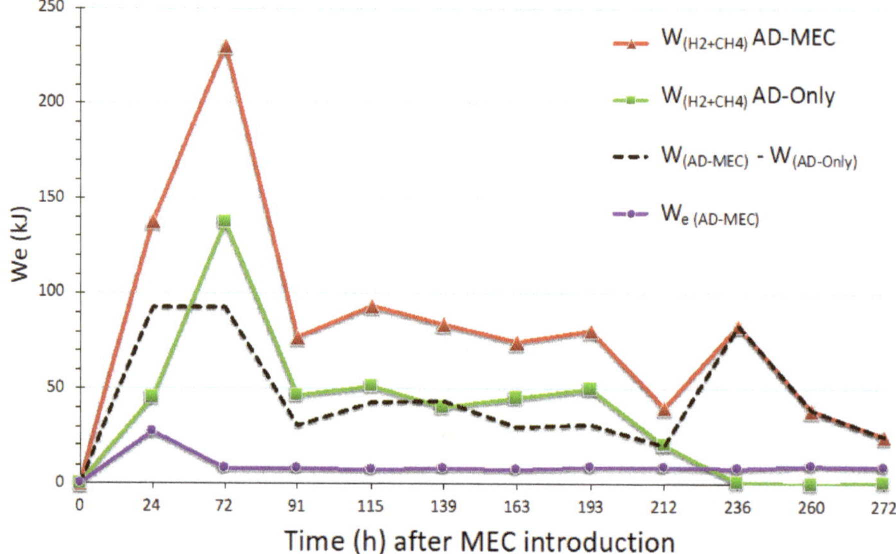

Figure 5. Energy from the CH_4 and H_2 produced by the AD-MEC and AD-only treatments from days 20–31 (272 cumulative h) after MEC introduction (days 0–20 not shown, as reactors acted as duplicate reactors).

The energy recovery efficiency, based on the extra energy produced ($W_{AD\text{-}MEC} - W_{AD\text{-}only}$) from the AD-MEC over the electrical energy needed to operate the MEC, ranged from 73.1% (ηE min.) to 324% (ηE max.), with an average increase of 170% over time (Figure 6). Huang et al. (2020) showed that coupling AD and MEC in the same chamber to treat food waste resulted in a 238% energy recovery efficiency when operated under negative pressure [47]. In the current study, the energy return was also greater than the input energy required to operate the MEC.

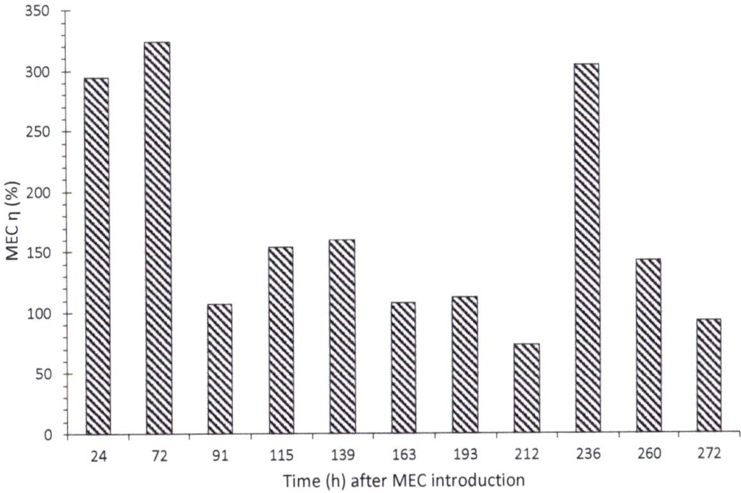

Figure 6. Electric energy recovery efficiency for MEC inclusion.

3.5. Biogas Utilization in a Fuel Cell or Combined Heat and Power (CHP) Generator

A combined heat and power (CHP) generator has a 30% electric conversion efficiency and 45% heat conversion efficiency [48], while a fuel cell has a 70% electric conversion efficiency [49]. The expected electricity production per m^3 digester volume using the produced biogas from the AD-MEC (during the last 11 days of digestion) in a fuel cell is 1.5 kwh/m^3/d, which is more than double the electricity production than AD-only (0.7 kwh/m^3/d), due to the 149% higher daily CH_4 and H_2 production in the AD-MEC system (Table 1). Similarly, the calculated electricity and heat production from a CHP generator was more than double with MEC inclusion compared to AD-only. Previous studies have found the potential electricity production for dairy manure digestion to be between 0.08 and 1.8 kWh/m^3/d over the entire digestion period [25,50]. By adding an MEC system to just the last 11 days of digestion, the dairy manure electric output went from the lower part of this range to the upper part of this range, while accounting for the electric input needed to operate the MEC. Overall, combining AD with an MEC greatly improved the biogas output during the last 11 days of digestion, allowing for a high electricity production when the biogas is used in a CHP or fuel cell (0.7 and 1.5 kWh/m^3/d, respectively).

Table 1. Biogas utilization in a fuel cell and combined heat and power (CHP) generator based on daily H_2 and CH_4 production per m^3 digester.

	H_2 Production	CH_4 Production	Fuel Cell Utilization	CHP Utilization	
Unit	m^3 H_2/m^3/d	m^3 CH_4/m^3/d	kWh/m^3/d	kWh/m^3/d	BTUs/m^3/d
AD-MEC	0.025	0.24	1.5	0.7	3388
AD-only	0	0.11	0.7	0.3	1515

4. Conclusions

An MEC was combined with AD in a single chamber to increase organic matter removal and energy production. The AD-MEC treatment produced 137.9% more CH_4 + H_2 in the produced biogas compared to the AD-only treatment. Furthermore, COD conversion efficiency was 14.5% higher in AD-MEC compared to AD-only. The efficiency of electrical energy recovery for MECs reached a maximum of 324%, with an average of 170% over the 11-day period. Incorporating AD with MEC could increase overall energy production from dairy manure digestion, even if added as a polishing step after 20 days of digestion.

Author Contributions: Conceptualization: A.H., F.W., S.L., L.Q., and Y.L.; Methodology: A.H. and F.W.; Software: A.H.; Validation: A.H. and F.W.; Formal analysis: A.H., F.W. and Y.L.; Investigation: A.H.; Resources: A.H., S.L., and L.Q.; Data curation: A.H., F.W.; Writing—original draft preparation: A.H.; Writing—review and editing, A.H., F.W., S.L., L.Q., and Y.L.; Visualization: A.H., F.W., L.Q., and Y.L.; Supervision: L.Q.; Project Administration: A.H., S.L., and L.Q.; Funding acquisition: A.H. and L.Q. All authors have read and agreed to the published version of the manuscript.

Funding: This research was funded by the China Scholarship Council (CSC) PhD program.

Acknowledgments: The first author acknowledges support from CSC.

Conflicts of Interest: The authors declare no conflict of interest.

References

1. Lu, L.; Xing, D.; Xie, T.; Ren, N.; Logan, B.E. Hydrogen production from proteins via electrohydrogenesis in microbial electrolysis cells. *Biosens. Bioelectron.* **2010**, *25*, 2690–2695. [CrossRef]
2. Wrana, N.; Sparling, R.; Cicek, N.; Levin, D.B. Hydrogen gas production in a microbial electrolysis cell by electrohydrogenesis. *J. Clean. Prod.* **2010**, *18*, S105–S111. [CrossRef]
3. Liu, H.; Grot, S.; Logan, B.E. Electrochemically assisted microbial production of hydrogen from acetate. *Environ. Sci. Technol.* **2005**, *39*, 4317–4320. [CrossRef] [PubMed]
4. Hassanein, A.; Witarsa, F.; Guo, X.; Yong, L.; Lansing, S.; Qiu, L. Next generation digestion: Complementing anaerobic digestion (AD) with a novel microbial electrolysis cell (MEC) design. *Int. J. Hydrogen Energy* **2017**, *42*, 28681–28689. [CrossRef]
5. Cheng, S.; Logan, B.E. Sustainable and efficient biohydrogen production via electrohydrogenesis. *Proc. Natl. Acad. Sci. USA* **2007**, *104*, 18871–18873. [CrossRef] [PubMed]
6. Gujer, W.; Zehnder, A.J.B. Conversion processes in anaerobic digestion. *Water Sci. Technol.* **1983**, *15*, 127–167. [CrossRef]
7. Gray, N.F. *Biology of Wastewater Treatment*; Oxford University Press: Oxford, UK, 1992; ISBN 0198563701.
8. Mudrack, K.; Kunst, S. *Biology of Sewage Treatment and Water Pollution Control*; Ellis Horwood Limited: Hertfordshire, UK, 1986; ISBN 0853129126.
9. Kim, J.; Park, C.; Kim, T.H.; Lee, M.; Kim, S.; Kim, S.W.; Lee, J. Effects of various pretreatments for enhanced anaerobic digestion with waste activated sludge. *J. Biosci. Bioeng.* **2003**, *95*, 271–275. [CrossRef]
10. Chae, K.J.; Choi, M.J.; Kim, K.Y.; Ajayi, F.F.; Park, W.; Kim, C.W.; Kim, I.S. Methanogenesis control by employing various environmental stress conditions in two-chambered microbial fuel cells. *Bioresour. Technol.* **2010**, *101*, 5350–5357. [CrossRef]
11. Hu, H.; Fan, Y.; Liu, H. Hydrogen production using single-chamber membrane-free microbial electrolysis cells. *Water Res.* **2008**, *42*, 4172–4178. [CrossRef]
12. Lee, H.S.; Torres, C.I.; Parameswaran, P.; Rittmann, B.E. Fate of H2 in an upflow single-chamber microbial electrolysis cell using a metal-catalyst-free cathode. *Environ. Sci. Technol.* **2009**, *43*, 7971–7976. [CrossRef]
13. Parameswaran, P.; Zhang, H.; Torres, C.I.; Rittmann, B.E.; Krajmalnik-Brown, R. Microbial community structure in a biofilm anode fed with a fermentable substrate: The significance of hydrogen scavengers. *Biotechnol. Bioeng.* **2010**, *105*, 69–78. [CrossRef] [PubMed]
14. Wang, A.; Liu, W.; Cheng, S.; Xing, D.; Zhou, J.; Logan, B.E. Source of methane and methods to control its formation in single chamber microbial electrolysis cells. *Int. J. Hydrogen Energy* **2009**, *34*, 3653–3658. [CrossRef]

15. Clauwaert, P.; Verstraete, W. Methanogenesis in membraneless microbial electrolysis cells. *Appl. Microbiol. Biotechnol.* **2009**, *82*, 829–836. [CrossRef] [PubMed]
16. Cusick, R.D.; Bryan, B.; Parker, D.S.; Merrill, M.D.; Mehanna, M.; Kiely, P.D.; Liu, G.; Logan, B.E. Performance of a pilot-scale continuous flow microbial electrolysis cell fed winery wastewater. *Appl. Microbiol. Biotechnol.* **2011**, *89*, 2053–2063. [CrossRef] [PubMed]
17. Wagner, R.C.; Regan, J.M.; Oh, S.E.; Zuo, Y.; Logan, B.E. Hydrogen and methane production from swine wastewater using microbial electrolysis cells. *Water Res.* **2009**, *43*, 1480–1488. [CrossRef]
18. Liu, W.; Huang, S.; Zhou, A.; Zhou, G.; Ren, N.; Wang, A.; Zhuang, G. Hydrogen generation in microbial electrolysis cell feeding with fermentation liquid of waste activated sludge. *Int. J. Hydrogen Energy* **2012**, *37*, 13859–13864. [CrossRef]
19. Logan, B.; Cheng, S.; Watson, V.; Estadt, G. Graphite fiber brush anodes for increased power production in air-cathode microbial fuel cells. *Environ. Sci. Technol.* **2007**, *41*, 3341–3346. [CrossRef]
20. Wei, J.; Liang, P.; Huang, X. Recent progress in electrodes for microbial fuel cells. *Bioresour. Technol.* **2011**, *102*, 9335–9344. [CrossRef]
21. Park, D.; Zeikus, J. Impact of electrode composition on electricity generation in a single-compartment fuel cell using Shewanella putrefaciens. *Appl. Microbiol. Biotechnol.* **2002**, *59*, 58–61. [CrossRef]
22. Lowy, D.A.; Tender, L.M.; Zeikus, J.G.; Park, D.H.; Lovley, D.R. Harvesting energy from the marine sediment-water interface II. Kinetic activity of anode materials. *Biosens. Bioelectron.* **2006**, *21*, 2058–2063. [CrossRef]
23. Call, D.; Logan, B.E. Hydrogen production in a single chamber microbial electrolysis cell lacking a membrane. *Environ. Sci. Technol.* **2008**, *42*, 3401–3406. [CrossRef]
24. Rader, G.K.; Logan, B.E. Multi-electrode continuous flow microbial electrolysis cell for biogas production from acetate. *Int. J. Hydrogen Energy* **2010**, *35*, 8848–8854. [CrossRef]
25. Achi, C.G.; Hassanein, A.; Lansing, S. Enhanced biogas production of cassava wastewater using zeolite and biochar additives and manure co-digestion. *Energies* **2020**, *13*, 491. [CrossRef]
26. American Public Health Association (APHA). *Standard Methods for the Examination of Water and Wastewater*, 21st ed.; APHA: Washington, DC, USA, 2005; ISBN 0875532357.
27. O'Dell, J.W. *The Determination of Chemical Oxygen Demand by Semi-Automated Colorimetry*; United States Environmental Protection Agency: Westlake, OH, USA, 1996.
28. Gajaraj, S.; Huang, Y.; Zheng, P.; Hu, Z. Methane production improvement and associated methanogenic assemblages in bioelectrochemically assisted anaerobic digestion. *Biochem. Eng. J.* **2017**, *117*, 105–112. [CrossRef]
29. Cai, W.; Han, T.; Guo, Z.; Varrone, C.; Wang, A.; Liu, W. Methane production enhancement by an independent cathode in integrated anaerobic reactor with microbial electrolysis. *Bioresour. Technol.* **2016**, *208*, 13–18. [CrossRef]
30. Asztalos, J.R.; Kim, Y. Enhanced digestion of waste activated sludge using microbial electrolysis cells at ambient temperature. *Water Res.* **2015**, *87*, 503–512. [CrossRef] [PubMed]
31. Logan, B.E.; Regan, J.M. Electricity-producing bacterial communities in microbial fuel cells. *Trends Microbiol.* **2006**, *14*, 512–518. [CrossRef] [PubMed]
32. Khater, D.Z.; El-Khatib, K.M.; Hassan, H.M. Microbial diversity structure in acetate single chamber microbial fuel cell for electricity generation. *J. Genet. Eng. Biotechnol.* **2017**, *15*, 127–137. [CrossRef]
33. Shornikova, O.N.; Kogan, E.V.; Sorokina, N.E.; Avdeev, V.V. The specific surface area and porous structure of graphite materials. *Russ. J. Phys. Chem. A* **2009**, *83*, 1022–1025. [CrossRef]
34. Zhang, Y.; Angelidaki, I. Counteracting ammonia inhibition during anaerobic digestion by recovery using submersible microbial desalination cell. *Biotechnol. Bioeng.* **2015**, *112*, 1478–1482. [CrossRef]
35. Xiao, B.; Chen, X.; Han, Y.; Liu, J.; Guo, X. Bioelectrochemical enhancement of the anaerobic digestion of thermal-alkaline pretreated sludge in microbial electrolysis cells. *Renew. Energy* **2018**, *115*, 1177–1183. [CrossRef]
36. Kadier, A.; Simayi, Y.; Abdeshahian, P.; Azman, N.F.; Chandrasekhar, K.; Kalil, M.S. A comprehensive review of microbial electrolysis cells (MEC) reactor designs and configurations for sustainable hydrogen gas production. *Alex. Eng. J.* **2016**, *55*, 427–443. [CrossRef]
37. An, J.; Lee, H.S. Implication of endogenous decay current and quantification of soluble microbial products (SMP) in microbial electrolysis cells. *RSC Adv.* **2013**, *3*, 14021–14028. [CrossRef]

38. Fu, Q.; Wang, D.; Li, X.; Yang, Q.; Xu, Q.; Ni, B.J.; Wang, Q.; Liu, X. Towards hydrogen production from waste activated sludge: Principles, challenges and perspectives. *Renew. Sustain. Energy Rev.* **2021**, *135*, 110283. [CrossRef]
39. Liu, W.; Cai, W.; Guo, Z.; Wang, L.; Yang, C.; Varrone, C.; Wang, A. Microbial electrolysis contribution to anaerobic digestion of waste activated sludge, leading to accelerated methane production. *Renew. Energy* **2016**, *91*, 334–339. [CrossRef]
40. Cai, W.; Liu, W.; Zhang, Z.; Feng, K.; Ren, G.; Pu, C.; Sun, H.; Li, J.; Deng, Y.; Wang, A. mcrA sequencing reveals the role of basophilic methanogens in a cathodic methanogenic community. *Water Res.* **2018**, *136*, 192–199. [CrossRef]
41. Jiang, Y.; Banks, C.; Zhang, Y.; Heaven, S.; Longhurst, P. Quantifying the percentage of methane formation via acetoclastic and syntrophic acetate oxidation pathways in anaerobic digesters. *Waste Manag.* **2018**, *71*, 749–756. [CrossRef]
42. Zakaria, B.S.; Ranjan Dhar, B. An intermittent power supply scheme to minimize electrical energy input in a microbial electrolysis cell assisted anaerobic digester. *Bioresour. Technol.* **2020**, 124109. [CrossRef]
43. Lohner, S.T.; Deutzmann, J.S.; Logan, B.E.; Leigh, J.; Spormann, A.M. Hydrogenase-independent uptake and metabolism of electrons by the archaeon Methanococcus maripaludis. *ISME J.* **2014**, *8*, 1673–1681. [CrossRef]
44. Villano, M.; Aulenta, F.; Ciucci, C.; Ferri, T.; Giuliano, A.; Majone, M. Bioelectrochemical reduction of CO2 to CH4 via direct and indirect extracellular electron transfer by a hydrogenophilic methanogenic culture. *Bioresour. Technol.* **2010**, *101*, 3085–3090. [CrossRef]
45. Yin, Q.; Zhu, X.; Zhan, G.; Bo, T.; Yang, Y.; Tao, Y.; He, X.; Li, D.; Yan, Z. Enhanced methane production in an anaerobic digestion and microbial electrolysis cell coupled system with co-cultivation of Geobacter and Methanosarcina. *J. Environ. Sci. (China)* **2016**, *42*, 210–214. [CrossRef] [PubMed]
46. Malvankar, N.S.; Tuominen, M.T.; Lovley, D.R. Biofilm conductivity is a decisive variable for high-current-density Geobacter sulfurreducens microbial fuel cells. *Energy Environ. Sci.* **2012**, *5*, 5790–5797. [CrossRef]
47. Huang, J.; Feng, H.; Huang, L.; Ying, X.; Shen, D.; Chen, T.; Shen, X.; Zhou, Y.; Xu, Y. Continuous hydrogen production from food waste by anaerobic digestion (AD) coupled single-chamber microbial electrolysis cell (MEC) under negative pressure. *Waste Manag.* **2020**, *103*, 61–66. [CrossRef]
48. Hampson, A.; Tidball, R.; Fucci, M.; Weston, R. *Combined Heat and Power (CHP) Technical Potential in the United States*; US Department of Energy: Washington, DC, USA, 2016. Available online: https://www.energy.gov/sites/prod/files/2016/04/f30/CHP%20Technical%20Potential%20Study%203-31-2016%20Final.pdf (accessed on 13 October 2020).
49. Chuahy, F.D.; Kokjohn, S.L. Solid oxide fuel cell and advanced combustion engine combined cycle: A pathway to 70% electrical efficiency. *Appl. Energy* **2019**, *235*, 391–408. [CrossRef]
50. Fan, X.; Li, Z.; Wang, T.; Yin, F.; Jin, X. Introduction to a Large-Scale Biogas Plant in a Dairy Farm. In Proceedings of the 2010 International Conference on Digital Manufacturing Automation, Changsha, China, 18–20 December 2010; Volume 1, pp. 863–866.

Article

Ferric Oxide-Containing Waterworks Sludge Reduces Emissions of Hydrogen Sulfide in Biogas Plants and the Needs for Virgin Chemicals

Tobias Persson, Kenneth M. Persson * and Jenny Åström

Sydvatten AB, Hyllie Stationstorg 21, SE-215 32 Malmö, Sweden; Tobias.Persson@Sydvatten.se (T.P.); Jenny.Astrom@Sydvatten.se (J.Å.)
* Correspondence: kenneth.persson@sydvatten.se

Abstract: Ferric oxide-containing waterworks sludge can be used to reduce the formation of hydrogen sulfide during anaerobic digestion. The ferric compound is reduced biochemically in the digester and forms insoluble pyrite in digester sludge. Virgin ferric chloride is often used to solve the hydrogen sulfide problem. Since 2013, Sydvatten AB has supplied a growing number of digestion plants in Sweden with ferric-containing dewatered waterworks sludge derived from the drinking water treatment plant Ringsjöverket to limit the formation of hydrogen sulfide. At the waterworks, ferric chloride is added to enhance the coagulation of organic matter from the source water. The sludge formed in this process is dewatered and landfilled, but also recycled in biogas production in order to decrease the hydrogen sulfide concentration. In this study, the use of sludge for hydrogen sulfide removal in digesters was technically and economically evaluated via case studies from 13 full-scale digesters in Sweden. Compared with the use of fresh ferric chloride, the operational costs are reduced by up to 50% by using sludge. The quality of the sludge is high and its content in metals is low or very low, especially when compared with the requirements of different certification standards for biosolid reuse applied in Sweden. The addition of waterworks sludge containing iron to a digester for the removal of dissolved hydrogen sulfide is a technically and economically good alternative when producing biogas. It is also one step closer to a circular economy, as replacing the use of virgin chemicals with the by-product waterworks sludge saves energy and materials and reduces the carbon footprint of the waterworks.

Keywords: biogas digestion; hydrogen sulfide; ferric oxide; waterworks sludge

Citation: Persson, T.; Persson, K.M.; Åström, J. Ferric Oxide-Containing Waterworks Sludge Reduces Emissions of Hydrogen Sulfide in Biogas Plants and the Needs for Virgin Chemicals. *Sustainability* **2021**, *13*, 7416. https://doi.org/10.3390/su13137416

Academic Editor: Shashi Kant Bhatia

Received: 24 May 2021
Accepted: 28 June 2021
Published: 2 July 2021

1. Introduction

Biogas production is gaining increasing attention as a source for replacing fossil-based fuels with renewable fuels in society. Biogas is typically produced in anaerobic digestion plants (AD), where different substrates rich in organics are digested by methanogenic bacteria. Most substrates also contain sulfur, which in anaerobic environments can be microbiologically reduced to hydrogen sulfide, which negatively affects the metabolic activity of the methanogens and eventually poisons the digester. Additionally, hydrogen sulfide is a technical issue in plants and downstream when biogas is used, since hydrogen sulfide corrodes pipes, generators and other equipment. It is also a health hazard, being toxic to humans. Improving the quality and quantity of biogas usually requires pre-treatment to maximize methane yields and/or post-treatment to remove hydrogen sulfide. This requires considerable energy consumption and higher costs; hence there are needs for better and more efficient measures to control hydrogen sulfide production [1].

One way to remove hydrogen sulfide as a gas is to add ferric salts to the substrate or to the digester. Ferric salts can be reduced to ferrous iron and form pyrite (FeS_2) as a precipitate. Often, ferric chloride solution is dosed into the reactor to achieve this removal effect on hydrogen sulfide. However, the addition of virgin ferric salts has an operational cost, and a water and carbon footprint. The profitability in, for instance, the Swedish biogas industry is relatively poor [2]. Swedish and European climate ambitions state that greenhouse gas emissions should be reduced to at least 55% below 1990 levels by 2030, and be climate-neutral by 2050. Policies to help transition society towards a circular economy later on suggest a reduction in waste generation and the reuse and recycling of materials and energy, as expressed in the EU New Green Deal, the agenda for sustainable growth. The EU's transition to a circular economy will reduce pressure on natural resources and will create sustainable growth and jobs. It is also a prerequisite to achieve the EU's 2050 climate neutrality target and to halt biodiversity loss. Could other ferric materials with lower costs, smaller climate footprints and better material use replace the need for virgin ferric chloride? If so, the operational costs would decrease, the climate footprint would be reduced, and more biogas plants could achieve positive results, thus contributing to the EU New Green Deal.

The segment that has the greatest untapped potential for biogas production in Sweden, but also the biggest economic challenges, is the agricultural sector. Reducing the hydrogen sulfide concentration during digestion is presently associated with significant costs and the handling of corrosive chemicals. For farm-based biogas plants, this is extra stressful because these plants are small, and have small financial margins and limited resources for handling hazardous chemicals. In addition, manure (especially pig manure) is rich in sulfur and contains concentrations that can be converted to several thousand ppm hydrogen sulfide during the digestion process. How much hydrogen sulfide is formed during the digestion process depends on the sulfur content of the substrate in the form of sulfate or as sulfur bound in amino acids [3,4]. High costs for the removal of hydrogen sulfide can mean the difference between a positive and a negative financial result at the end of the year for a farm facility, and thus are also something that limits the expansion of biogas production in agriculture in Sweden. The removal of hydrogen sulfide also represents a significant cost in co-digestion plants. Requirements are higher for the separation of hydrogen sulfide in these plants, since the biogas produced is in principle exclusively upgraded to vehicle gas quality. In the case of biogas generated from sewage sludge in wastewater treatment plants, primarily those based on biological phosphorus separation have issues with high hydrogen sulfide concentrations during digestion. For these plants, a reduction in hydrogen sulfide is also associated with costs that make the biogas business less profitable, which is why alternative solutions can be of interest. Before upgrading the biogas, all sulfur must be removed, unless the upgrade is performed with a water scrubber, or in some cases an amine scrubber, as a few hundred ppm can be accepted. When the biogas is to be used for power/heat production, the requirements are usually around 50–200 ppm, but by lowering the concentration further, the service life can be increased and the need for maintenance on the engine/turbine used for power/heat production can be reduced.

The addition of an iron source, may it be iron chloride, iron oxides or waterworks sludge rich in iron salts, binds the hydrogen sulfide in the slurry in the digestion chamber, and reduces the possibility that the hydrogen sulfide can inhibit biogas production [3]. The addition of ferric salts can also increase the availability of trace metals that the microorganisms need, and thus increase the efficiency of the biogas process [5]. The addition of air and oxygen reduces the hydrogen sulfide concentration in the gas phase, but does not resolve the problem of hydrogen sulfide inhibiting the microorganisms in biogas production to the same extent. Furthermore, the use of oxygen/air in methane streams is associated with some risks, and it is important that biogas producers leave a sufficient margin to the lower explosion limit for biogas. It is not possible to use air if the biogas is to be upgraded to vehicle quality, as this requires that the oxygen in the air first be separated from the air nitrogen in an external process [6]. Ferric chloride and ferric oxide have similar properties when it comes to binding sulfide, with the difference that iron oxide is less reactive and less corrosive and thus easier to handle. Regardless of the method used, the reduction of hydrogen sulfide is associated with significant costs for the biogas producer, with the exception of those plants that use only air. A Swedish feasibility study for biogas production at farms [7] showed that the cost of hydrogen sulfide reduction was around EUR 0.01–0.02 per Nm^3 biogas at farm biogas plants. The cost is higher for plants where the hydrogen sulfide level must be kept below 100 ppm in the produced biogas.

Since 1997, Sydvatten AB has utilized ferric chloride as a coagulant in drinking water production at Ringsjöverket, the waterworks in Stehag, south Sweden. The coagulant forms a sludge that is gravimetrically removed from the sedimentation step in the waterworks. The waterworks sludge is dewatered in two steps and landfilled in an area previously used for peat extraction. Sydvatten AB has a sustainability plan laid out by the board of directors in 2018, stating, among other items, that resources should be utilized as efficiently as possible and that energy and material should be reused and recycled to the greatest possible extent [8]. The board of directors has stated that the company must be climate-neutral by 2030, and the work on defining how climate neutrality can be reached and what measures must be taken in the organization to achieve climate neutrality has been reviewed in the Climate Accounts Report 2020 [9].

In 2016, tests were performed to investigate if the reuse of dewatered waterworks sludge could be applied in anaerobic digesters in the biogas industry in Sweden [10]. A growing number of biogas plants using varying sulfur-containing substrates means a growing need for efficient hydrogen sulfide management. To minimize the amounts of sludge deposited and to increase the recycling of materials is beneficial for society and reduces the costs and carbon footprint in the digester. The sludge contains mostly ferric oxide in various forms that originate from chemical precipitation with iron chloride in the waterworks. The purpose of this study is to technically and economically evaluate the use of the sludge for hydrogen sulfide reduction and to discuss to what extent the reuse of ferric waterworks sludge can contribute to the company reaching climate neutrality by 2030. A technical evaluation of the methods employed to add sludge to digesters and which specific dose of sludge should be added to digesters is also presented. We present some accounts from the field of the quantities of ferric compounds required to reduce the hydrogen sulfide concentration in different biogas plants.

The residual solids from the biogas production should be of such quality that they can be brought back to arable land as organic fertilizers when using ferric waterworks sludge as a hydrogen sulfide measure in the digester. Efficient material use requires these measures in a sustainable society. In Sweden, two different certification standards are used, depending on the origin of the substrate in biogas plants. If the substrate comes from a wastewater treatment plant, REVAQ is applied [11]. This is the national standard for the quality control of residuals from wastewater treatment plants and has been used since 2008. If the substrate originates from other sources, such as manure or food waste, the SPCR 120 standard is used instead [12]. This certification standard has been developed by the solid waste industry in Sweden and has been used since 1999.

In substrates containing sulfur and rich in organic material, the anaerobic microbial metabolism generates sulfide and hydrogen sulfide, depending on the pH. If iron is present, some iron is reduced microbiologically to ferrous iron. Pyrite (FeS_2) is a highly insoluble sulfide that can be formed in anaerobic conditions in the presence of sulfide ions. Waterworks sludge from drinking water treatment plants utilizing ferric salts for coagulation contains large amounts of ferric oxide. Mixing such sludge into the digester will cause the ferric ion (Fe^{3+}) to be reduced in the anaerobic environment to ferrous iron (Fe^{2+}), which binds sulfide ions to form pyrite. To dose ferric compounds into the digester is a method that can facilitate the removal of hydrogen sulfide from the biogas. It has been reported that around 2–4% of influent S enters the digesters, which could be removed sufficiently by a dosage of 1.1 mg/L of Fe into the raw wastewater. A higher dry matter content was also observed in the dewatered cake as an additional secondary benefit when changing from alum dosage to iron dosage for phosphorous removal [13]. A drop in hydrogen sulfide emission from full-scale ADs at a large-scale municipal wastewater treatment plant could be achieved when dosing ferric chloride. The ferric salt was applied in the range of 24–105 mg $FeCl_3$/L into the feeding line and the sludge thickener unit. The hydrogen sulfide emission was reduced by 4 mg/L with the direct dosing into an AD, but this emission was reduced by only 1.3 mg/L in non-dosed ADs. The formation of hydrogen sulfide could be correlated to the volatile primary sludge solid loading rates, based on data from a 17-month study period [14].

The waste iron powder produced by laser cutting machines in the steel and iron industry was mixed with dairy manure at a concentration between 2.0 and 20.0 g/L in digestion batch experiments and between 1.0 and 4.0 g/L in bench experiments. For batch experiments, the hydrogen sulfide concentration could be reduced by up to 93% at a dosage of waste iron powder of 2.0 g/L. If the waste iron powder concentration was higher than 8.0 g/L, the reduction was more than 99%. Waste iron powder did not have a significant effect on methane yield in the batch and bench experiments, but the hydrolysis rate constant was almost doubled and the lag-phase period halved in test digesters compared to control digesters without iron dosage. In bench experiments, the H_2S concentration was reduced by 89% at 2.0 g/L, and by 50% at 1.0 g/L, without harming the digestion process [15].

Fe_2O_3 and TiO_2 nanoparticles at four different concentrations in two different combinations, from 20 to 500 mg/L, were used for the mitigation of hydrogen sulfide emission during the anaerobic digestion of cattle manure in a batch system. The H_2S production was 2.13–2.64 times lower than in the control. Additionally, biogas and CH_4 production were 1.09–1.191 times higher than those of the control [16]. Titanium is relatively costly, and in another study, the researchers investigated whether directly adding waste iron powder and iron oxide nanoparticles into batch digesters could offer a more cost-efficient solution to hydrogen sulfide generation. By adding iron in the form of microscale iron powder at concentrations of 100 mg/L to 1000 mg/L, the methane yield could be improved by up to 57%. The equivalent dosages of iron nanoparticles improved the yield by up to 21%. The highest iron powder dose (1000 mg/L) achieved the maximum improvement in the rate of hydrolysis, which was 1.25 times higher than in the control reactions. A high dosage of iron powder also decreased the rate of hydrogen sulfide production by up to 77% compared with the reference. The direct mixing of microscale iron powder was proposed as a practical and economical means of supporting the production of biogas from dairy manure [17].

The addition of iron-rich drinking water sludge directly into the urban domestic wastewater system was tested to reduce the content of dissolved sulfide in sewer systems, to aid phosphate removal in wastewater treatment, and to reduce hydrogen sulfide in the anaerobic digester. It was tested using two laboratory-scale urban wastewater systems, one as an experimental system and the other as a control, each comprising sewer reactors, a sequencing batch reactor (SBR) for wastewater treatment, sludge thickeners, and anaerobic digestion reactors. The experimental system received in-sewer drinking water sludge corresponding to 10 mg Fe/L, while the control had none. The addition of ferric sludge reduced the hydrogen sulfide concentration in the wastewater by 3.5 mg S/L as compared with the control. The phosphate concentration decreased by 3.6 mg P/L after biological wastewater treatment in the experimental SBR. In the experimental anaerobic digester, the sulfide concentration decreased by 16 mg S/L compared with the reference. Drinking water sludge dosing also enhanced the settleability of the mixed liquid suspended sludge and the dewaterability of the anaerobically digested sludge. The cake solids concentration increased from 16% to 19%. Additionally, the chemical oxygen demand (COD) and total suspended solids (TSS) concentrations in the wastewater were increased, but did not affect normal operation. The authors concluded that the addition of iron-rich drinking water sludge could be employed in the urban wastewater system, achieving multiple benefits [18].

Just over 2.1 TWh of biogas was produced in Sweden in 2019. Swedish biogas production increased by 3.3% in 2019, to a total of 2111 GWh (Table 1). Biogas production increased at all plant types except industrial plants and gasification plants in 2019. The largest increase was at digestion plants (+68 GWh), which also accounted for most of the increase in the last decade. A total of 49% of the biogas was produced in co-digestion plants and 35% at sewage treatment plants. There are a total of 280 biogas production facilities in Sweden [19]. The biogas is mainly produced from various types of waste and residual products such as sewage sludge, food waste, manure and waste from the food industry and slaughterhouses. Increasing quantities of biogas are produced from manure. A total of 71 plants use fertilizer as a substrate, and the amount of manure that is digested has increased by 9% to 1.1 million tons. In total, around 2.8 million tons of digestate (wet weight) were produced at Swedish biogas plants in 2019, of which 2.4 million tons (87%) were used as fertilizer in agriculture. From co-digestion plants and farm plants, all digestate (bio fertilizer) was used as fertilizer. From the sewage treatment plants, 41% of the digestate (digestate sludge) was used as fertilizer. Just under two-thirds of the biogas is upgraded. The long-term trend whereby an increasing amount of biogas is being upgraded continues, after a temporary decline in 2018. The upgraded biogas is used as vehicle gas or fed into the gas network. Of the biogas produced, 64% is upgraded (1351 GWh) and 19% is used for heat production (Table 2). Direct electricity production continues to decline. The share of biogas that goes into flaring is a total of 11% of production, showing a definite increase up to 2018. Flaring has to be carried out during the start-up phases of digesters, and occasionally when operational problems occur. In 2019, a large new digester was commissioned, and the start-up issues took some time to solve [19].

Table 1. Volume of biogas production and number of plants in Sweden in 2019 per plant type and change since 2018 [19].

Plant Type	Number of Plants	Production (GWh)	Share (%)	Change Since 2018 (%)
Sludge from wastewater treatment plants	135	738	35	+2
Co-digesters	36	1031	49	+7
Farm units	50	58	3	+4
Industrial plants	7	142	6	−1
Landfill gas plants	52	142	7	+1
Gasification plants	0	0	0	−100
Sum	280	2111	100	+3.3

Table 2. Use of produced biogas in Sweden 2019 with change since 2018 [19].

Area	Use (GWh)	Share (%)	Change Since 2018 (%)
Upgrading	1351	64	4
Heat	397	19	−1
Electricity	38	2	−10
Industrial use	52	2	0
Other uses	23	1	−15
Flaring	235	11	11
Losses and lack of data	15	1	2
Sum	2111	100	3.3

Of the upgraded biogas, 539 GWh was injected directly into the gas distribution network in south-west Sweden and in the regional network in Stockholm. In 2019, the total biogas use increased by 7%, and the import was estimated at around 1.8 TWh, meaning the total biogas use in Sweden in 2019 was 4 TWh. The biogas market is growing in Sweden. Since 2015, it doubled, but the Swedish production only grew by a total of 9% during the same period [19]. Profitability in the Swedish biogas industry is relatively poor, and many biogas producers are struggling to achieve positive results. The segment that has the greatest untapped potential for biogas production in Sweden, but also the biggest economic challenges, is the agricultural sector. In order for there to be growth in this segment, it is necessary to be able to report profitability for the business. Reducing the hydrogen sulfide concentration during digestion is today associated with significant costs and the handling of corrosive chemicals. For farm-based biogas plants, this is extra stressful, because these plants are small, and have small financial margins and limited resources for handling hazardous chemicals. In addition, manure (especially pig manure) contains sulfur, which can be converted to several thousand ppm of hydrogen sulfide during the digestion process. How much hydrogen sulfide is formed during the digestion process depends on the sulfur content of the substrate in the form of sulfate or as sulfur bound in amino acids [3,4]. The high costs of hydrogen sulfide reduction can mean the difference between a positive and a negative financial result at the end of the year for a farm facility, and this is thus also something that limits the expansion of biogas production in agriculture in Sweden today.

The reduction of hydrogen sulfide is also a significant cost for co-digestion plants. Here, the requirements for the separation of hydrogen sulfide are higher than at farm-based biogas plants that produce power/heat, since the biogas produced at digestion plants is in principle exclusively upgraded to vehicle gas quality (see Table 2). In the case of wastewater treatment plants, it is primarily plants that perform biological phosphorus separation that experience high hydrogen sulfide concentrations during digestion. For these plants, a reduction in hydrogen sulfide is also associated with costs that make the biogas business less profitable, which is why alternative solutions can be of interest. Sulfur hydrogen is corrosive, and must be removed before the biogas is upgraded to vehicle fuel or used for power/heat production. Before upgrading biogas, all sulfur must be removed, unless the upgrade is performed with a water scrubber, or in some cases an amine scrubber, as a few hundred ppm can be accepted. When the biogas is to be used for power/heat production, the requirements are usually around 50–200 ppm, but by lowering the concentration further, the service life can be increased and the need for maintenance on the engine/turbine used for power/heat production can be reduced. The addition of iron chloride, iron oxides or waterworks sludge from the iron coagulation steps binds the hydrogen sulfide in the slurry in the digestion chamber, and reduces the probability of the hydrogen sulfide inhibiting biogas production [3].

2. Materials and Methods

Since 2013, iron-containing sludge derived from drinking water production at Sydvatten's waterworks in Stehag has been offered to biogas production plants in southern Sweden for hydrogen sulfide control. Sydvatten's interest is to minimize and eventually avoid the landfilling of waterworks sludge and find pathways to reusing the sludge in other applications. A survey of the properties of the waterworks sludge and how it has been used for counteracting hydrogen sulfide formation during biogas production has previously been reported [10]. The sludge contains mostly iron in various forms that originate from chemical precipitation with iron chloride in the waterworks.

Dewatered waterworks sludge was collected three times in 2016 and analyzed with reference to metal content at an accredited lab, AlControl AB. Sludge was collected from three different dewatering batches and mixed prior to analysis. Thirteen biogas producers from different sites in south Sweden who use waterworks sludge at full-scale for hydrogen sulfide removal were asked to share their experiences from these facilities, which have been collected and compiled below under different categories. Experiences concerning waterworks sludge transportation, transport cost, operational and maintenance costs for storage, the dosing and cleaning of the equipment used in the handling of waterworks sludge at the biogas plant, the practical dosage and use of waterworks sludge in the digester, the effects of storage conditions due to storage time and ambient temperature, and general operational observations of conditions when the waterworks sludge was dosed into the digester and mixed with substrate, were recorded in the interview series. All interviews were carried out through direct visits to the plants and through interviews with plant operators and managers.

3. Results

At Sydvatten's Ringsjöverket waterworks in Stehag, approximately 9000 tons of sludge with 15% total solids (TS) are produced annually (see Table 3). The sludge is formed in the chemical precipitate, which is the first part of the waterworks process. The raw water comes from Lake Bolmen in Småland, and the organic content of Bolmen's water is virtually inert, i.e., it will not contribute to the biogas production in the digestion chamber. Today, ferric chloride is used as a coagulation chemical to coagulate organic material in the water. The water pH is corrected with lye before the addition of the coagulant. The sludge settles in lamella sedimentation and is then thickened in a gravity thickener after the addition of iron chloride, lime water and polymer, so as to reach a dry content of about 2%. The sludge is then pumped into a sludge handling plant where it is pressed in sieve belt presses after the addition of additional polymer and iron chloride to a dry content of about 15%. The sludge is finally landfilled in a closed peat extraction area. Data on the overall generation of sludge and the fraction used in biogas plants are presented in Table 3.

Table 3. Waterworks sludge generation and fraction used for sulfide removal in biogas plants (Sydvatten, internal statistics).

Year	Total Waterworks Sludge Generation (tons)		Fraction to Biogas Plants for Hydrogen Sulfide Removal
	Wet Weight	**Total Solids**	**(%)**
2012	10,388	1558	0%
2012	10,388	1558	0%
2013	10,709	1606	5%
2014	9378	1407	10%
2015	9521	1428	16%
2016	7907	1186	27%
2017	6739	1011	40%
2018	8730	1310	34%
2019	7682	1152	56%
2020	9584	1438	56%

There are operating situations at the waterworks when no iron-containing sludge is produced. This occurs especially if there is a landslide in the tunnel that runs between Lake Bolmen in Småland and Ringsjöverket. This has historically occurred on three occasions over the past 30 years, and then the tunnel has had to be drained and renovated. On these occasions, water from Ringsjön is used at the waterworks and then an aluminum-based precipitation chemical is used instead of iron chloride. The sludge formed then cannot be used in the biogas industry. These malfunctions usually take between one and two years to rectify. It is therefore important for biogas producers who choose to use the waterworks sludge to be able to switch quickly to an alternative solution, such as ferric chloride, during such a period.

The waterworks sludge from Ringsjöverket has a black-brown appearance (see Figure 1), and if it has been in contact with air for a while, small black iron crystals form on the surface. The sludge is water-soluble and has a slight iron odor. Its density is around 1.1 kg/dm^3, and the sludge is slightly acidic, with a pH value around 4.2. The sludge consists mostly of iron compounds (about 30% iron and >40% iron oxides) and various organic compounds (about 25% TOC). In the waterworks, before the sludge is separated, only iron chloride and lye are added. Then, a small amount of the drinking water-grade polymer Magnafloc LT22S-DWI is added together with additional iron chloride in both the thickener and the screen belt press to facilitate dewatering [10].

Figure 1. Waterworks sludge (**left**) and manure storage (**right**). Photo: Annika Nyberg.

A detailed metal analysis has been carried out on the dewatered sludge from Ringsjöverket on three different occasions. Results for the most relevant metals are reported in Table 4. Table 4 compares the analysis results with the limit values that exist within Avfall Sverige's certification rules for biofertilizers, SPCR 120 [12] and the Swedish Environmental Protection Agency's general guideline values for contaminated land for sensitive land use, published in 2009 [20]. It is clear that the metal content of the waterworks sludge is below these levels by a very good margin for virtually all metals. Only the limit value for arsenic in the sludge is in the same order of magnitude as the Swedish Environmental Protection Agency's guideline value for contaminated soil in sensitive land use. Revaq is a certification system that has the aim of reducing the flow of hazardous substances to treatment plants and creating a sustainable return of plant nutrients by spreading sludge from wastewater treatment plants on arable land [11]. The heavy metals that hold the greatest interest in Revaq and that require the most frequent analysis intervals within Revaq are lead, cadmium, copper, chromium, nickel, zinc, mercury, silver and tin, and analysis results from all of these are included in Table 4. Even in this case, the concentrations are low compared to the guideline values that exist. Table 4 also shows that the iron content is relatively constant during the year, with only small variations. This is important for biogas producers to be able to use a similar dosage for different sludge deliveries.

Table 4. Composition of waterworks sludge from Ringsjöverket compared with guideline values from Swedish EPA [20] and the Swedish Waste Association standard SPCR 120 [12] for permissible metal content for biosolids reuse.

Metal	Guideline Value from Swedish EPA for Contaminated Land	SPCR 120	12 October 2015	11 April 2016	27 June 2016	Unit
Antimony	12		<2.1	<2.1	<2.1	mg/kg TS
Arsenic	10		7.2	5.1	8.6	mg/kg TS
Barium	200		40	<23	45	mg/kg TS
Cadmium	1.5	1	<0.11	<0.11	0.13	mg/kg TS
Chromium	80	100	12	18	15	mg/kg TS
Cobalt	15		4.3	5	4.7	mg/kg TS
Copper	80	600	11	17	15	mg/kg TS
Iron			30	29	28	% of TS
Lead	50	100	6.5	5.3	11	mg/kg TS
Mercury	0.25	1	<0.051	<0.051	0.052	mg/kg TS
Molybdenum	40		4.1	4.5	4.1	mg/kg TS
Nickel	40	50	5.5	5.3	6.7	mg/kg TS
Silver			<0.51	<0.51	<0.52	mg/kg TS
Tin			<0.51	0.7	0.82	mg/kg TS
Vanadium	100		47	47	64	mg/kg TS
Zinc	250	800	49	50	65	mg/kg TS

In 2016, waterworks sludge was accepted as an approved additive to the digestion process within the certification system SPCR 120. Therefore, it is now possible for the biogas plants that have this certification to use the waterworks sludge to reduce the hydrogen sulfide concentration in the biogas. When it comes to certification according to Revaq, the use of the waterworks sludge should only affect the cadmium (Cd)/phosphorus (P) ratio marginally, because the cadmium content is very low. However, the phosphorus content is also low, which means that even a low cadmium concentration can have a negative effect on the Cd/P ratio. The size of the sludge feed is also affected because this is regulated not only by the metal concentration, but also by the amount of a certain metal that may be laid per unit area. This applies in particular to the metals lead, cadmium, chromium, nickel, zinc and mercury. However, the content of these metals is very low, and should only have a minor impact. The rules for the Revaq certification system state, however, that "The certificate holder shall not receive such material that is deemed to adversely affect the quality of sludge, through low nutrient content or high content of contaminants. The 60 trace elements must always be analyzed before receiving a new type of material". This can be a problem for the use of waterworks sludge in Revaq-certified treatment plants, but it is not really different from using ferric chloride to reduce the hydrogen sulfide concentration, as it also contains some other heavy metals.

During the project, 13 biogas producers that use waterworks sludge at the full scale were contacted. The experiences at these facilities have been collected and compiled below under different categories. Some of the results have already been published in a report in Swedish [10].

3.1. Transport Cost

The transport cost varies between different biogas producers depending on the distance and how the transport is performed. Some use trucks with trailers and others without. Some drive a few kilometers while others drive up to 600 km. In addition to the actual transport of the sludge, the transporter must also spend time cleaning the platform afterwards. The cost reported by the biogas producers in different parts of the country varies between EUR 15 and 40/ton of waterworks sludge, with 15% TS.

3.2. Other Costs

In addition to transport, biogas producers face costs for the storage, dosing and cleaning of the equipment used. For most users of waterworks sludge for hydrogen sulfide control, these costs are considered to be marginal and estimated at somewhere between EUR 2 and 10/ton wet waterworks sludge.

Most of the biogas producers who currently use waterworks sludge have previously used iron chloride that they bought from a chemical supplier. The cost of virgin iron chloride is up tp twice as high as the total handling costs of the waterworks sludge. The location of the plant matters. A large transport cost reduces the net savings of operational costs. According to Broberg [7], the cost of hydrogen sulfide reduction with iron chloride and iron oxide is around EUR 0.01–0.02 per Nm^3 in farm biogas plants. The cost of using waterworks sludge ends up in the same order of magnitude if the cost of handling and transport is estimated at approximately EUR 40 per ton. However, several biogas producers have reported substantially lower transport and handling costs, and in addition, they reduced the hydrogen sulfide concentration to lower levels when using the sludge compared to when they used ferric chloride. This supports the conclusion that cost savings of up to 50% are possible, but that this depends on the transport cost. The reason for the further reduction in the hydrogen sulfide concentration with sludge is that they think they can afford to add a surplus of iron to the digester, since the marginal material cost is lower, and thus they can then control the dosage towards a lower hydrogen sulfide concentration in the generated biogas. A lower residual hydrogen sulfide concentration in the biogas increases the life of the power/heating unit and reduces its maintenance needs and costs. If the substrate used has lower sulfate content, the amount of sludge needed per volume of biogas produced is lower. Other benefits that the operators of the biogas production plants have observed is the easier handling of waterworks sludge when dosing compared to the corrosive ferric chloride solution and the corrosive damage to the equipment that this can lead to. Some of the farm biogas producers have previously used only aeration or aeration in combination with iron additive. Then, the economic gain achieved with the transition to waterworks sludge is smaller because the addition of air is associated with very marginal costs.

3.3. Use of Waterworks Sludge in the Digester

The number of plants that utilize waterworks sludge from Ringsjöverket has increased since the test started in 2013. In 2020, a total of 24 different plants collected iron-containing waterworks sludge for hydrogen sulfide removal in the digesters. Based on information from seven biogas plants where only manure is used as the substrate, one ton of wet waterworks sludge (with a TS of 15%) is sufficient to produce an average of 2700 Nm^3 of biogas if the hydrogen sulfide concentration in the biogas is to be reduced below 100 ppm. This corresponds to about 0.2–0.5% of the amount of substrate added, expressed as dry matter. The stoichiometric relation between sulfur and iron could be observed at another biogas plant where the residual hydrogen sulfide content was allowed to be higher. This plant was designed to produce biogas with a residual hydrogen sulfide concentration below 300 ppm. An addition of one ton of waterworks sludge to the digester was enough to produce 8000 Nm^3 of biogas with <300 ppm H_2S.

In co-digestion plants where a mixture of manure, starch, food waste and slaughterhouse waste is applied, one ton of wet waterworks sludge was sufficient to reduce the hydrogen sulfide concentration in a significantly larger volume of biogas, since the mix of substrate contains less sulfur than pure manure. Five co-digestion plants surveyed in this study dosed less than half the amount of waterworks sludge into the substrate compared with the manure-based biogas plants, and could still generate biogas with less than 100 ppm hydrogen sulfide. If the proportion of manure dominates in the substrate, the required addition of waterworks sludge remains high, since that kind of substrate is similar to pure manure. The iron in waterworks sludge is less available compared to the addition of pure ferric chloride solutions to control the hydrogen sulfide concentration. When comparing the addition of iron from waterworks sludge with ferric chloride solution, the total amount of iron added to the substrate had to be increased 2.5 to 3 times if added as waterworks sludge in order to achieve a similar effect on hydrogen sulfide removal, compared with the dosing of ferric chloride solution.

3.4. Impact of Temperature and Storage of Waterworks Sludge

Out of 13 surveyed plants, 1 had experienced a slight loss of efficiency resulting from waterworks sludge dosing in the summer. When stored for a long time in the summer, the waterworks sludge lost some of its function, since iron crystals were formed on the sludge surface when it dried in the sun. The iron in the crystals was less available for the microorganisms in the digester. Only one of the producers surveyed identified this as problematic. The storage of sludge in a shaded environment and the modest addition of moisture to the sludge could mediate this issue.

3.5. Observed Operational Conditions When Mixing Waterworks Sludge with Substrate and Feeding the Mix into Digesters

The mixing and feeding of waterworks sludge into the digester was generally a carefree process. Very few problems have been experienced in connection with the handling of the waterworks sludge in biogas plants. The exception was for biogas plants utilizing solid substrates. In these plants, the substrate is mixed with waterworks sludge and fed into the digestion chamber with a screw. The screw is designed for handling dry materials. If the substrate becomes too wet after mixing with the sludge, it slides backwards and stops following the screw. With less feed into the digester, the production decreases. The solution to the problem is to mix the sludge with drier materials and preferably also with longer straw in the substrate. In one of these plants, it was observed that some of the waterworks sludge remained at the bottom of the digestion chamber when it was opened. No action has been taken, but this suggests that the dissolution of the sludge is slow when it is fed together with solid material.

In most of the plants that use liquid substrates, the waterworks sludge is scooped into a mixing well. In these wells the pH is often quite low and the stirring is vigorous. A low pH facilitates the dissolution of the sludge as the solubility of the iron increases with decreasing pH. Those who use this type of mixture have not experienced any problems with dissolving the sludge, nor have they seen any residues of undissolved sludge in the mixing well when it has been drained. For wastewater treatment plants, operators have expressed concerns that the addition of waterworks sludge could affect the drainage properties of the digestate. The opposite effect is indicated by literature data, finding that the drainage of biosolids after the addition of waterworks sludge is improved [13,18].

3.6. *Transportation of Waterworks Sludge into the Digestion Plant*

The waterworks sludge is slightly adhesive and may get stuck on the flatbed when transported. For this reason, various measures have been taken to make handling and cleaning easier. Many people have added straw or sawdust to the flatbed before loading the sludge to make cleaning easier. Another possibility is to spray the flatbed with rapeseed oil or similar prior to loading. There is also a risk in cold climates that the sludge gets stuck on the platform due to freezing. In wintertime, it may be necessary to transport the sludge in closed containers.

4. Discussion

Assuming that the waterworks sludge is virtually inert and does not degrade in the digestion chamber, the entire dry content of the added waterworks sludge will pass through the digestion chamber and be present in the residual biosolids. Based on the information collected from the examined biogas plants included in the study, the waterworks sludge contributes to an increased amount of digestate; that is, about 1–3% based on dry matter, i.e., about 1–3% of TS. The digestate in most plants has a TS content of around 5%, while the waterworks sludge has a TS content of around 15%. The volume increase in the digestate to be handled due to the addition of waterworks sludge is 1% at the most; the TS content in the digestion increases by about 0.1–0.3%, while the total amount of metals in the digestate increases by approximately 1–3% (see calculation in Table 5). Table 5 also refers to the quality requirements according to the biofertilizer certification system, SPCR 120 [12]. This standard states that the proportion of each of the metals (lead, cadmium, copper, chromium, mercury and zinc) may not exceed 15% of the total amount of the metal in the certified biofertilizer. In addition, the nickel content in the biofertilizer from the waterworks sludge must not exceed 6 mg/kg biofertilizer (wet weight). In 2019, around 2.8 million tons of digestate (wet weight) was produced in Sweden, of which 87% was used as fertilizer in agriculture [19]. From farm plants and co-digestion plants, 2.13 million tons, virtually all digestate (biofertilizer), was used as fertilizer. From the wastewater treatment plants, 0.25 million tons (41% of all the sludge) was used as fertilizer certified in the Revaq system. The remaining amount was used mainly as construction material or for the final coverage of landfills [19]. No biosolids are generated at landfill gas plants.

The biogas production from the other plants that also generate biosolids corresponds to a total of 1969 GWh [17]. These plants simultaneously produced 2.8 million tons of biosolids, or around 700 kWh/ton of biofertilizer. Assuming that 1 Nm^3 CH_4 corresponds to 9.81 kWh [21] and that the biogas contains 65% methane [22], this corresponds to a production of 110 Nm^3 of biogas per ton of biofertilizer. According to the values for the 13 plants using waterworks sludge from Ringsjöverket surveyed above, around 1 ton of waterworks sludge is added per 8750 Nm^3 of biogas that has been produced. This corresponds to adding 1 ton of waterworks sludge per 108 tons of biofertilizer produced in the plant, excluding the added waterworks sludge, or 9.2 kg of waterworks sludge per ton of biofertilizer produced. Table 5 shows how adding waterworks sludge affects the concentration of metals in the biofertilizer. The starting point for the calculation is the average value of the metal contents in the biofertilizer for the 18 plants that were certified within SPCR 120 in 2014 [23]. Nickel is not included as there is no risk that the content of nickel can reach up to 6 mg/kg wet weight given the concentration of nickel present in the waterworks sludge. Among other heavy metals, lead and chromium make up the largest parts of the total metal concentration, at 7 and 6%, respectively. However, this is below the limit of 15% specified according to SPCR 120. Therefore, this is also not considered to be a problem for the use of waterworks sludge. With the addition of the waterworks sludge, which is assumed to be inert, the S content of the biofertilizer increases. Assuming the same waterworks sludge dose as above, the TS in the biofertilizer increases from 3.9 to 4.0%. An effect of this is that the concentration of metals, stated as mg/kg TS, decreases for, e.g., copper and zinc, while the concentration increases slightly for lead and chromium (see Table 5).

Table 5. Average concentration of heavy metals in biofertilizer for the facilities that were certified in 2014 [23] before and after an estimated addition of waterworks sludge. The TS in the biofertilizer is 3.9% before the addition of waterworks sludge. The concentration in the sludge is the average of analyzed data according to Table 3.

	Unit	Pb	Cd	Cu	Cr	Hg	Zn
Concentration in biofertilizer	mg/kg TS	3.6	0.4	89	8.3	0.06	292
Concentration in biofertilizer	mg/kg DS	0.14	0.016	3.47	0.32	0.002	11.4
Concentration in waterworks sludge	mg/kg	1.15	0.019	2.15	2.25	0.01	8.27
Content in 9.2 g waterworks sludge	mg	0.011	0.0002	0.020	0.041	0.00007	0.150
Total amount in 1 kg bio fertilizer + 9.2 g waterworks sludge	mg	0.15	0.016	3.49	0.34	0.002	11.5
Fraction of metals from waterworks sludge	%	7.0	1.1	0.6	6.0	3.0	0.7
Total concentration in biofertilizer produced with waterworks sludge	mg/kg TS	3.7	0.4	86	8.4	0.06	281

Since about half of the waterworks sludge consists of various organic compounds from lake-source water, the waterworks sludge contributes to the increased organic content in soil where the digestate is spread as fertilizer. In many Swedish soils, the organic content is low, which is why this is a welcome contribution to improving the soil's properties. The organic content in soils improves the physical, chemical and biological properties of the soil, such as its water holding capacity, nutrient content, buffer capacity, and the activity of soil organisms.

To reuse material is beneficial for climate and society, and reduces the carbon footprint. According to the evaluation in Sydvatten's Climate Account Report 2020, the production of virgin iron chloride generates about 0.395 kg CO_2 emission per kg $FeCl_3$. Since Sydvatten used 3132 tons of $FeCl_3$ in 2020 for drinking water treatment, the reuse of 56% sludge by replacing virgin ferric chloride with waterworks sludge would eliminate 740 tons of carbon dioxide, which is about 17% of all the carbon dioxide that was emitted by the company in 2020 [9], and is well in accordance with EU's Circular Economy Plan [24].

5. Conclusions

Waterworks sludge that contains iron works very effectively as an additive to reduce the hydrogen sulfide concentration in biogas production. This has been demonstrated in the 13 full-scale biogas plants surveyed in this project. According to the costs reported for the transport and handling of the sludge, there is potential to save up to 50% compared to a scenario in which these plants use virgin ferric chloride instead. In the manure-based biogas plants, one ton of waterworks sludge with 15% TS is sufficient to reduce the hydrogen sulfide concentration to below 100 ppm in 2–3000 Nm^3 biogas. In the digestion plants that participated in the study, the same amount of sludge was enough for more than twice as much gas—about 8000 Nm^3. In both cases, the exact figure depends on the substrate composition and the level at which the hydrogen sulfide concentration is reduced. The content of heavy metals in the waterworks sludge is well below the concentrations used in

Avfall Sverige's certification system, SPCR 120, for the reuse of biosolids from solid waste plants. The reuse of 56% of the sludge by replacing virgin ferric chloride with waterworks sludge saves 740 tons of carbon dioxide, corresponding to about 17% of all carbon dioxide emitted by Sydvatten in 2020.

Author Contributions: Conceptualization, T.P. and K.M.P.; methodology, T.P., investigation, T.P., K.M.P. and J.Å.; writing, T.P., K.M.P. and J.Å. All authors have read and agreed to the published version of the manuscript.

Funding: This research received no external funding.

Acknowledgments: Lars Månsson and Irene Bohn are acknowledged for sampling and other technical support in the project.

Conflicts of Interest: The authors declare no conflict of interest.

References

1. Wellinger, A.; Murphy, J.D.; Baxter, D. *The Biogas Handbook. Science, Production and Applications Woodhead Publishing*; Woodhead Publishing Series in Energy no. 52; Elsevier: Amsterdam, The Netherlands, 2013; 512p.
2. Ahlberg-Eliasson, K.; Nadeau, E.; Levén, L.; Schnürer, A. Production efficiency of Swedish farm-scale biogas plants. *Biomass Bioenergy* **2017**, *97*, 27–37. [CrossRef]
3. Schnürer, A.; Jarvis, Å. Mikrobiologisk Handbok för Biogasanläggningar. SGC-Rapport 207. 2009. Available online: http://www.sgc.se/ckfinder/userfiles/files/SGC207.pdf (accessed on 11 May 2021).
4. Möestedt, J.; Påledal Nillson, S.; Schnürer, A. The effect of substrate and operational parameters on the abundance of supphate reducing bacteria in industrial anaerobic biogas digester. *Bioresour. Technol.* **2013**, *132*, 327–332. [CrossRef]
5. Yekta, S.S. Chemical Speciation of Sulfur and Metals in Biogas Reactors. Implication for Cobolt and Nickel Bio-Uptake Processes. Ph.D. Thesis, The Department of Thematical Studies–Environmental Change Linköping University, Linköping, Sweden, 2014.
6. Bauer, F.; Hulteberg, C.; Persson, T.; Tamm, D. Biogas upgrading–Review of commercial technologies. *SGC Rapp.* **2013**, *2013*, 270. Available online: http://www.sgc.se/ckfinder/userfiles/files/SGC270.pdf (accessed on 11 May 2021).
7. Broberg, A. Metoder för Svavelvätereducering, Hushållningsskapets Förbund 2013. 2013. Available online: http://www.bioenergiportalen.se/attachments/42/691.pdf (accessed on 11 May 2021).
8. Sydvatten 2021A: Sustainability Report. 2020. Available online: https://sydvatten.se/app/uploads/2021/04/Hallbarhetsredovisning_2020_Sydvatten.pdf (accessed on 11 May 2021). (In Swedish)
9. Sydvatten 2021B: Climate Account Report. 2020. Available online: https://sydvatten.se/app/uploads/2021/04/Klimatbokslut_2020_Sydvatten.pdf (accessed on 11 May 2021). (In Swedish)
10. Tobias, P.; Lars, M.; Irene, B. Vattenverksslam Reducerar Biogasens Svavelväte. Rapport 2017, 344. Energiforsk, Stockholm. Available online: https://energiforskmedia.blob.core.windows.net/media/22106/vattenverksslam-reducerar-biogasens-svavelvate-energiforskrapport-2017-344.pdf (accessed on 14 May 2021).
11. Svenskt Vatten 2020: REVAQ-Rules for Certification. Available online: https://www.svensktvatten.se/globalassets/avlopp-och-miljo/uppstromsarbete-och-kretslopp/revaq-certifiering/revaq-regler-2021---utgava-7.0-gul.pdf (accessed on 11 May 2021).
12. Avfall Sverige 2021: SPCR 120. Certification Rules. Available online: https://www.avfallsverige.se/kunskapsbanken/certifierad-atervinning/certifieringsregler/ (accessed on 11 May 2021). (In Swedish)
13. Parker, W.; Celmer, R.D.; Bicudo, J.; Law, P. Assessment of the use of mainstream iron addition for phosphorous control on H2S content of biogas from anaerobic digestion of sludges. *Water Environ. Res.* **2020**, *92*, 338–346. [CrossRef] [PubMed]
14. Erdirencelebi, D.; Kucukhemek, M. Control of hydrogen sulphide in full-scale anaerobic digesters using iron (III) chloride: Performance, origin and effects. *Water SA* **2018**, *44*, 176–183. Available online: https://search.ebscohost.com/login.aspx?direct=true&db=a9h&AN=129677686&site=eds-live&scope=site (accessed on 11 May 2021). [CrossRef]
15. Andriamanohiarisoamanana, F.J.; Shirai, T.; Yamashiro, T.; Yasui, S.; Iwasaki, M.; Ihara, I.; Nishida, T.; Tangtaweewipat, S.; Umetsu, K. Valorizing waste iron powder in biogas production: Hydrogen sulfide control and process performances. *J. Environ. Manag.* **2018**, *208*, 134–141. Available online: https://search.ebscohost.com/login.aspx?direct=true&db=edselp&AN=S030147971731174X&site=eds-live&scope=site (accessed on 11 May 2021). [CrossRef] [PubMed]
16. Farghali, M.; Andriamanohiarisoamanana, F.J.; Ahmed, M.M.; Kotb, S.; Yamashiro, T.; Iwasaki, M.; Umetsu, K. Impacts of iron oxide and titanium dioxide nanoparticles on biogas production: Hydrogen sulfide mitigation, process stability, and prospective challenges. *J. Environ. Manag.* **2019**, *240*, 160–167. Available online: https://search.ebscohost.com/login.aspx?direct=true&db=edselp&AN=S0301479719303925&site=eds-live&scope=site (accessed on 11 May 2021). [CrossRef]
17. Farghali, M.; Andriamanohiarisoamanana, F.J.; Ahmed, M.M.; Kotb, S.; Yamamoto, Y.; Iwasaki, M.; Yamashiro, T.; Umetsu, K. Prospects for biogas production and H 2 S control from the anaerobic digestion of cattle manure: The influence of microscale waste iron powder and iron oxide nanoparticles. *Waste Manag.* **2020**, *101*, 141–149. Available online: https://search.ebscohost.com/login.aspx?direct=true&db=cmedm&AN=31610475&site=eds-live&scope=site (accessed on 11 May 2021). [CrossRef] [PubMed]

18. Rebosura, J.M.; Salehin, S.; Pikaar, I.; Kulandaivelu, J.; Jiang, G.; Keller, J.; Sharma, K.; Yuan, Z. Effects of in-sewer dosing of iron-rich drinking water sludge on wastewater collection and treatment systems. *Water Res.* **2020**, *171*, 115396. Available online: https://search.ebscohost.com/login.aspx?direct=true&db=edselp&AN=S0043135419311704&site=eds-live&scope=site (accessed on 11 May 2021). [CrossRef] [PubMed]
19. Klackenberg, L. Biomethane in Sweden—Market Overview Andpolicies, Swedish Gas Association 16 March 2021. 2021. Available online: https://www.energigas.se/media/boujhdr1/biomethane-in-sweden-210316-slutlig.pdf (accessed on 14 May 2021).
20. Naturvårdsverket. Riktvärden för förorenad mark. Modellbeskrivning och vägledning. In *Rapport 5976*; Naturvårdsverket (Swedish Environmental Protection Agency): Stockholm, Sweden, 2009.
21. Berglund, M.; och Börjesson, P. *Energianalys av Biogassystem. Rapport nr. 44*; Institution för Teknik Och Samhälle, Avdelningen för Miljö- Och Energisystem; Lunds Tekniska Högskola: Lund, Sweden, 2003.
22. Nelsson, C. *Varierande Gaskvalitet–Litteraturstudie. SGC-Rapport 209*; Svenskt Gasteknisk Center: Malmö, Sweden, 2009. Available online: http://www.sgc.se/ckfinder/userfiles/files/SGC209.pdf (accessed on 11 May 2021).
23. Persson, E. Årsrapport 2014 Certifierad återvinning, SPCR 120. *Avfall Sveriges Utvecklingssatsning Rapport 2015:20*. Available online: https://www.avfallsverige.se/kunskapsbanken/rapporter/rapportera/article/arsrapport-2014-certifierad-atervinning-spcr-120/ (accessed on 1 July 2021).
24. EU 2020: COM/2020/98 final. Communication from the Commission to the European Parliament, the Council, the European Economic and Social Committee and the Committee of the Regions: A New Circular Economy Action Plan for a Cleaner and More Competitive Europe. Available online: https://eur-lex.europa.eu/legal-content/EN/TXT/?qid=1583933814386&uri=COM:2020:98:FIN (accessed on 11 May 2021).

Review

Lattice Boltzmann Method in Modeling Biofilm Formation, Growth and Detachment

Mojtaba Aghajani Delavar and Junye Wang *

Faculty of Science and Technology, Athabasca University, Athabasca, AB T9S 3A3, Canada; maghajanidelavar@athabascau.ca

* Correspondence: junyew@athabascau.ca; Tel.: +1-7803-944-883

Abstract: Biofilms are a complex and heterogeneous aggregation of multiple populations of microorganisms linked together by their excretion of extracellular polymer substances (EPS). Biofilms can cause many serious problems, such as chronic infections, food contamination and equipment corrosion, although they can be useful for constructive purposes, such as in wastewater treatment, heavy metal removal from hazardous waste sites, biofuel production, power generation through microbial fuel cells and microbially enhanced oil recovery; however, biofilm formation and growth are complex due to interactions among physicochemical and biological processes under operational and environmental conditions. Advanced numerical modeling techniques using the lattice Boltzmann method (LBM) are enabling the prediction of biofilm formation and growth and microbial community structures. This study is the first attempt to perform a general review on major contributions to LBM-based biofilm models, ranging from pioneering efforts to more recent progress. We present our understanding of the modeling of biofilm formation, growth and detachment using LBM-based models and present the fundamental aspects of various LBM-based biofilm models. We describe how the LBM couples with cellular automata (CA) and individual-based model (IbM) approaches and discuss their applications in assessing the spatiotemporal distribution of biofilms and their associated parameters and evaluating bioconversion efficiency. Finally, we discuss the main features and drawbacks of LBM-based biofilm models from ecological and biotechnological perspectives and identify current knowledge gaps and future research priorities.

Keywords: biofilm; lattice Boltzmann method; cellular automata; individual-based model

Citation: Delavar, M.A.; Wang, J. Lattice Boltzmann Method in Modeling Biofilm Formation, Growth and Detachment. *Sustainability* **2021**, *13*, 7968. https://doi.org/10.3390/su13147968

Academic Editor: Shashi Kant Bhatia

Received: 15 June 2021
Accepted: 10 July 2021
Published: 16 July 2021

1. Introduction

Bacteria mainly live together, forming complex and heterogeneous aggregations of structured communities connected together by their excretion of extracellular polymer substances (EPS). Such aggregates of microorganisms are called biofilms. Biofilms can consist of bacteria and various non-cellular materials such as mineral crystals, corrosion particles, clay or silt particles, or blood components, depending on the environment in which the biofilm has developed. Biofilms often contain one or more microbial species, which adhere to each other; to surfaces (living or non-living); and to solid–liquid, liquid–air, liquid–liquid and solid–air interfaces [1]. Biofilms can cause many industrial and biological problems, such as chronic infections, food contamination, biofouling and equipment corrosion, although they have been applied for a variety of industrial applications, such as in biohydrogen production, fermentative food processing, power generation through microbial fuel cells, biological wastewater treatment and microbial-enhanced oil recovery [2–15]. As such, it is important to understand the mechanisms of biofilm formation, growth and detachment, as well as their interactions with their surrounding environments, the social interactions between single-species and multispecies biofilms and the potential applications of biofilms for environmental protection and enhanced biotechnology. Extensive efforts have been made to understand the mechanisms of biofilm growth and their interactions with the affective signals [3,16–24]. Due to the very complex interactions among physicochemical and biological processes, direct measurement of the impacts of bioactivity on transport dynamics in 3D structures is very difficult due to sample opacity [25,26]; thus, many numerical models have been developed to examine biofilm growth, which can be categorized into discrete or particle-based models and continuum models [27,28].

1.1. Mathematical Models of Biofilms

Biofilm modeling dates back to the 1970s. Biofilms were originally described as uniform biomasses of single microorganism species [29,30]. The first-generation biofilm models can be implemented quickly and easily but they cannot capture certain details of biofilm structures. Later, stratified biofilms were dynamically modeled [31]. These one dimensional (1D) stratified models were evolved to describe multi-substrate and multi-species interactions within the biofilm as the second-generation models; however, the second-generation models were not capable of describing the characteristic structural heterogeneity of biofilms. As such, discrete models of biofilm growth were developed in the 1990s and have continued until the current day as the third-generation mathematical models. Cellular automata (CA) or individual-based models (IbM) are two of the discrete model types, which have been used for simulations of inherently stochastic and individual cells and for cell spreading at the local level [32]. The rules used to model the interactions at the local level can be determined not only through biological principles, but also through mathematical and physical frameworks using CA or IbM approaches. Hydrodynamics and multispecies transport are solved using the finite difference method (FDM), finite volume method (FVM), or finite element method (FEM). As bottom-up models, the third-generation biofilm models allow the description of the sophisticated two- (2D) and three-dimensional (3D) structures and morphologies of microbial biofilms. They can incorporate not only substrate transport and transformations [33,34], but also hydrodynamics [35] and population dynamics [36]. The large-scale morphology dynamics results from the interactions of the small-scale individual cells and the environment; however, the third-generation biofilm models have difficulty in accounting for complex non-equilibrium interface dynamics, multiphase flow with phase changes and complex geometries, such as porous media.

1.2. History of Lattice Boltzmann Equation in Modeling of Biofilms

The lattice Boltzmann method (LBM)-based models are more recent in the field of biofilm modeling research than others using FDM or FVM. LBM approaches are particle-based, bottom-up models that are considered suitable for tracing the dynamics and properties of individual bacterial cells. Conversely, the behavior of individual cells is not directly tracked in continuum models and the biofilm is modeled using volume- or mass-averaged behaviors of different functional groups. The first attempt to model biofilm growth using LBM can be dated back to 1999 by Picioreanu et al. [35,37]. In the 2000s, many researchers used the LBM approach to study biofilm growth [34,38–41]. In LBM-CA models, the CA describes the spreading possess for a limited number of directions and with ad hoc or probabilistic rules [42,43]. In the LBM-based biofilm models, biofilm growth is simulated through coupling with either IbM (e.g., LBM-IbM) [28,44] or with CA (e.g., LBM-CA) [3,19,35,37]. In recent years, LBM-based models have been extended to the modeling of several key processes, such as multispecies competition and cooperation [3], the thermal effects on biofilm growth [19,20], the effects of pH values on biofilm growth [21] and heterogeneous permeability in biofilms [28]. The LBM-based models are very effective in modeling complex multilateral interactions among biofilm growth, transport phenomenon and hydrodynamics, allowing better and more detailed description of processes in different industrial or academic applications. Here, we call the LBM-based biofilm models as the fourth-generation models. Despite remarkable advances in modeling biofilm formation and growth, the LBM-based biofilm models have not yet been reviewed.

This review aims to synthesizes our understanding in modeling the biofilm formation, growth and detachment using LBM-based models. We present the fundamentals of various LBM-based biofilm models and how the LBM couples with CA or IbM with their applications for the simulation of spatiotemporal distribution of biofilms and bioconversion in biotechnologies and bioengineering. Finally, we highlight the value of an ecological and biotechnological perspective and identify current knowledge gaps and future research priorities.

2. Multiscale Biofilm Formation Processes and Lattice Boltzmann Method

Biofilm growth, formation and detachment are multiscale processes in nature [45]. Biofilm growth links cell and population scales and interacts with hydrodynamic and nutrient. Their multiscale nature requires to incorporate more intracellular states, behaviors and interplay among biofilm growth, reaction-diffusion and hydrodynamics in quantitative modeling of biofilms.

2.1. Multiscale Biofilm Processes

Biofilms are ubiquitous in nature and are persistently sticked to both biotic and abiotic surfaces ranging from the human tooth or lung to a rock submerged in a stream and/or reactor wall. Therefore, many studies of biofilms have been performed due to their importance in infectious disease processes in clinical and public health [46] and bioconversion in biotechnology [19,20]. Medical devices may be colonized by biofilms, resulting in measurable rates of device-associated infections. Bacterial biofilms are mostly composed of multiple populations of different bacterial species in hierarchical structure of a microbial biofilm [47] (Figure 1). Each population needs to be adapted to either environmental fluctuations, such as nutrient and water availability, temperature, moisture and pH value, or metabolism and migration of other species. Therefore, the surviving of any given population in a multispecies biofilm greatly relies on the three universal types of processes occurring in a biofilm system: transformations (e.g., substrate utilization), transport (e.g., convection and diffusion) and biofilm formation and growth.

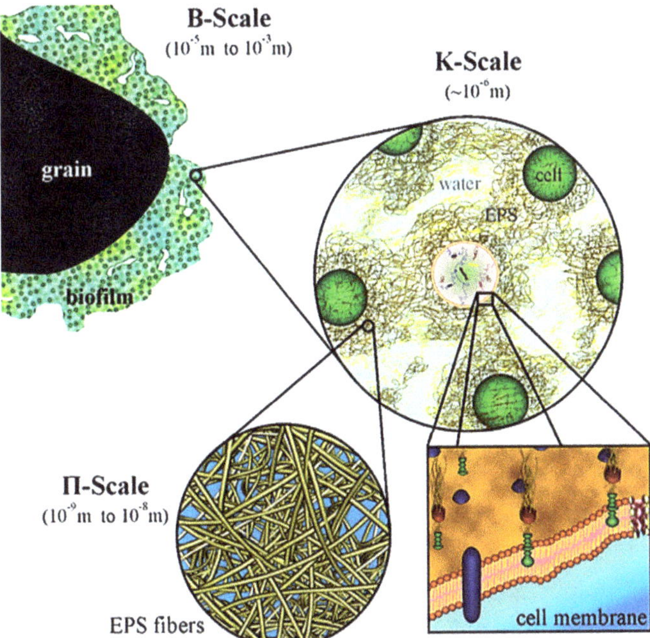

Figure 1. Schematic representation of the hierarchical structure of a microbial biofilm (Reprinted from [47] with permission from Elsevier).

Figure 2 shows the hierarchical characteristic time scale for microbial biofilm growth and fluid flow processes that occur in a biofilm community [48]. The order of magnitude of the characteristic times can be represented by dimensionless forms [48]:

$$\tau_{gr} = \frac{1}{\mu} \gg \tau_{dif} = \frac{L_F^2}{D} \quad (1)$$

where τ_{gr} is the characteristic time for biomass growth, τ_{dif} is the time of substrate transport by diffusion and L_F is the biofilm thickness. The difference of characteristic times could be large as 10 orders of magnitude, although a specific time scale can change from minutes to months, relying on the processes of most interest. Biofilm growth has a time scale > 10^5 s (~1 day). In contrast, hydrodynamic changes occur at a time scale usually less than 1 s. This means that hydrodynamic conditions are changed quickly but the change of other biofilm components is so slow that they can be presumed as a constant state. Thus, this provides the possibility to reduce drastically the computational effort without sacrificing accuracy if multiscale modeling approaches are used.

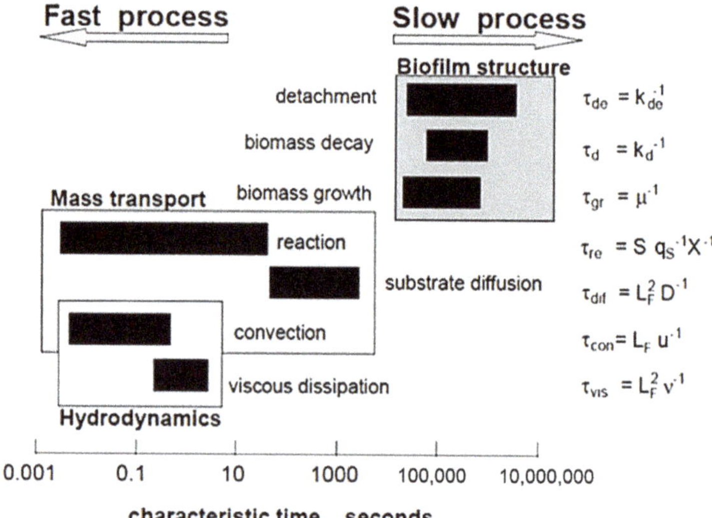

Figure 2. Characteristic times of different processes occurring in reactive transport and biofilm systems (Reprinted from [48] with permission from John Wiley and Sons).

There are several numerical methods to simulate transport phenomenon from macroscopic continuum media to microscopic scheme (Figure 3). In the macroscopic methods, the domain is treated as a continuum media such as Navier–Stokes equation which is used in fluid mechanics and thermodynamics. On the other hand, in microscopic methods of molecular dynamics (MD), the domain is considered as small particles such as atoms, molecules or parcels. Both methods can produce acceptable outcomes when the system involves enough large number of particles.

Molecular dynamics is based on the Lagrangian specification of the flow field in which an individual fluid parcel or molecule moves through space and time according to Newton's law and an intermolecular potential. In CA models, the biomass in a grid is represented using volume averaged variables of properties (density or concentration) and is represented as percentage of whole area (2D) or volume (3D). The set of CA rules is used to describe the interaction of cells with each other and its neighborhood in the lattice network. Each rule designates one of the main processes occurring in biofilm, such as spreading direction, biofilm growth and decay and biofilm distribution. Cell movement is limited in the finite number of lattice directions. In contrast, the IbM is a type of multiscale models to represent the interactions among individual cells, population or community [43,49]. A bacterial cell is assumed as a sphere which is characterized by essential variables, such as volume, mass and density. These spherical cells can locate anywhere in continuous 3D space and act independently under a set of rules, analogous to behaviors of individual bacterial cells, such as biofilm growth, cell division and production of metabolites. IbM allows off-lattice cell movement on a continuous set of directions and distances. This type of IbMs overcome the drawbacks deriving from the discrete rules to biofilm spreading in the CA approach [34]. In biofilm modeling, both CA and IbM do not simulate reactive transport of a solute species and fluid flow. When movement of particles is controlled by Newton's law, CA can be used for simulation of hydrodynamics and reactive transport. Such a CA algorithm is called as lattice gas cellular automata (LGCA) [50].

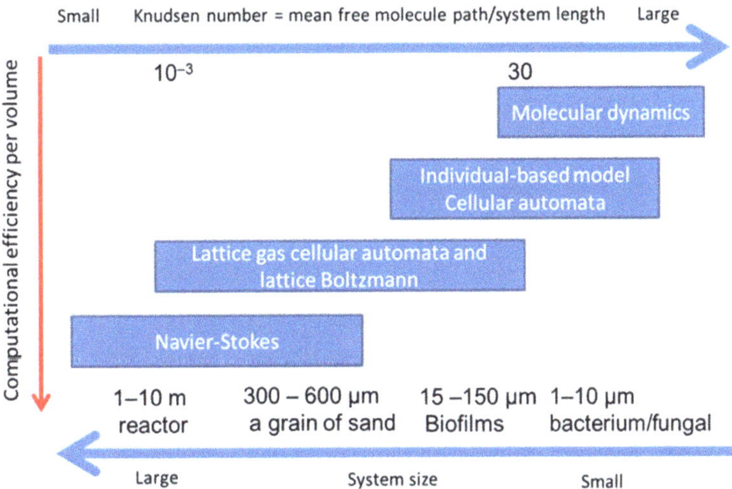

Figure 3. Various approaches to hydrodynamics and reactive transport together with their preferred range of applicability.

LBM originated from classical statistical physics and lattice gas automata (LGCA) [50]. In LBM, there are two main practical modifications: (1) The distribution functions of particles replace the tracing of each particle. This allows a group/parcel of atoms or molecules to represent individual particles and significantly reduces simulation scale from microscopic to mesoscopic scale, that considerably saves the computational time. (2) The degree of freedom of particles movement is reduced to restrict the movement of particles to a finite number of directions. LBM treats flows in terms of fictive parcels of particles which reside on a mesh and conduct translation according to collision steps entailing overall fluid-like behavior. The statistical noise of LGCA was eliminated due to using a continuous distribution function and the collision operator within Bhatnagar–Gross–Krook (LBGK) model [51]. LBM can recover macro governing equations of transport phenomenon, including momentum, energy and mass transfer using correctly choosing the equilibrium distribution functions after implementation of the Chapman–Enskog multiscale expansion [52–54]. This makes LBM more efficient, stable and flexible in hydrodynamics modeling.

Figure 3 shows four groups of methods which can describe the conservation of mass, momentum and energy at computational grids. These methods have their respective strengths at different Knudsen numbers. The Knudsen number is the ratio of the mean free molecule path to the representative system length, which is a characteristic length scale representing spatial resolution, such as bacterium and the sand size [55]. Navier–Stokes approaches solve continuum-based partial differential equations using FDM or FVM at a larger spatial scale, while MD, CA or IbM approaches are suitable processes at microscale.

LBM and LGCA are at mesoscale between macroscale of Navier–Stokes approaches and microscale of MD. Therefore, LBM is more suitable for simulations of reactive transport and fluid flow with biofilm growth at meso or macroscale.

2.2. Lattice Boltzmann Equation

Compared to traditional computational fluid dynamics (CFD) using FDM or FVM, LBM, due to its fundamental kinetic nature, has a number of key benefits, such as easier parallelization, handling of pressure field (no need to solve the Poisson equation that is computationally expensive), complex boundary conditions and geometries and interface interactions. These advantages make LBM an ideal scale-bridging scheme to integrate simplified kinetic models with hydrodynamics in the simulation of microscopic/mesoscopic processes and biofilm growth.

In biofilm modeling, the system must be governed by the mass, energy and momentum conservations with their initial and boundary conditions. The substrate conservation is described by the reactive transport equation. The flow field is expressed by the Navier–Stokes equation and the continuity equation. The set of conservation equations is mathematically complicated and difficult to be handled analytically. Thus, the set of conservation equations have been solved using numerical methods, such as FDM and LBM.

In lattice Boltzmann method, as a mesoscopic approach, the movement of groups of fluid particles are modeled to capture macroscopic quantities such as velocity during two processes, streaming and collision. This modeling is established using the distribution function, $f(x, c, t)$, which is defined as the number of particles in time t, with velocity ranging from c to $c + dc$, located in the position between x and $x + dx$. The distribution function stands as a replacement of each single particle in molecular dynamics which leads to substantial saving in computational cost and making simulations feasible for larger computational domain, compared to ones for molecular dynamics.

In LBM, the domain is discretized to uniform Cartesian cells (often squares in 2D or cubes in 3D). The movement of particles (group of atoms or molecules) is limited to a number of specific directions, depending on LBM schemes. The DnQm scheme is widely used for classifying the different LBM methods is. In the DnQm scheme, n represents the dimension of model (2 for two-dimensional and 3 for three-dimensional) and m does the number of specific streaming directions. For example, D2Q9 represents two-dimensional model with nine velocity directions. Table 1 shows several commonly-used 2D and 3D LBM schemes. For D2Q9, the model equations are expressed as:

$$f_k(x + c_k\Delta t, t + \Delta t) = f_k(x,t) + \frac{\Delta t}{\tau}\left[f_k^{eq}(x,t) - f_k(x,t)\right] + \Delta t F_k \tag{2}$$

$$f_k^{eq}(x,t) = \omega_k \rho \left[1 + \frac{c_k.U}{c_s^2} + 0.5\frac{(c_k.U)^2}{c_s^4} - 0.5\frac{U.U}{c_s^2}\right] \tag{3}$$

where $\omega = 1/\tau$ stands for collision frequency, τ is for relaxation factor and f_k^{eq} represents the local equilibrium distribution function. Subscript k denotes the streaming direction of LBM model. The set of LBM equations can recover the continuity of mass and momentum (Navier–Stokes) equations by using the Chapman–Enskog multiscale expansion [56,57]:

$$\nabla.U = 0 \tag{4}$$

$$\rho\frac{\partial U}{\partial t} + \rho U.\nabla U = -\nabla p + \mu \nabla^2 U \tag{5}$$

To consider all fluid flow, temperature field and species transport, the conservations of energy and species concentration in the domain can be described as the below reactive transport equation and temperature equation:

$$\frac{\partial T}{\partial t} + U.\nabla T = \alpha \nabla^2 T + \frac{q_g}{\rho C_p} \tag{6}$$

where T is local temperature, α is thermal diffusion coefficient, q_g is the heat source.

Table 1. Common Lattice Boltzmann Models.

Model	Schematic	Velocities	Weighting Factors
D2Q5	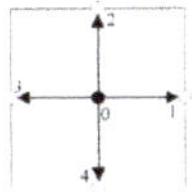	$c_k = \begin{cases} (0,0), & k=0 \\ (\pm1,0) & k=1,3 \\ (0,\pm1), & k=2,4 \end{cases}$	$\omega_k = \begin{cases} 0, & k=0 \\ \frac{1}{4}, & k=1-4 \end{cases}$
D2Q9	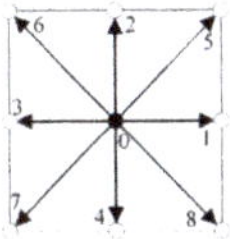	$c_k = \begin{cases} (0,0), & k=0 \\ (\pm1,0),(0,\pm1), & k=1-4 \\ (\pm1,\pm1), & k=5-8 \end{cases}$	$\omega_k = \begin{cases} \frac{4}{9}, & k=0 \\ \frac{1}{9}, & k=1-4 \\ \frac{1}{36}, & k=5-8 \end{cases}$
D3Q15	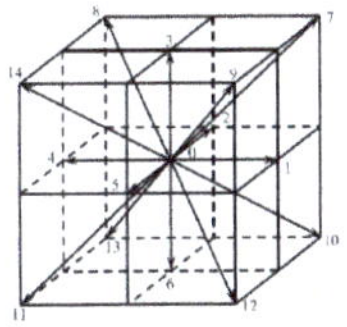	$c_k = \begin{cases} (0,0,0), & k=0 \\ (\pm1,0,0),(0,\pm1,0),(0,\pm1,0), & k=1-6 \\ (\pm1,\pm1,\pm1), & k=7-14 \end{cases}$	$\omega_k = \begin{cases} \frac{2}{9}, & k=0 \\ \frac{1}{9}, & k=1-6 \\ \frac{1}{72}, & k=7-14 \end{cases}$
D3Q19	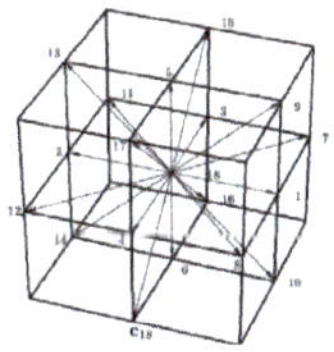	$c_k = \begin{cases} (0,0,0), & k=0 \\ (\pm1,0,0),(0,\pm1,0),(0,\pm1,0), & k=1-6 \\ (\pm1,\pm1,0),\ (\pm1,0,\pm1),(0,\pm1,\pm1) & k=7-18 \end{cases}$	$\omega_k = \begin{cases} \frac{1}{3}, & k=0 \\ \frac{1}{18}, & k=1-6 \\ \frac{1}{36}, & k=7-18 \end{cases}$

Corresponding distribution function for each parameter is required if LBM is applied to solve scalar parameters such as concentration and temperature in the domain. Thus, different distribution functions, g and g_l, are defined for every scalar parameter, respectively [19,20].

The g is the distribution function used for temperature calculation:

$$g_k(x + c_k\Delta t, t + \Delta t) = g_k(x,t) + \frac{\Delta t}{\tau_g}\left[g_k^{eq}(x,t) - g_k(x,t)\right] + \Delta t\omega_k S_T \quad (7)$$

where g_k^{eq} is the corresponding equilibrium distribution function, Δt is time step that should be adjusted by macroscopic time step and S_T is the heat source. For the temperature as a scalar variable, the equilibrium distribution function is determined by [3,19,54,58–60]:

$$g_k^{eq}(x,t) = \omega_k\rho\left[1 + \frac{c_k.U}{c_s^2}\right] \quad (8)$$

The source term, S_T, is a function of heat source, q_g, as:

$$S_T = \frac{q_g}{\rho C} \quad (9)$$

As there are more than one species in biofilm growth systems, the concentration field of multispecies reactive transport can be described as

$$\rho U.\nabla C_l = \rho D_l^{eff}\nabla^2 C_l + S_l \quad (10)$$

where C_p is the specific heat capacity, C_l represents the concentration fraction of species l, S_l is the mass generation rate for the species per unit volume and time and D_l^{eff} is the effective diffusion coefficient of the lth species. Subscript l denotes the lth species.

The concentration field of each species (as a scalar parameter which is similar as temperature) can be simulated using corresponding distribution functions, g_l, defined as below [3,19–21,59]:

$$g_{lk}(x + c_k\Delta t, t + \Delta t) = g_{lk}(x,t) + \frac{\Delta t}{\tau_l}\left[g_{lk}^{eq}(x,t) - g_{lk}(x,t)\right] + \Delta t\omega_k S_l \quad (11)$$

$$g_{lk}^{eq}(x,t) = \omega_k C_l \left[1 + \frac{c_k.U}{c_s^2}\right] \quad (12)$$

where g_{lk} is the distribution function of the lth species.

Having calculated distribution functions, the temperature and species concentrations are calculated as:

$$T = \sum_k g_k \quad (13)$$

$$C_l = \sum_k g_{lk} \quad (14)$$

Although each biofilm simulation has its own known and unknown parameters that affect the simulation process, a general simulation procedure would generally be applicable to LBM simulations, shown in Figure 4 [19]:

1. Geometry definition: as the first step, the computational domain is defined regarding to the bioreactor geometry.
2. There are usually three types of zones in the domain as fluid, solid, biofilm that are updated in this step due to initial geometry or zones calculated in previous time step.
3. Solve the hydrodynamic equations (Equations (2) and (3)) to capture local velocity vectors.
4. Solve energy/temperature equations (Equations (7) and (8)) to find local temperature and update fluid and species properties, if they are considered temperature dependent.
5. Solve the species transport equations (Equations (11) and (12)), that gives local species concentration and consumption by bacteria regarding to local fluid velocity, temperature and concentrations of nutrients.
6. Biomass calculation (its concentration, detachment and shrinkage) in each grid in order to determine the biofilm growth and spread of growth and spread.
7. Update biofilm pattern, affecting flow connection channels.
8. Return to Step (2) for the next time step, calculations continue up to desired total simulation time.

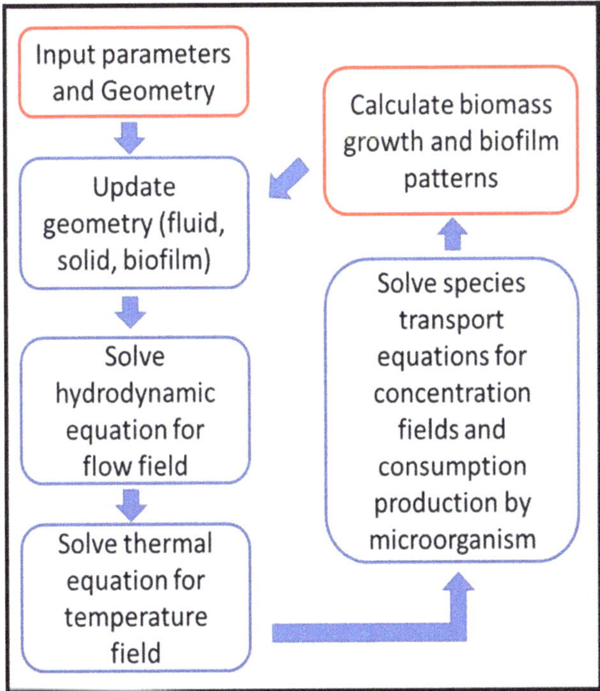

Figure 4. LBM simulation flow chart (Reprinted from [19] with permission from American Chemical Society).

3. Applications of LBM-Based Biofilm Models to Biotechnology and Bioengineering

LBM-based biofilm models are based on the idea that flow field, substrate fields and temperature field can be solved by LBM while biofilm growth can be characterized using IbM or CA. Therefore, an LBM-based approach is essentially a hybrid scheme. Because modeling biofilm growth, detachment, attachment and shrinkage have two approaches: IbM and CA, an LBM model can couple with one of them to solve interactions among biofilm growth, flow and reactive transport and temperature. These hybrid models have shown their capability in representing the biofilm structure heterogeneity. In this section, we classify the LBM-based biofilm models as three types: LBM with no biofilm pattern, LBM coupled with CA (LBM-CA) and LBM coupled with IbM (LBM-IbM).

3.1. LBM for Biological Reaction with no Biofilm Growth

LBM can describe biological reaction without tracing biofilm growth. In such LBM model, flow and reactive transport, such as temperature and substrate concentration, are solved using LBM. Substrate consumptions are calculated by using zero, the first order or Monod kinetics but biofilm growth patterns will not be traced. Graf von der Schulenburg et al. [61] developed an LBM model coupled with a directed random walk algorithm to study the hydrodynamics of flow induced by biofilm in a randomly packed bed. They investigated the irregular transport dynamics due to effects of biofilms and examined the effects of observation time on the displacement distributions (propagators). They reported that increasing observation time caused propagators to have a transition from a pre-asymptotic to a Gaussian-shaped distribution. In both situations with and without biofilm growth, this transition was observed; however, it was very significantly delayed due to the significant development of essentially stagnant regions if the biofilm was present.

Zhang et al. [62] used lattice Boltzmann method to simulate fluid flow and mass transfer and their effects on an herbicide-degrading bacterium growth rate in porous media. They investigated the effects of the heterogeneity of pore structures and transverse mixing on the growth of a pure culture (*Delftia acidovorans*) degrading (R)-2-(2, 4-dichlorophenoxy) propionate (R-2, 4-DP). Figure 5 shows reaction rate contours in homogeneous and aggregate pore networks. Although they experimentally observed biofilm growth, they used LBM to simulate the reaction rate without biofilm growth. The results could contribute to enhance the knowledge of biofilm growth in complex porous media affected by the coupling between fluid flow, species mass transfer and bioreaction rates.

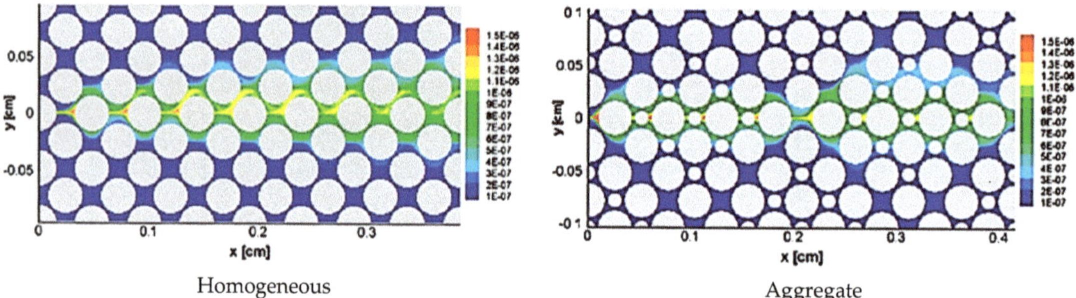

Figure 5. Reaction rate contours in mol.m^{-3}.s^{-1}, in homogeneous and aggregate pore networks, for reaction rate constants (Da = 1.39) (Reprinted from [62] from with permission American Chemical Society).

Hydrogen is considered as high energy density and renewable energy source and has an enormous potential to replace fossil fuels. Hydrogen can be generated through some microbial processes such dark and photo fermentation that are eco-friendly production technologies having a huge potential to be a leading future-interest energy technology [21]. Yang et al. [63] numerically investigated hydrogen production by photosynthetic bacteria (PSB) in a biofilm bioreactor using LBM. They investigated the biochemical reaction in an attached thin PSB biofilm to a circular cylinder considering the curved boundaries. The substrate consumption efficiency and the hydrogen yield were examined for different Reynolds and Sherwood numbers that represented different fluid velocity and mass transfer characteristics. The simulated results were validated against the drag and lift coefficients and concentration profiles of theoretical/numerical data available in the literature. They found that the concentrations of both the substrate and products decrease with increasing Reynolds number. However, they did not include biofilm growth. Liao et al. [64,65] simulated biohydrogen production through photo fermentation in photobioreactor. They used LBM to simulate fluid flow and species transport in the bioreactor and evaluated the impacts of accumulation mode of particles, fluid velocity, porosity of immobilized granule and illumination intensity on biohydrogen production performance in terms of the substrate consumption efficiency and hydrogen yield. Figure 6 shows biohydrogen concentration fields around the immobilized granule [65]. However, they did not consider biofilm growth as they calculated photo fermentative biohydrogen production assuming a biofilm region in the domain without considering biofilm growth.

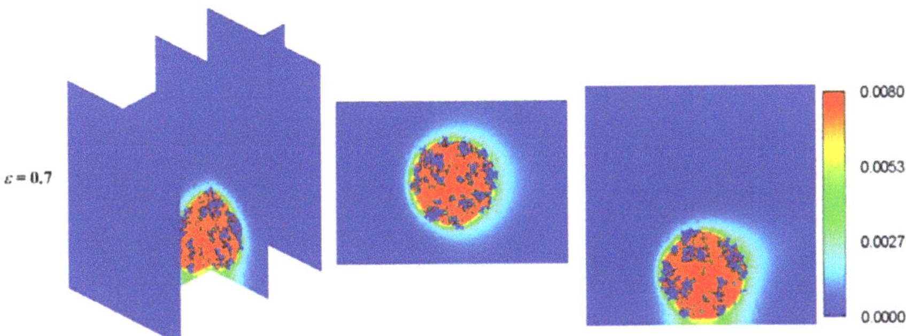

Figure 6. Concentration fields of biohydrogen concentration around the immobilized granule (Reprinted from [64] with permission from Elsevier).

3.2. LBM-CA Based Biofilm Models

LBM-based models are more recent in time in the biofilm modeling research than others using FDM or FVM. A summary of all LBM-CA based models is shown in Table 2. It is common that CA is discretized spatially as same grid of rectangular elements as LBM. Each grid element has four first-order neighbors and another four second-order neighbors in the 2D square space discretization. A grid cell is allowed to be filled up to a predetermined maximum value and a simple rule is employed to locate the extra biomass in a new compartment. As one of the first studies in the file of biofilm growth using LBM-CA, Picioreanu et al. [35,37] presented a two- and three-dimensional model for biofilm growth considering both convective and diffusive mass transfer and cellular automata approach for the growth calculation. A two- and three dimensional solute transport were solved using LBM when convection term was considered, or the FDM if without convection term. A CA approach was used for the biofilm growth calculation. This platform simulated bacterial growth, biomass spreading and biofilm detachment for multispecies and multi substrate biofilms. Figure 7 shows their simulated two-species biofilm growth on a spherical carrier. Their results showed that the LBM platform is a useful tool for multidimensional biofilm modeling.

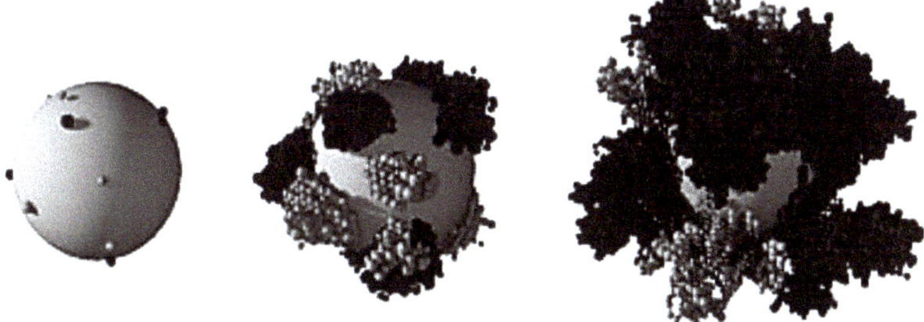

Figure 7. A two-species biofilm simulated growth during time grown on a spherical carrier. Two bacteria are shown as dark and light-colored balls (Reprinted from [35] with permission from the copyright holders, IWA Publishing).

Biofilm growth in porous media has many important applications, such as, bioremediation of contaminated sites, bio-barriers development in groundwater protection, biodegradation, microbially-enhanced oil recovery, dynamic membrane filters and biofilters of wastewater treatment [21,28,40,41,66,67]. Due to the highly heterogeneous biofilm growth within porous media, this can cause a decrease of pore space, leading to a reduction in porosity and permeability of the system and increasing hydrodynamic depression. Therefore, understanding biofilm growth in porous media is of interests to achieve an efficiency of industrial applications matter.

Table 2. Characteristics of LBM-CA based biofilm models.

Ref.	Biofilm Growth	Dimension	Flow	Species	Temperature	pH	Consideration
Picioreanu et al. [35,37]	CA-Herbert-Pitt	2D&3D	LBM	LBM	-	-	Reynolds number, Thiele number
Picioreanu et al. [48]	CA-Beeftink	2D	LBM	LBM	-	-	
Picioreanu et al. [68]	CA-Beeftink	2D	LBM	LBM	-	-	
Eberl et al. [34]	CA-Monod kinetics	3D	LBM	HOC/CDS	-	-	
Knutson et al. [38]	CA-Dual Monod	2D	LBM	Volume of Fluid Approach	-	-	
Knutson et al. [39]	CA-Dual Monod	2D	LBM	Volume of Fluid Approach	-	-	
Tang et al. [69]	CA-Monod kinetics	2D	LBM	Finite Difference Method	-	-	
Benioug et al. [70]	CA-Monod kinetics	2D	LBM	Volume of Fluid Approach	-	-	Considering Damköhler and Péclet numbers and dimensionless shear stress
Benioug et al. [71]	CA-Monod and Haldane kinetics	2D	LBM	Volume of Fluid Approach	-	-	Considering Damköhler and Péclet numbers and dimensionless shear stress
Delavar and Wang [3]	CA-Haldane and Monod Kinetics	2D	LBM	LBM	-	-	Reynolds, Prandtl and Schmidt numbers and variable diffusion coefficient due to biofilm growth were considered.
Delavar and Wang [19]	CA	2D	LBM	LBM	+	-	Variable properties and source term due to the local temperature and/or concentration change.
Delavar and Wang [20]	CA	2D	LBM	LBM	+	-	Modified boundary conditions, shear stress calculation and extra biomass transfer. Considering variable properties.
Delavar and Wang [21]	CA	2D	LBM	LBM	-	+	Including spatiotemporal effects of pH change due to inlet and local acid production and concentration on the bioprocess.

Two-dimensional models were developed for simulations of fluid flow, substrate mass transfer, biomass growth and spreading around an irregular biofilm surface [48,68]. In their model, both diffusion and convection terms of the substrate mass transport were considered in the liquid phase and only diffusion term in the biofilm-gel matrix. They evaluated effects of convection and diffusion on biofilm heterogeneity under different mass transfer regimes. Their results showed that the maximum substrate flux to the biofilm was significantly affected by both internal and external mass transfer rates. However, it was observed that the inoculation density did not affect the maximum substrate flux to the biofilm. Eberl et al. [34] used LBM to simulate fluid flow using a combination of standard central difference scheme (CDS), high order compact (HOC) method for substrate concentration and CA for simulation of biofilm growth. They indicated that the solution algorithms were computationally expensive.

Knutson et al. [38,39] developed a hybrid model in which two-dimensional pore-scale LBM numerical model was used for flow field calculation, FVM for substrate transport and CA for biodegradation and biofilm growth. They simulated transvers mixing of species and biofilm growth within a 2D porous medium. Their results showed a qualitative agreement between numerical results and experimental ones. They also examined impact of some parameters, including reaction rate and biofilm shear strength, on biofilm growth and distribution in the domain. It was found that shear strength of new biomass and solute degradation rates are the most important factors that control biofilm growth.

Tang et al. [69] developed a pore-scale biofilm model to study biofilm growth in porous media. LBM was used for simulations of fluid flow using lattice Boltzmann method, FDM for the concentration of species and CA for biofilm growth. They claimed that they have included the shrinkage procedure in biofilm growth for the first time. They found that an insufficient grid accuracy in the narrow liquid paths could cause about 90% reduction in computational time due to improvement of the numerical instability. Figure 8 shows their results for biofilm distribution in pore-space for both numerical and experimental investigations: (a) numerical and (b) experimental [69].

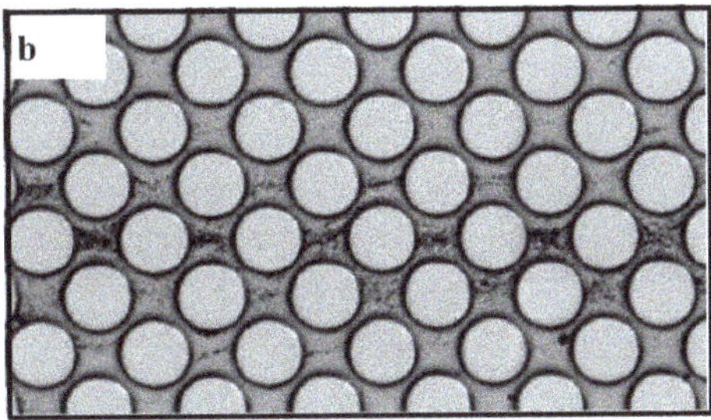

Figure 8. Biofilm distribution in pore-space for both investigations: (**a**) numerical and (**b**) experimental (Reprinted from [69] with permission from John Wiley and Sons).

Benioug et al. [70] presented a two-dimensional LBM-CA pore scale numerical model to investigate biofilm growth in porous media. They simulated fluid flow using an immersed boundary LBM and substrate transport was captured using a volume of fluid type approach. They implemented a combined model of CA algorithm with immersed boundary methods to account for cell attachment and detachment mechanisms for the spreading and distribution of biomass in the two-dimensional domain. The dimensionless parameters, Damköhler and Péclet numbers and dimensionless shear stress, were used to describe biofilm growth effects on macroscopic properties of the porous media. Their results showed the capability of the coupled model to predict appropriately the interactions among fluid flow, species transport and bacterial growth under different hydrostatics and hydrodynamics conditions. Benioug et al. [71] examined the interaction between biofilm growth and non-aqueous-liquid (NAPL) dissolution/biodegradation in both abiotic and biotic conditions to assess the capability of the model. Although the FVM was used to solve solute transport in the domain, the fluid flow field was simulated using immersed boundary lattice Boltzmann method (IB-LB). Biomass growth was investigated using a CA which was same as their previous study [70]. Figure 9 shows the evolution of simulated biofilm growth on porous media [71]. They studied how the hydrodynamic regimes and spatial distribution of NAPL blobs affect the distribution rate under different blob size and Peclet numbers in abiotic conditions.

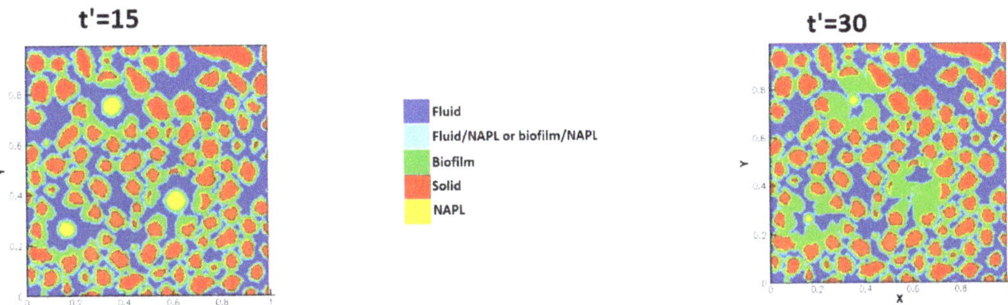

Figure 9. Evolution of the transport processes and biofilm growth (coloured green) at Da =10 (Reprinted from [71] with permission from Elsevier).

In recent years, Delavar and Wang [3,19–21] perform a series of studies using LBM-CA to investigate impact of different working parameters and signals on microorganisms' cooperation and competition with their industrial applications. Biofilms are mostly composed of multiple bacteria. As complex consortia of different microbial species, these species need to cooperate/compete with their neighbors for nutrient and space to survive. However, it is an imperative challenge to simulate the competition and the growth of multiple microorganisms in biofilms due to complex and multilateral interactions amid fluid, solid and bio-interfaces.

Due to the importance of the competitive growth, Delavar and Wang [3] investigated effects of aerobic nitrite and ammonium oxidizers in a micro bioreactor using an LBM-CA platform. If the biomass density in a lattice cell exceeds a maximum value, amount of the exceeded biomass is transferred to its neighboring grid cell. The selection of the new location is random or specific to find an appropriate neighbor. Unlike other LBM-CA models, if there are no empty neighbors, different spreading strategies were be applied by Delavar and Wang [3,19] to find a transfer direction of biomass according to concentration gradients. Newborn cells would be located to its neighbors with the maximum concentration gradient. The platform showed the ability to simulate bacterial growth, biomass spreading and biofilm detachment in multispecies and multi substrate biofilms. The effects of some parameters, such as the ratio of nutrient transfer rate to the biofilm and bacterial growth rate, were also investigated under different Reynolds number and Thiele modulus. Figure 10 shows biofilm concentration and aerobic nitrite and ammonium oxidizers bacteria in the microbioreactor at different simulation time. Their results revealed that the inlet nutrient concentrations had substantial effects on biofilm growth in terms of maximum biofilm concentration, microorganisms' concentrations, growth pattern and time. This showed that the LBM-CA framework provides a powerful tool to improve our knowledge of dynamic multilateral behavior of many complex microbial consortia systems.

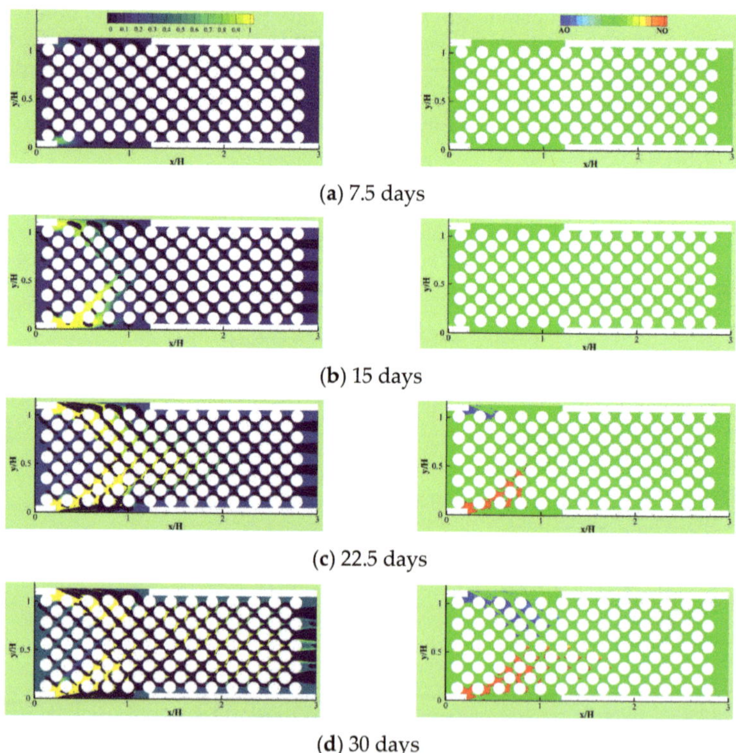

Figure 10. Biofilm concentration (left) and cells occupied with aerobic nitrite oxidizer (red) or ammonium oxidizer (blue) at different simulation time: (**a**) 7.5 days, (**b**) 15 days, (**c**) 22.5 days and (**d**) 30 days (Reprinted from [3] with permission from John Wiley and Sons).

Delavar and Wang [19] developed their LBM-CA further to account for thermal effects on competitive biofilm growth in a microbioreactor. Two microbial species, aerobic nitrite and ammonium oxidizers, were considered in three geometrical configurations of the microbioreactor. Figure 11 illustrates aerobic nitrite oxidizer (red) and ammonium oxidizer (blue) patterns in different models in a reactor. In all geometries, two heating blocks were inserted to find and compare the impacts of changing structures and temperatures on competitive biofilm growth. It was observed that the heating blocks temperatures and locations had significant effects on both the biofilm growth rate and its pattern.

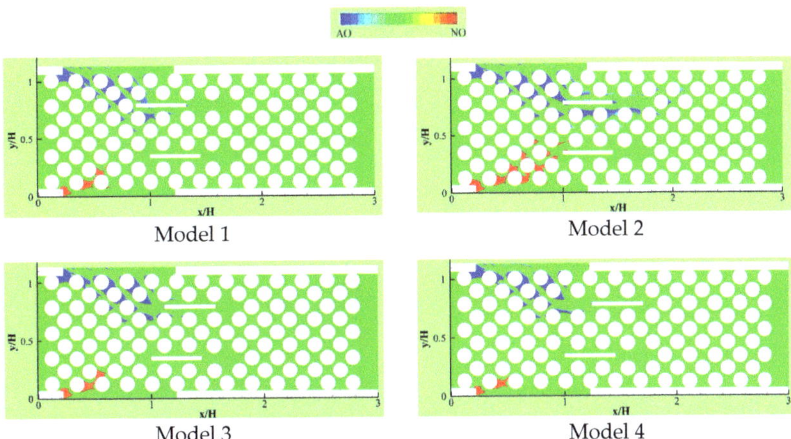

Figure 11. Aerobic nitrite oxidizer (red) or ammonium oxidizer (blue) patterns in different places of heated blocks for the changed heating arrangement in the bioreactor (Reprinted from [19] with permission from American Chemical Society).

A detailed development of a thermal LBM-CA platform was presented with significant modifications in boundary condition implementation, variable properties, detachment and extra biomass transfer [20]. The effects of dimensionless number (e.g., Reynolds, Prandtl and Schmidt numbers) were analyzed. Their results showed that changing temperature has considerable effects on not only the biofilm growth rate but also the maximum biofilm concentration in the reactor. For example, it was observed that time to reach maximum concentration reduced to 5 days (high temperature case) from 30 days (low temperature case).

Delavar and Wang [21] developed an LBM-CA platform to account for fermentative biohydrogen production. The platform was successfully used to study dark fermentation in a microbioreactor. Figure 12 shows calculated pH contours in the microbioreactor at different acid concentration at inlet. Their results demonstrated that pH is one of signals affecting the performance of dark fermentation and. Acids are one of the main byproducts of many dark fermentation processes. A change of local acid production and inlet acid concentration could significantly affect local pH and dark fermentative biohydrogen production and extraction rates.

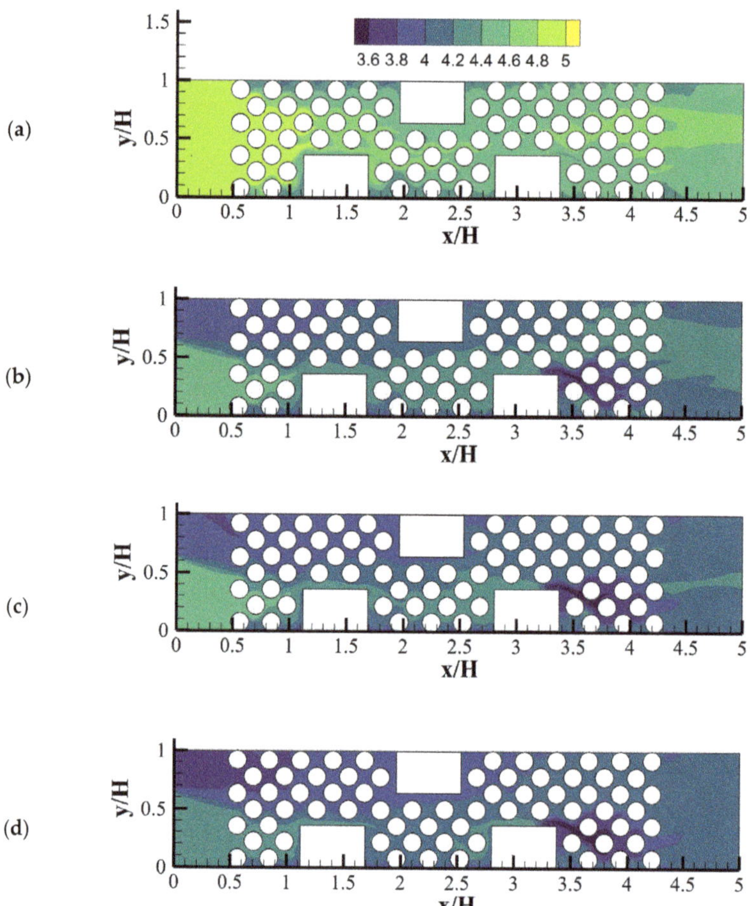

Figure 12. *pH* contours for different inlet acetic acid concentration through Inlet 2 when $pH_i = 5$: (**a**) $C_{i\text{-}2\text{-}acetic} = 0$, (**b**) $C_{i\text{-}2\text{-}acetic} = 10$, (**c**) $C_{i\text{-}2\text{-}acetic} = 20$ and (**d**) $C_{i\text{-}2\text{-}acetic} = 30$ (mole/m^3) (Reprinted from [21] with permission from Elsevier).

3.3. LBM-IbM Based Biofilm Models

In LBM-IbM models, hydrodynamics, temperature and substrate transport are solved using LBM and biofilm growth, formation and detachment are simulated using IbM. Thus, LBM-IbM can account for large phenotypic heterogeneity in nature resulting from environmental differences. A summary of all LBM-IbM based models is shown in Table 3. LBM-IbM platforms have been used to study biofilm growth in different porous structures [44,72–74]. Graf von der Schulenburg et al. [44] developed a three-dimensional LBM-IbM model to examine multilateral interactions among fluid flow, nutrient mass transport and biofilm growth in porous media. Their results revealed that the permeability of porous media was highly affected by biofilm growth and showed the necessity of 3D simulations to capture more accurate results, compared to 2D models. Pintelon et al. [72] modified an LBM-IbM model to enable it to consider non-zero permeability of biofilms structures. They examined permeability reduction in porous medium due to the biofilm growth with a range of biofilm strengths. Their results showed that given stronger biofilms under constant pressure drop, less biomass deposition and energy input are required to reduce the system permeability. However, for weaker biofilms, it was observed that constant volumetric flow rate was more efficient to decline the permeability than constant pressure drop. This demonstrated that the significant influence of biofilm permeability on overall biofilm growth rate at large biofilm concentrations. Biofilm growth in sponge carrier media (which has a large surface area to volume ratio) was carried out by So et al. [74] using an LBM-IbM model. The simulated oxygen and biomass distributions inside both the porous network and biofilm were examined in mobbing bed biofilm reactors (MBBR). Their results showed that at low detachment rates more clogging happened in outer pores that limited biofilm growth on the surface region of the sponge.

Table 3. Characteristics of LBM-IbM based biofilm models.

Ref.	Biofilm Growth	Dimension	Flow	Species	Temperature	pH	Consideration
Graf von der Schulenburg et al. [44]	IbM-Monod kinetics	3D	LBM	LBM	-	-	
Pintelon et al. [72]	IbM-Monod kinetics	2D	LBM	LBM	-	-	Flow shear to simulate the biomass detachment
Pintelon et al. [75]	IbM-Monod kinetics	3D	LBM	LBM	-	-	
Creber et al. [76]	IbM-Monod kinetics	3D	LBM	LBM	-	-	
Pintelon et al. [73]	IbM-Monod kinetics	3D	LBM	LBM	-	-	
So et al. [74]	IbM-Monod kinetics	2D	LBM	LBM	-	-	
Tian and Wang [28]	IbM-Monod kinetics	2D	LBM	LBM	-	-	Multiscale Permeability in biofilms

The fouling of biofilms in reverse osmosis (RO) membrane is one of the important issues in industrial production of high purity water. In good condition, RO membrane can remove 99% of mono- and divalent ions. However, biofilm growth can dwindle the fluid flow and system efficiency if biofilm fouling happens. Pintelon et al. [75] developed a 3D LBM-IbM model to simulate biofouling in the RO membrane. Monod kinetics was used for description of biofilm growth. Their simulated results agreed well with experimental data obtained with magnetic resonance imaging (MRI) techniques. They examined the effects of biofilm cohesive strength on the membrane pressure drop. It is found that weaker biomass has lower effects on pressure drop. Some cleaning strategies such as forward or backward flushing and electro-chemical treatment could be used to remove foulants, in RO membranes. Creber et al. [76] simulated membrane cleaning and reduction in the biofilm cohesive strength using an extension of LBM-IbM model proposed by Pintelon et al. [75]. They examined biofilm accumulation and chemical removal in RO membranes. The biofilm distribution and water flow field in the membrane, before and after cleaning, were captured. and their simulated biomass changes were in good agreement with experimental data.

In LBM-IbM models by Graf von der Schulenburg et al. [44], Pintelon et al. [72] and Pintelon et al. [73], the conventional LBGK was used to recover continuity and standard Navier–Stokes equations. Because Navier–Stokes equation is limited to simulate flows in porous media due to neglecting the variable porosity and permeability of unresolved microporosity in solid matrix and biofilms, the permeability of biofilm was considered indirectly through the manipulation of the LBGK relaxation parameter as the pseudo-viscosity for flow within biofilm-occupied simulation lattice. However, many porous materials, such as biofilms and soils, include multiscale pores or porosities that can contribute to the macro-permeability of porous media. Due to importance of the unresolved intra-aggregate pores in porous media, Tian and Wang [28] presented an LBM-IbM model which recovered the Darcy–Brinkman–Forchheimer equation. They used a multilayer bounce-back treatment developed by Tian and Wang [77] to account for multiscale porosities in biofilms and solid matrix. The coupled LBM-IbM model was implemented to study interactions among oxygen, bioclogging, chemical oxygen demand (COD), biofilm growth and multiscale permeability of the porous media and biofilms. Figure 13 shows the porosity distribution and velocity vectors and the biofilm colony distribution in the domain. Their results showed a very heterogeneous biofilm growth on the surface of the solid matrix and pores. In addition, they found a biofilm porosity threshold, beyond which no obvious effects on the flow rate and COD removal were observed. This demonstrated the capability of the model to study biofilm growth, clogging and contaminants degradation under various operational conditions in porous media. The model is potentially applicable to investigate processes in wastewater treatment.

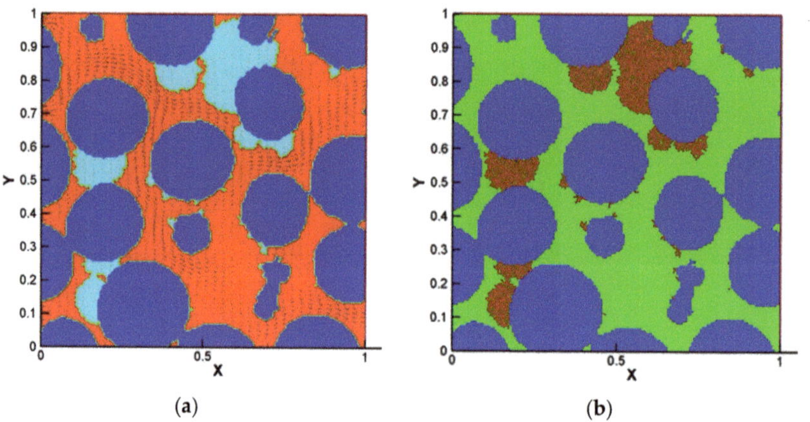

Figure 13. (**a**) The porosity distribution and velocity vectors, (**b**) the biofilm colony distribution in the domain, at 38×10^4 time steps (Reprinted from [28] with permission from John Wiley and Sons).

4. Current Trends and Upcoming Challenges

LBM-based models are more recent in time in the biofilm modeling research than other biofilm models. Both LBM-CA and LBM-IbM approaches represent powerful tools in modeling biofilm growths. Compared to the FDM or FVM, as a meso-scale approach, LBMs coupled with IbM or CA have advantages to incorporate non-equilibrium interface processes and events and adopt distinct rules for cell spreading by different bacteria and, thus, can capture the micro-scale interactions among individual cells, local substrate variability and macroscopic structure of biofilms. Because geometries generated by computed x-ray tomography scan, such as porous media, can be directly inputted into LBM solver, this allows very complex geometries to be simulated [28]. Furthermore, because LBM solver uses two basic steps: stream and collision, this makes the code efficient and effective to be parallelized and implemented [57,78].

4.1. *Advantages and Disadvantages of Different LBM-Based Models*

The LBM-IbM model seems to be very appealing to microbiologists since it allows individual variability and tracks individual cells as the fundamental units [28,44]. The use of IbM would become necessary if the biofilm heterogeneity resulting from the intracellular interaction and their feedbacks between the individuals and the population needs to be simulated. In IbM, the collective action of individual cells reflects population or community level properties [79].

For LBM-CA model, a CA approach can use the same lattice grid as LBM. A set of rules is used for biomass spreading. As a CA grid is much bigger than an individual cell size, a CA grid is usually occupied by several cells. Therefore, CA does not resolve individual cells. Instead, CA uses a predetermined maximum biomass in a grid. Under the maximum biomass, biomass in the grid is represented by its percentage [3,19]. Any newborn cell in a grid is transferred to its neighbour grids in the biofilm front and the cell shoving with one another is not required.

Despite the aforementioned advantages, LBM-based models have disadvantages. First, because LBM-based models require to calculate distribution functions and then recover the macro variables, such as velocity, temperature and concentration, LBM-based models require more computer memory and computational time than FDM [57]. Particularly, LBM-IbM is very computationally intensive. Because IbM traces every individual cell and their interaction. Thus, at every time step, shoving and spreading of cells require many iterations to find appreciated position of every cell. The more the cell number, the more iterations will require. Thus, LBM-IbM models have been limited at small-scale systems. However, if the larger sizes of cells (e.g., 10 to 20 μm in diameter) or a parcel of cells are used, this will allow a larger system to be simulated while still keeping the shoving and spreading rules for cell redistribution. Furthermore, because individual cells are assumed as hard spheres and a predetermined shoving parameter is used to simulate the direction of the biomass movement. Constant biofilm porosity and the elastic contact between cells can also cause uncertainties.

In CA approach, spreading successfully movements are limited in a number of directions. The rules of spreading in CA are sometimes arbitrarily formulated. This might lead to it being aesthetically driven, rather than physically motivated [42].

4.2. *Upcoming Challenges and Future Research Perspectives*

Biofilm modeling is driven by the scientific and practical interest to understand, control and engineer biofilms in different applications, such as bioconversion, biofuel production, wastewater treatment and prevention of biofouling. With the impressive advances in genetic and biochemical technologies, it has been realized that the biofilms could not be defined as simply bacteria fastened to a hydrated surface. For example, antibiotic tolerance in biofilms causes the inability of antibiotic treatment in suppressing bacterial infections. Bacterial populations produce persisters that exhibit multidrug tolerance and survive treatment by all antimicrobials but the tolerance of persister cells is different from inherited, reversible antimicrobial resistance. Persister cells are greatly enhanced in biofilms. This can lead to difficulty to deal with biofilm-related diseases [80]. Quorum sensing involves perceiving and responding to extracellular signals of bacterial communities for competition and cooperation [81]. Furthermore, nutrient and microbial heterogeneity, quorum sensing and environmental conditions all affect the phenotypic distinctiveness of the individual stages of biofilm development. Individuals of the same species can clearly compete with each other, but they need cooperation in the multicellular-level. LBM-based models require to incorporate these features and behaviors of the complex biofilm communities as a possible way in controlling and removing biofilms of industrial and clinical concern [82].

Although IbMs have simulated signal transduction mechanisms in chemotaxis in the new field of synthetic biology [83] and the integration of intracellular mechanisms also enables the use of the rapidly increasing amount of metagenomic data, LBM-IbM or LBM-CA models have not been reported to simulate these complex features. It is still a challenge to account for different cells, such as active and inert cells, quorum sensing and antibiotics. Furthermore, biofilm models need to expand their capacities beyond the illustrative processes and simple geometries because many bioreactors are complex in industrial applications. However, simplification and parameterization of the general biofilm models are inevitable to achieve a specific goal in industrial applications. Therefore, a comprehensive model of biofilm reactors may not require including as many parameters as possible, but rather, to assess the level of significance of various parameters. This can make it feasible to capture the main features in the description of the different biofilm processes as the necessary data can be extracted for the reactor design and operation.

5. Conclusions

In this paper, we revisit a family of the LBM-based biofilm models chronologically. We present the fundamentals of LBM-based biofilm modeling first and two main integrated approaches: the integrated LBM with CA or IbM. Then, we focus on progresses of LBM-CA and LBM-IbM approaches and their applications. The LBM-based models are very promising to handle complex multilateral interactions which occur in a bioreactor and have successfully simulated many complicated phenomena in the biofilm's dynamics and interactions between biofilm growth and hydrodynamics in different industrial or academic applications. However, long-term challenges exist in modeling biofilms when bridging different spatiotemporal scales, such as multispecies competition and cooperation, community assembly, cell-to-cell signaling, quorum sensing, antimicrobial resistance and cellular motility. Addressing these challenges would inevitably need a combined approach that not only selectively couples multiple relevant models by adding their strengths, but also the incorporation of new findings. LBM-based models doubtlessly offer new ways in understanding the dynamics of biofilms and controlling the biofilm formation and growth in public health, biotechnology and bioengineering and microbiology.

Author Contributions: Conceptualization, M.A.D. and J.W.; methodology, M.A.D.; formal analysis, M.A.D. and J.W.; data curation, M.A.D. and J.W.; writing—original draft preparation, M.A.D.; writing—review and editing, M.A.D. and J.W.; supervision, J.W.; project administration, J.W.; funding acquisition, J.W. All authors have read and agreed to the published version of the manuscript.

Funding: This work was supported by Natural Sciences and Engineering Research Council (No. RGPIN-2019-07071).

Institutional Review Board Statement: Not applicable.

Informed Consent Statement: Not applicable.

Conflicts of Interest: The authors declare no conflict of interest.

References

1. Flemming, H.C.; Wingender, J.; Szewzyk, U.; Steinberg, P.; Rice, S.A.; Kjelleberg, S. Biofilms: An emergent form of bacterial life. *Nat. Rev. Microbiol.* **2016**, *14*, 563–575. [CrossRef] [PubMed]
2. Gunes, B. A critical review on biofilm-based reactor systems for enhanced syngas fermentation processes. *Renew. Sustain. Energy Rev.* **2021**, *143*, 110950. [CrossRef]
3. Delavar, M.A.; Wang, J. Pore-scale modeling of competition and cooperation of multispecies biofilms for nutrients in changing environments. *AIChE J.* **2020**, *66*, e16919.
4. Wong, A.C.L. Biofilms in food processing environments. *J. Dairy Sci.* **1998**, *81*, 2765–2770. [CrossRef]
5. Alvarez-Ordóñez, A.; Coughlan, L.M.; Briandet, R.; Cotter, P.D. Biofilms in food processing environments: Challenges and opportunities. *Annu. Rev. Food Sci. Technol.* **2019**, *10*, 173–195. [CrossRef]
6. Zhao, X.; Zhao, F.; Wang, J.; Zhong, N. Biofilm formation and control strategies of foodborne pathogens: Food safety perspectives. *RSC Adv.* **2017**, *7*, 36670–36683. [CrossRef]
7. Xu, Z.; Wang, J.; Xia, L.; Zhang, Y. A biofouling thermal resistance model with a growth term of surface biofilm on heat transfer surface. *Int. J. Therm. Sci.* **2021**, *161*, 106699. [CrossRef]
8. de Vries, H.J.; Kleibusch, E.; Hermes, G.D.; van den Brink, P.; Plugge, C.M. Biofouling control: The impact of biofilm dispersal and membrane flushing. *Water Res.* **2021**, *198*, 117163. [CrossRef]
9. Akao, P.K.; Singh, B.; Kaur, P.; Sor, A.; Avni, A.; Dhir, A.; Verma, S.; Kapoor, S.; Phutela, U.G.; Satpute, S.; et al. Coupled microalgal–bacterial biofilm for enhanced wastewater treatment without energy investment. *J. Water Process Eng.* **2021**, *41*, 102029. [CrossRef]
10. Mallikarjuna, C.; Dash, R.R. A review on hydrodynamic parameters and biofilm characteristics of inverse fluidized bed bioreactors for treating industrial wastewater. *J. Environ. Chem. Eng.* **2020**, *8*, 104233. [CrossRef]
11. di Biase, A.; Kowalski, M.S.; Devlin, T.R.; Oleszkiewicz, J.A. Moving bed biofilm reactor technology in municipal wastewater treatment: A review. *J. Environ. Manag.* **2019**, *247*, 849–866. [CrossRef]
12. Swain, A.K.; Sahoo, A.; Jena, H.M.; Patra, H. Industrial wastewater treatment by aerobic inverse fluidized bed biofilm reactors (AIFBBRs): A review. *J. Water Process Eng.* **2018**, *23*, 61–74. [CrossRef]
13. Obileke, K.; Onyeaka, H.; Meyer, E.L.; Nwokolo, N. Microbial fuel cells, a renewable energy technology for bio-electricity generation: A mini-review. *Electrochem. Commun.* **2021**, *125*, 107003. [CrossRef]
14. Saravanan, A.; Kumar, P.S.; Vardhan, K.H.; Jeevanantham, S.; Karishma, S.B.; Yaashikaa, P.R.; Vellaichamy, P. A review on systematic approach for microbial enhanced oil recovery technologies: Opportunities and challenges. *J. Clean. Prod.* **2020**, *258*, 120777. [CrossRef]
15. Gaol, C.L.; Ganzer, L.; Mukherjee, S.; Alkan, H. Parameters govern microbial enhanced oil recovery (MEOR) performance in real-structure micromodels. *J. Pet. Sci. Eng.* **2021**, *205*, 108814. [CrossRef]

16. Roy, P.K.; Ha, A.J.W.; Mizan, M.F.R.; Hossain, M.I.; Ashrafudoulla, M.; Toushik, S.H.; Nahar, S.; Kim, Y.K.; Ha, S.D. Effects of Environmental Conditions (temperature, pH, and glucose) on Biofilm Formation of Salmonella enterica serotype Kentucky and Virulence Gene Expression. *Poult. Sci.* **2021**, *100*, 101209. [CrossRef]
17. Kim, W.J.; Kim, S.H.; Kang, D.H. Thermal and non-thermal treatment effects on Staphylococcus aureus biofilms formed at different temperatures and maturation periods. *Food Res. Int.* **2020**, *137*, 109432. [CrossRef] [PubMed]
18. Shojaei, B.; Khazaee, I. 1-D transient microbial fuel cell simulation considering biofilm growth and temperature variation. *Int. J. Therm. Sci.* **2021**, *162*, 106801. [CrossRef]
19. Delavar, M.A.; Wang, J. Modeling combined effects of temperature and structure on competition and growth of multispecies biofilms in microbioreactors. *Ind. Eng. Chem. Res.* **2020**, *59*, 16122–16135. [CrossRef]
20. Delavar, M.A.; Wang, J. Modeling coupled temperature and transport effects on biofilm growth using thermal lattice Boltzmann model. *AIChE J.* **2021**, *67*, e17122. [CrossRef]
21. Delavar, M.A.; Wang, J. Numerical investigation of pH control on dark fermentation and hydrogen production in a microbioreactor. *Fuel* **2021**, *292*, 120355. [CrossRef]
22. Matheus, M.C.; Ekenberg, M.; Bassin, J.P.; Dezotti, M.W.; Piculell, M. High loaded moving bed biofilm reactors treating pulp & paper industry wastewater: Effect of hydraulic retention time, filling degree and nutrients availability on performance, biomass fractions and nutrients utilization. *J. Environ. Chem. Eng.* **2021**, *9*, 104944.
23. Chaiwong, C.; Koottatep, T.; Polprasert, C. Development of kinetic models for organic and nutrient removal in biofilm photo-bioreactor for treatment of domestic wastewater. *Environ. Technol. Innov.* **2021**, *23*, 101547. [CrossRef]
24. Chen, X.; Chen, X.; Zhao, Y.; Zhou, H.; Xiong, X.; Wu, C. Effects of microplastic biofilms on nutrient cycling in simulated freshwater systems. *Sci. Total Environ.* **2020**, *719*, 137276. [CrossRef]
25. Iltis, G.C.; Armstrong, R.T.; Jansik, D.P.; Wood, B.D.; Wildenschild, D. Imaging biofilm architecture within porous media using synchrotron-based X-ray computed microtomography. *Water Resour. Res.* **2011**, *47*, W02601. [CrossRef]
26. Zhang, Y.; Xu, A.; Lv, X.; Wang, Q.; Feng, C.; Lin, J. Non-Invasive Measurement, Mathematical Simulation and In Situ Detection of Biofilm Evolution in Porous Media: A Review. *Appl. Sci.* **2021**, *11*, 1391. [CrossRef]
27. Mattei, M.R.; Frunzo, L.; D'Acunto, B.; Pechaud, Y.; Pirozzi, F.; Esposito, G. Continuum and discrete approach in modeling biofilm development and structure: A review. *J. Math. Biol.* **2018**, *76*, 945–1003. [CrossRef]
28. Tian, Z.; Wang, J. Lattice Boltzmann simulation of biofilm clogging and chemical oxygen demand removal in porous media. *AIChE J.* **2019**, *65*, e16661. [CrossRef]
29. Williamson, K.; McCarty, P. A model of substrate utilization by bacterial films. *J. Water Pollut. Control Fed.* **1976**, *48*, 9–24.
30. Harremoes, P. The significance of pore diffusion to filter denitrification. *J. Water Pollut. Control Fed.* **1976**, *48*, 377–388.
31. Wanner, O.; Gujer, W. A multispecies biofilm model. *Biotechnol. Bioeng.* **1986**, *28*, 314–328. [CrossRef]
32. Laspidou, C.S.; Kungolos, A.; Samaras, P. Cellular-automata and individual-based approaches for the modeling of biofilm structures: Pros and cons. *Desalination* **2010**, *250*, 390–394. [CrossRef]
33. Picioreanu, C.; van Loosdrecht, M.C.M.; Heijnen, J.J. Mathematical modeling of biofilm structure with a hybrid differential-discrete cellular automaton approach. *Biotechnol. Bioeng.* **1998**, *58*, 101–116. [CrossRef]
34. Eberl, H.; Picioreanu, C.; van Loosdrecht, M. Modeling geometrical heterogeneity in biofilms. In *High Performance Computing Systems and Applications*; Pollard, A., Mewhort, D.J.K., Weaver, D.F., Eds.; Springer: Boston, MA, USA, 2002; pp. 497–512.
35. Picioreanu, C.; van Loosdrecht, M.; Heijnen, J. Discrete-Differential modelling of biofilm structure. *Water Sci. Technol.* **1999**, *59*, 115–122. [CrossRef]
36. Noguera, D.R.; Pizarro, G.; Stahl, D.A.; Rittmann, B.E. Simulation of multispecies biofilm development in three dimensions. *Water Sci. Technol.* **1999**, *39*, 123–130. [CrossRef]
37. Picioreanu, C.; van Loosdrecht, M.; Heijnen, J. Multidimensional Modeling of Biofilm Structure. In *Microbial Biosystems: New Frontiers*; Bell, C.R., Brylinsky, M., Johnson-Green, P., Eds.; Atlantic Canada Society for Microbial Ecology: Halifax, NS, Canada, 1999.
38. Knutson, C.; Werth, C.J.; Valocchi, A.J.; Travis, B.J. Modeling biofilm morphology along a transverse mixing zone in porous media at the pore scale. In *Developments in Water Science*; Miller, C.T., Farthing, M.W., Gray, W.G., Pinder, G.F., Eds.; Elsevier: Amsterdam, The Netherlands, 2004; Volume 55, pp. 61–69.
39. Knutson, C.E.; Werth, C.J.; Valocchi, A.J. Pore-scale simulation of biomass growth along the transverse mixing zone of a model two-dimensional porous medium. *Water Resour. Res.* **2005**, *41*, W07007. [CrossRef]
40. Singh, R.; Paul, D.; Jain, R.K. Biofilms: Implications in bioremediation. *Trends Microbiol.* **2006**, *14*, 389–397. [CrossRef] [PubMed]
41. Seo, Y.; Lee, W.H.; Sorial, G.; Bishop, P.L. The application of a mulch biofilm barrier for surfactant enhanced polycyclic aromatic hydrocarbon bioremediation. *Environ. Pollut.* **2009**, *157*, 95–101. [CrossRef] [PubMed]
42. Eberl, H.J.; Parker, D.F.; van Loosdrecht, M.C.M. A new deterministic spatio-temporal continuum model for biofilm development. *J. Theor. Med.* **2001**, *3*, 161–175. [CrossRef]
43. Picioreanu, C.; Kreft, J.U.; van Loosdrecht, M.C.M. Particle-Based multidimensional multispecies biofilm model. *Appl. Environ. Microbiol.* **2004**, *70*, 3024–3040. [CrossRef]
44. Graf von der Schulenburg, D.A.; Pintelon, T.R.R.; Picioreanu, C.; Van Loosdrecht, M.C.M.; Johns, M.L. Three-dimensional simulations of biofilm growth in porous media. *AIChE J.* **2009**, *55*, 494–504. [CrossRef]

45. Picioreanu, C.; van Loosdrecht, M. Use of mathematical modelling to study biofilm development and morphology. In *Biofilms in Medicine, Industry and Environmental Biotechnology: Characteristics, Analysis and Control*; Lens, P., Moran, A.P., Mahony, T., Stoodley, P., O'Flaherty, V., Eds.; IWA Publishing: London, UK, 2003; pp. 413–437.
46. Donlan, R.M. Biofilms: Microbial Life on Surfaces. *Emerg. Infect. Dis.* **2002**, *8*, 881–890. [CrossRef]
47. Kapellos, G.E.; Alexiou, T.S.; Payatakes, A.C. A multiscale theoretical model for diffusive mass transfer in cellular biological media. *Math. Biosci.* **2007**, *210*, 177–237. [CrossRef]
48. Picioreanu, C.; Van Loosdrecht, M.C.; Heijnen, J.J. Effect of diffusive and convective substrate transport on biofilm structure formation: A two-dimensional modeling study. *Biotechnol. Bioeng.* **2000**, *69*, 504–515. [CrossRef]
49. Kreft, J.U.; Booth, G.; Wimpenny, J.W.T. BacSim, a simulator for individual-based modelling of bacterial colony growth. *Microbiology* **1998**, *144*, 3275–3287. [CrossRef]
50. Frisch, U.; d'Humieres, D.; Hasslacher, B.; Lallemand, P.; Pomeau, Y.; Rivet, J.P. Lattice gas hydrodynamics in two and three dimensions. *Complex Syst.* **1987**, *1*, 649–707.
51. Bhatnagar, P.L.; Gross, E.P.; Krook, M. A model for collision processes in gases. I. Small amplitude processes in charged and neutral one-component systems. *Phys. Rev.* **1954**, *94*, 511. [CrossRef]
52. Qian, Y.H.; d'Humières, D.; Lallemand, P. Lattice BGK models for Navier-Stokes equation. *EPL Europhys. Lett.* **1992**, *17*, 479. [CrossRef]
53. Chen, H.; Chen, S.; Matthaeus, W.H. Recovery of the Navier-Stokes equations using a lattice-gas Boltzmann method. *Phys. Rev. A* **1992**, *45*, R5339. [CrossRef]
54. Mohamad, A.A. *Fundamentals and Engineering Applications with Computer Codes*; Springer: London, UK, 2011.
55. Raabe, D. Overview of the lattice Boltzmann method for nano-and microscale fluid dynamics in materials science and engineering. *Model. Simul. Mater. Sci. Eng.* **2004**, *12*, R13. [CrossRef]
56. Chen, S.; Doolen, G.D. Lattice Boltzmann method for fluid flows. *Annu. Rev. Fluid Mech.* **1998**, *30*, 329–364. [CrossRef]
57. Wang, J.; Zhang, X.; Bengough, A.G.; Crawford, J.W. Domain-decomposition method for parallel lattice Boltzmann simulation of incompressible flow in porous media. *Phys. Rev. E* **2005**, *72*, 016706. [CrossRef] [PubMed]
58. Mezrhab, A.; Jami, M.; Abid, C.; Bouzidi, M.H.; Lallemand, P. Lattice-Boltzmann modelling of natural convection in an inclined square enclosure with partitions attached to its cold wall. *Int. J. Heat Fluid Flow* **2006**, *27*, 456–465. [CrossRef]
59. Delavar, M.A.; Farhadi, M.; Sedighi, K. Numerical simulation of direct methanol fuel cells using lattice Boltzmann method. *Int. J. Hydrogen Energy* **2010**, *35*, 9306–9317. [CrossRef]
60. Ajarostaghi, S.S.M.; Delavar, M.A.; Poncet, S. Thermal mixing, cooling and entropy generation in a micromixer with a porous zone by the lattice Boltzmann method. *J. Therm. Anal. Calorim.* **2019**, *140*, 1–19. [CrossRef]
61. Graf von der Schulenburg, D.A.; Gladden, L.F.; Johns, M.L. Modelling biofilm-modified hydrodynamics in 3D. *Water Sci. Technol.* **2007**, *55*, 275–281. [CrossRef]
62. Zhang, C.; Kang, Q.; Wang, X.; Zilles, J.L.; Müller, R.H.; Werth, C.J. Effects of pore-scale heterogeneity and transverse mixing on bacterial growth in porous media. *Environ. Sci. Technol.* **2010**, *44*, 3085–3092. [CrossRef]
63. Yang, Y.; Liao, Q.; Zhu, X.; Wang, H.; Wu, R.; Lee, D.J. Lattice Boltzmann simulation of substrate flow past a cylinder with PSB biofilm for bio-hydrogen production. *Int. J. Hydrogen Energy* **2011**, *36*, 14031–14040. [CrossRef]
64. Liao, Q.; Yang, Y.X.; Zhu, X.; Chen, R.; Fu, Q. Pore-scale lattice Boltzmann simulation of flow and mass transfer in bioreactor with an immobilized granule for biohydrogen production. *Sci. Bull.* **2017**, *62*, 22–30. [CrossRef]
65. Liao, Q.; Yang, Y.X.; Zhu, X.; Chen, R.; Fu, Q. A simulation on flow and mass transfer in a packed bed photobioreactor for hydrogen production. *Int. J. Heat Mass Transf.* **2017**, *109*, 1132–1142. [CrossRef]
66. Al-Bader, D.; Kansour, M.K.; Rayan, R.; Radwan, S.S. Biofilm comprising phototrophic, diazotrophic, and hydrocarbon-utilizing bacteria: A promising consortium in the bioremediation of aquatic hydrocarbon pollutants. *Environ. Sci. Pollut. Res.* **2013**, *20*, 3252–3262. [CrossRef]
67. Huang, T.; Xu, J.; Cai, D. Efficiency of active barriers attaching biofilm as sediment capping to eliminate the internal nitrogen in eutrophic lake and canal. *J. Environ. Sci.* **2011**, *23*, 738–743. [CrossRef]
68. Picioreanu, C.; Van Loosdrecht, M.C.; Heijnen, J.J. Two-dimensional model of biofilm detachment caused by internal stress from liquid flow. *Biotechnol. Bioeng.* **2001**, *72*, 205–218. [CrossRef]
69. Tang, Y.; Valocchi, A.J.; Werth, C.J.; Liu, H. An improved pore-scale biofilm model and comparison with a microfluidic flow cell experiment. *Water Resour. Res.* **2013**, *49*, 8370–8382. [CrossRef]
70. Benioug, M.; Golfier, F.; Oltéan, C.; Buès, M.A.; Bahar, T.; Cuny, J. An immersed boundary-lattice Boltzmann model for biofilm growth in porous media. *Adv. Water Resour.* **2017**, *107*, 65–82. [CrossRef]
71. Benioug, M.; Golfier, F.; Fischer, P.; Oltean, C.; Buès, M.A.; Yang, X. Interaction between biofilm growth and NAPL remediation: A pore-scale study. *Adv. Water Resour.* **2019**, *125*, 82–97. [CrossRef]
72. Pintelon, T.R.; Graf von der Schulenburg, D.A.G.; Johns, M.L. Towards optimum permeability reduction in porous media using biofilm growth simulations. *Biotechnol. Bioeng.* **2009**, *103*, 767–779. [CrossRef] [PubMed]
73. Pintelon, T.R.; Picioreanu, C.; van Loosdrecht, M.C.; Johns, M.L. The effect of biofilm permeability on bio-clogging of porous media. *Biotechnol. Bioeng.* **2012**, *109*, 1031–1042. [CrossRef] [PubMed]
74. So, M.; Naka, D.; Goel, R.; Terashima, M.; Yasui, H. Modelling clogging and biofilm detachment in sponge carrier media. *Water Sci. Technol.* **2014**, *69*, 1298–1303. [CrossRef] [PubMed]

75. Pintelon, T.R.; Creber, S.A.; Graf von der Schulenburg, D.A.G.; Johns, M.L. Validation of 3D simulations of reverse osmosis membrane biofouling. *Biotechnol. Bioeng.* **2010**, *106*, 677–689. [CrossRef]
76. Creber, S.A.; Pintelon, T.R.R.; Von Der Schulenburg, D.G.; Vrouwenvelder, J.S.; Van Loosdrecht, M.C.M.; Johns, M.L. Magnetic resonance imaging and 3D simulation studies of biofilm accumulation and cleaning on reverse osmosis membranes. *Food Bioprod. Process.* **2010**, *88*, 401–408. [CrossRef]
77. Tian, Z.; Wang, J. Lattice Boltzmann simulation of CO_2 reactive transport in network fractured media. *Water Resour. Res.* **2017**, *53*, 7366–7381. [CrossRef]
78. Wang, J.; Zhang, X.; Bengough, A.G.; Crawford, J.W. Performance evaluation of the cell-based algorithms for domain decomposition in flow simulation. *Int. J. Numer. Methods Heat Fluid Flow* **2008**, *18*, 656–672. [CrossRef]
79. Hellweger, R.L.; Clegg, R.J.; Clark, J.R.; Plugge, C.M.; Kreft, J.U. Advancing microbial sciences by individual-based modelling. *Nat. Rev. Microbiol.* **2016**, *14*, 461–471. [CrossRef]
80. Brauner, A.; Fridman, O.; Gefen, O.; Balaban, N.Q. Distinguishing between resistance, tolerance and persistence to antibiotic treatment. *Nat. Rev. Microbiol.* **2016**, *14*, 320–330. [CrossRef]
81. Dalwadi, M.P.; Pearce, P. Emergent robustness of bacterial quorum sensing in fluid flow. *Proc. Natl. Acad. Sci. USA* **2021**, *118*, e2022312118. [CrossRef]
82. Cogan, N.G.; Harro, J.M.; Stoodley, P.; Shirtliff, M.E. Predictive computer models for biofilm detachment properties in pseudomonas aeruginosa. *mBio* **2016**, *7*, e00815-16. [CrossRef] [PubMed]
83. Emonet, T.; Cluzel, P. Relationship between cellular response and behavioral variability in bacterial chemotaxis. *Proc. Natl. Acad. Sci. USA* **2008**, *105*, 3304–3309. [CrossRef] [PubMed]

Article

Energy Utilization Efficiency of China Considering Carbon Emissions—Based on Provincial Panel Data

Ge Huang [1], Wei Pan [2,*], Cheng Hu [1], Wu-Lin Pan [1] and Wan-Qiang Dai [1]

1 School of Economic and Management, Wuhan University, Wuhan 430072, China; huanggekaku@whu.edu.cn (G.H.); 2017201050215@whu.edu.cn (C.H.); panwulin@whu.edu.cn (W.-L.P.); dwqiang28@163.com (W.-Q.D.)
2 School of Applied Economics, Renmin University of China, Beijing 100872, China
* Correspondence: mrpanwei2020@ruc.edu.cn

Abstract: With the development of the economy, environmental pollution caused by energy consumption has become increasingly prominent. Improving the efficiency of energy utilization is an important way to solve this problem. Firstly, we used a data envelopment analysis (DEA) model to calculate the energy utilization efficiency of China's provinces and regions from the perspective of environmental constraints, including four inputs—labor force, capital stock, energy consumption and carbon emission—and one output, GDP. Secondly, an entity fixed effect model of panel data was built to investigate the influence of openness, urbanization, marketization and industrial structure on energy utilization efficiency in the process of economic structure change. The results indicate that China's energy efficiency shows a trend of first stabilizing and then declining from 2007 to 2017. Meanwhile, the comprehensive energy efficiency of all provinces and regions is not very ideal. Only Beijing, Shanghai and Guangdong constitute the forefront of China's energy efficiency. The lack of pure technical efficiency in most provinces is the main reason for the low comprehensive efficiency, but there are also obvious differences among provinces and regions. In addition, urbanization, openness and industrial structure have a negative impact on energy efficiency, while marketization has a significant positive impact on energy efficiency. Finally, based on the regional differences, some suggestions were put forward to improve China's energy utilization efficiency.

Keywords: DEA; regional difference; energy utilization efficiency; carbon emission

Citation: Huang, G.; Pan, W.; Hu, C.; Pan, W.-L.; Dai, W.-Q. Energy Utilization Efficiency of China Considering Carbon Emissions—Based on Provincial Panel Data. *Sustainability* **2021**, *13*, 877. https://doi.org/10.3390/su13020877

Academic Editor: Tomonobu Senjyu
Received: 7 December 2020
Accepted: 29 December 2020
Published: 16 January 2021

1. Introduction

With the rapid development of the economy in recent years, China's energy production and consumption have been growing rapidly, but environmental pollution has also been significantly aggravated. Moreover, the restriction of energy on the economy and the environment is becoming more and more obvious, which has attracted the attention of many scholars. Therefore, in order to realize the coordinated development of energy–economy–environment, the 13th Five-Year Plan (2016–2020) has emphasized that China should adhere to the energy development strategy of "four revolutions and one cooperation" and deepen the energy revolution. In addition, energy consumption per unit of GDP and carbon dioxide emission are required to be reduced by 15% and 18%, respectively. Energy efficiency is generally measured by energy intensity (energy consumption per 10,000 yuan of GDP). The higher the energy intensity per unit of GDP, the lower the energy efficiency. As can be seen from Figure 1, since 1980, with the improvement of China's energy science and technology innovation ability, the rapid development of energy technology and equipment, and the optimization of the energy system driven by automation, intellectualization and digitization, China's energy efficiency has undergone great changes, and the energy intensity has generally declined. It can be seen that China's energy utilization efficiency drops sharply every 10 years, and the overall trend is a stepwise decline. However, the energy intensity dropped to 0.55 tons of standard coal per

10,000 yuan in 2018, which was not reduced by 15% on the basis of 2015. There, it is necessary to analyze China's energy efficiency in detail to develop more effective strategies for green development.

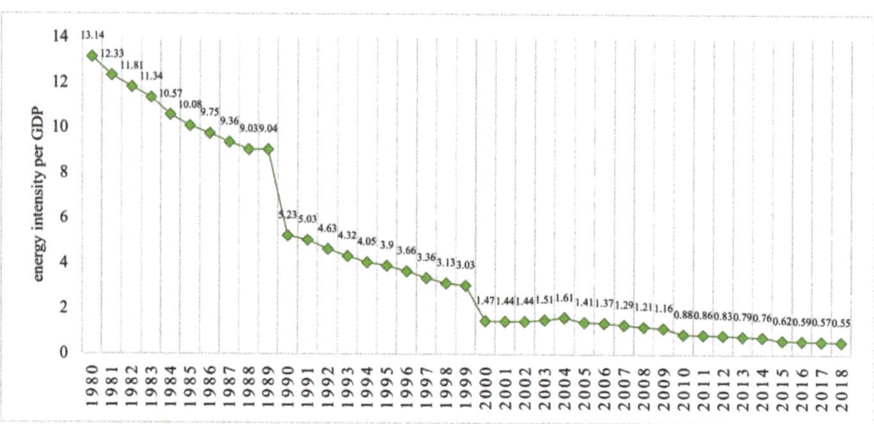

Figure 1. Energy intensity in China from 1980 to 2018. Data source: China economic network.

Many scholars have used various methods to study energy utilization efficiency. Patterson pointed out that energy efficiency itself is a general term [1], which can be measured by a variety of quantitative indicators The measuring methods can be divided into single-factor energy efficiency and total-factor energy efficiency. The former is the reciprocal of energy intensity. Compared with total-factor, the single-factor method is calculated in a simple way, and the alternative effects between various input elements are not taken into account. Total-factor energy efficiency considers the interaction between energy consumption and other factors of production and has more than one input and output indicator. Among various methods, data envelopment analysis (DEA) is widely used to calculate energy efficiency. George Halkos and Kleoniki Natalia Petrou analyzed the efficiency of the 28 EU member states in 2008, 2010, 2012 and 2014 using DEA and directional distance function, among which Germany, Ireland and the United Kingdom were the most efficient countries [2]. Zhongshan Yang, Xiaoxue Wei calculated the energy utilization efficiency of 26 major cities in China by building a multi input-output DEA model [3]. Tengfei Huo, Miaohan Tang et al. used the DEA model and the TFEE algorithm to calculate the actual total factor energy efficiency (TFEE) of the construction industry [4]. Cheng et al. measured the TFEE of 30 provinces in China from 1997 to 2016 based on the nonradial directional distance function of the DEA model and found that TFEE had significant regional heterogeneity [5]. The extended DEA method is also very common in the energy field. Geng et al. proposed a novel DEA model based on the affinity propagation (AP) clustering algorithm (AP-DEA) [6]. Iftikhar et al. used the network DEA method to measure the economic efficiency and distributive efficiency of major economies [7]. Fernández et al. evaluated the productivity and energy efficiency of existing industrial gases facilities through DEA and the Malmquist index [8]. Wen et al. combined a multiregional input-output model with DEA to evaluate the energy efficiency of China's construction industry at the provincial level [9].

From the above literature, it can be found that a variety of input and output factors should be considered when the DEA method is used to calculate energy utilization efficiency. In addition, we also need to pay attention to which social and environmental variables will affect the efficiency value. Scholars generally agree that technological progress and economic structure are two key factors. Zhu et al. held that the rationalization and advancement of industrial structure had a positive impact on the efficiency of green development [10]. The results of Wang et al. showed that the development of 21 industries would decrease the national carbon intensity in China's 28 industries [11]. Xiong, Ma and Ji supported that industrial structure efficiency was a determinative factor to provincial industrial energy efficiency [12]. In addition to industrial structure, openness, marketization and urbanization could also affect energy efficiency. Zhao and Lin constructed a simultaneous equation model and demonstrated positive feedback between foreign trade and energy efficiency in the textile industry [13]. Koengkan conducted a survey on the impact of trade openness on energy consumption in four Andean community countries, and found that economic growth and trade openness had a positive impact on energy consumption [14]. Peng et al. analyzed the overall textile industry energy efficiency gap between China and the United States, and argued that the main factors influencing the energy efficiency of the chemical fiber industry in China included economic structure, energy structure, industry scale and technology [15]. Wang, Shi and Zhang explored the influencing factors of energy efficiency using the Tobit regression model and found that market concentration and foreign direct investment had significantly positive effects on industrial energy efficiency [16]. Zhao et al. used a three-stage DEA model, and the results showed that economic and energy structure, urbanization rate and R&D investment would affect energy efficiency [17]. Morfeldt and Silveira believed that the government's behavior did not contribute to the improvement of the energy efficiency of the industry [18]. Li, Fang and He found that the overall impact of urbanization on energy efficiency was negative in China [19].

We find that previous studies cover countries, regions and industries, but lack analysis of regional differences. Meanwhile, most scholars only consider the economic output measured by GDP, and the choice of output indicators is often single. In addition, most studies focus on the relationship between technological progress, industrial structure and energy efficiency, with insufficient attention paid to openness, marketization and urbanization. This paper differs from the past researches as follows:

(1) We incorporate carbon emissions into the DEA model. In calculating the energy efficiency of regions and four major economic zones, it is considered an undesirable output to measure environmental pollution.
(2) Based on the provincial panel data from 2007 to 2017, the entity fixed effect models in China and its four major economic zones are constructed.
(3) We explore the impact of openness, marketization and urbanization on energy efficiency in the period of economic structure transformation, and put forward some suggestions to improve efficiency.

2. Energy Utilization Efficiency in China

2.1. DEA Method and Index Selection

DEA is proposed by Charnes [20], which is a non-parametric econometrics method based on operational research theory and linear programming technology. Many scholars have measured energy efficiency by the DEA method because of its distinct advantages in evaluating relative efficiency of decision-making units with multiple input and output indicators. The DEA model can be divided into two types: input-oriented and output-oriented. Due to the different situations in different provinces, we choose the input-oriented and changeable scale of the BCC model as the original DEA model (1) to calculate energy utilization efficiency, where n represents the number of decision-making units (DMU), U represents input, V represents output, and the relative efficiency of DMU_0 is a_0.

$$\min a_0 \text{S. t.} \begin{cases} \sum\limits_{i=1}^{n} \lambda_i x_{iu} \leq a_0 x_{0u}, u = 1,2,\cdots,U \\ \sum\limits_{i=1}^{n} \lambda_i y_{iv} \geq y_{0v}, v = 1,2,\cdots,V \\ \lambda_1 + \lambda_2 + \cdots + \lambda_n = 1, i = 1,2,\ldots n \\ \lambda_i \geq 0, i = 1,2,\ldots n \end{cases} \quad (1)$$

In the previous studies, most scholars only considered expected output measured by GDP. However, the use of energy will inevitably bring undesirable outputs, such as environmental pollution, which should also be considered in efficiency evaluation. In this paper, we use carbon emissions to measure environmental pollution. For economic and environmental systems, carbon emissions are unexpected output. In the environmental efficiency evaluation model, there are many processing methods to deal with the undesirable outputs, among which "regarding unexpected output as input" is one of the main ideas. This measure conforms to the basic idea of the DEA model, that is, we expect to get more expected output with less input. Therefore, we regard carbon emissions as an input variable. In addition, under the framework of total-factor production function, labor, capital and energy consumption are regarded as input variables, and GDP is regarded as an output variable.

2.2. Data Sources and Processing

The panel data of 30 provinces in China from 2007 to 2017 are selected as the evaluation units for energy utilization efficiency (considering the availability and completeness of data, Shanxi, Guizhou and Tibet are not included, and the data of Chongqing are incorporated into Sichuan for accounting). The sources and processing of input and output data are as follows:

(1) **Energy consumption.** The total energy consumption of each province is used as the energy consumption index, and the data are from China Energy Statistical Yearbook in 2018.

(2) **Labor.** The number of employees in each province is selected to measure the labor input. The number of current employees = (the number of employees at the end of the previous period + the number of employees at the end of the current period)/2. The data are from CEInet Statistics Database.

(3) **Capital stock.** Perpetual inventory method is adopted to estimate the actual capital stock of each province every year. The calculation method is as follows: $K_{it} = K_{i,t-1}(1 - \delta_{it}) + I_{it}$, where K_{it} is the capital stock of province i in year t, I_{it} is the actual investment of province i in year t, δ is economic depreciation rate. I_{it} is calculated according to the total investment and the fixed asset investment price index of each province. The data are from China Statistical Yearbook, δ takes 5%. The initial value of capital stock $K_0 = \frac{I_0}{\delta+g}$ where I_0 is the actual fixed assets investment in initial year, g is the annual average growth rate of investment in sample period. All data are calculated according to the constant price in 2007.

(4) **Carbon emissions.** It is regarded as an unexpected environmental output indicator for measuring energy utilization efficiency. The energy consumption (coal, coke, crude oil, gasoline, kerosene, diesel oil, fuel oil and natural gas) of each province is multiplied by the standard coal coefficient to obtain the standard consumption. Then multiply it by the carbon emission coefficient, and we can get the carbon emission of energy consumption in each province during 2007–2017. The data are from China Energy Statistics Yearbook 2018.

(5) **GDP.** It is used as an economic output index for measuring energy utilization efficiency. We get the GDP data (calculated by current price) of each province during 2007–2017 from China Statistical Yearbook. Meanwhile, we convert those data by the price in 2007.

2.3. Analysis of Energy Utilization Efficiency

2.3.1. Analysis of Provincial Energy Utilization Efficiency

We use DEAP2.1 software to calculate the energy efficiency of 30 provinces in China during 2007–2017. The results are shown in Table 1. The DEA model can be used to analyze the comprehensive efficiency, pure technical efficiency and scale efficiency. Pure technical efficiency (vrste) refers to the minimum factor cost of input under the maximum output of the same scale, which is calculated under the assumption of variable returns to scale. It can be used to measure the extent to which input factors are caused by pure technical inefficiency. Scale efficiency (scale) refers to the ratio between the output of technical efficiency on the production boundary and its optimal scale output under certain input conditions. The larger the scale efficiency is, the closer the production scale of the production unit is to the optimal production scale. Comprehensive efficiency (crste) refers to the minimum factor cost of required input under the maximum output, which is calculated under the assumption of constant return to scale. The quantitative relationship among them is that comprehensive efficiency = pure technical efficiency × scale efficiency.

First of all, according to Figure 2, the comprehensive energy utilization efficiency of provinces and cities in 2007–2017 is not very ideal. Only Beijing, Shanghai and Guangdong have a comprehensive efficiency value of 1, which constitutes the forefront of China's energy utilization efficiency. Secondly, the average energy efficiency of Tianjin, Zhejiang and Jiangsu reaches 0.8 in 2007–2017, which is relatively effective. Thirdly, the average energy efficiency of 12 provinces, such as Gansu, Yunnan, Ningxia and Qinghai, is below 0.6, so there is an urgent need to improve energy efficiency and find reasonable ways to save energy and reduce emissions. From a horizontal perspective, the energy utilization efficiency of all provinces and cities in China declined from 2007 to 2017. Improving energy efficiency is an urgent problem.

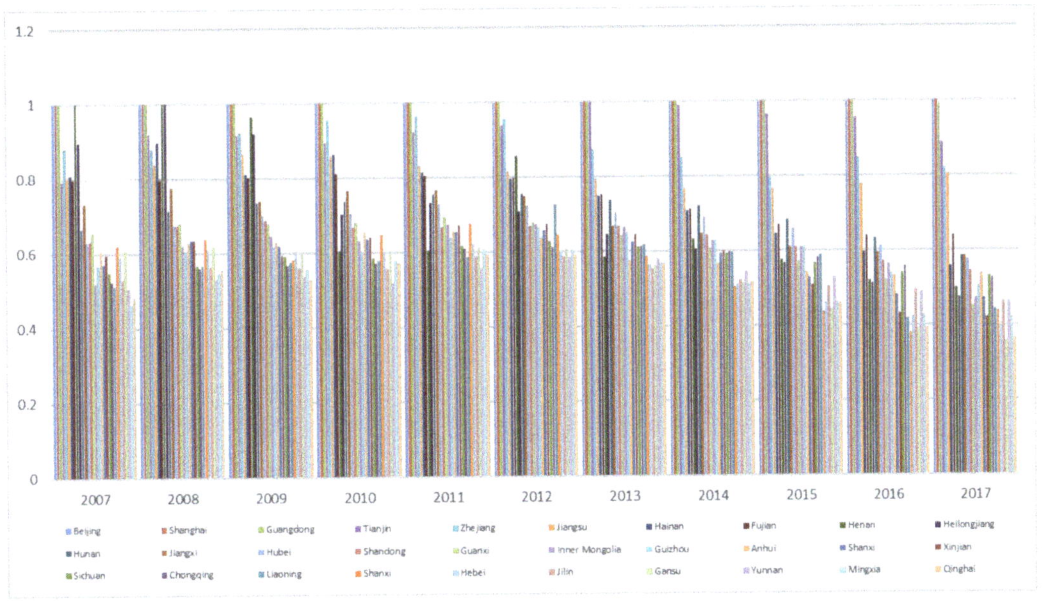

Figure 2. Comprehensive efficiency of China's regions during 2007–2017.

Table 1. Energy efficiency results of DEA in different provinces of China from 2007 to 2017.

		Beijing	Tianjing	Hebei	Shanxi	Inner Mongolia	Liaoning	Jilin	Heilong jiang
	crste	1.00	0.79	0.59	0.62	0.52	0.51	0.53	0.90
2007	vrste	1.00	0.93	0.59	0.65	0.52	0.63	0.53	0.93
	scale	1.00	0.85	1.00	0.95	1.00	0.81	0.99	0.97
	crste	1.00	0.92	0.61	0.64	0.62	0.57	0.56	1.00
2008	vrste	1.00	1.00	0.61	0.67	0.62	0.71	0.57	1.00
	scale	1.00	0.92	1.00	0.95	1.00	0.80	0.99	1.00
	crste	1.00	0.91	0.61	0.58	0.64	0.58	0.56	0.92
2009	vrste	1.00	1.00	0.61	0.61	0.66	0.73	0.57	0.93
	scale	1.00	0.91	1.00	0.96	0.98	0.79	0.99	0.99
	crste	1.00	0.90	0.60	0.65	0.63	0.58	0.56	0.70
2010	vrste	1.00	1.00	0.61	0.67	0.66	0.75	0.56	0.73
	scale	1.00	0.90	0.98	0.96	0.96	0.77	1.00	0.96
	crste	1.00	0.92	0.62	0.68	0.68	0.59	0.59	0.73
2011	vrste	1.00	1.00	0.64	0.70	0.72	0.77	0.59	0.76
	scale	1.00	0.92	0.97	0.96	0.94	0.76	1.00	0.96
	crste	1.00	0.94	0.60	0.65	0.67	0.73	0.59	0.71
2012	vrste	1.00	1.00	0.63	0.67	0.72	0.82	0.60	0.73
	scale	1.00	0.94	0.96	0.96	0.93	0.89	0.99	0.96
	crste	1.00	1.00	0.56	0.59	0.66	0.62	0.55	0.65
2013	vrste	1.00	1.00	0.61	0.61	0.72	0.77	0.57	0.67
	scale	1.00	1.00	0.93	0.97	0.92	0.80	0.97	0.97
	crste	1.00	0.99	0.51	0.50	0.63	0.60	0.52	0.60
2014	vrste	1.00	1.00	0.57	0.52	0.68	0.73	0.55	0.63
	scale	1.00	0.99	0.90	0.97	0.93	0.82	0.95	0.96
	crste	1.00	0.96	0.46	0.44	0.61	0.59	0.50	0.57
2015	vrste	1.00	1.00	0.53	0.44	0.65	0.70	0.54	0.57
	scale	1.00	0.96	0.86	1.00	0.94	0.84	0.94	0.99
	crste	1.00	0.92	0.62	0.68	0.68	0.59	0.59	0.73
2016	vrste	1.00	1.00	0.64	0.70	0.72	0.77	0.59	0.76
	scale	1.00	0.92	0.97	0.96	0.94	0.76	1.00	0.96
	crste	1.00	0.89	0.40	0.44	0.47	0.44	0.46	0.48
2017	vrste	1.00	0.98	0.50	0.45	0.48	0.50	0.48	0.48
	scale	1.00	0.91	0.79	0.98	0.98	0.89	0.96	0.99
	mean	1.00	0.92	0.56	0.59	0.62	0.58	0.55	0.73

		Shanghai	Jiangsu	Zhejiang	Anhui	Fujian	Jiangxi	Shandong	Henan	Hubei	Hunan
	crste	1.00	0.81	0.88	0.61	0.80	0.73	0.63	1.00	0.63	0.67
2007	vrste	1.00	0.90	0.89	0.62	0.80	0.74	0.75	1.00	0.64	0.68
	scale	1.00	0.89	1.00	0.98	1.00	0.99	0.84	1.00	0.98	0.98
	crste	1.00	0.84	0.88	0.63	0.80	0.77	0.67	1.00	0.67	0.72
2008	vrste	1.00	0.96	0.89	0.65	0.80	0.78	0.79	1.00	0.69	0.73
	scale	1.00	0.87	0.99	0.97	1.00	0.99	0.85	1.00	0.98	0.98
	crste	1.00	0.86	0.92	0.63	0.80	0.74	0.69	0.97	0.70	0.73
2009	vrste	1.00	1.00	0.94	0.66	0.80	0.75	0.82	0.98	0.71	0.75
	scale	1.00	0.86	0.98	0.95	1.00	0.99	0.84	0.98	0.98	0.98
	crste	1.00	0.86	0.95	0.65	0.81	0.77	0.67	0.61	0.71	0.74
2010	vrste	1.00	1.00	0.97	0.69	0.81	0.77	0.79	0.62	0.72	0.75
	scale	1.00	0.86	0.99	0.95	1.00	1.00	0.85	0.97	0.99	0.98
	crste	1.00	0.83	0.96	0.66	0.81	0.77	0.67	0.60	0.73	0.75
2011	vrste	1.00	1.00	0.98	0.72	0.81	0.77	0.78	0.63	0.74	0.77
	scale	1.00	0.83	0.98	0.92	1.00	1.00	0.85	0.96	0.99	0.98
	crste	1.00	0.82	0.96	0.64	0.80	0.75	0.67	0.86	0.72	0.76
2012	vrste	1.00	1.00	0.97	0.71	0.81	0.75	0.79	0.87	0.73	0.77
	scale	1.00	0.82	0.98	0.90	0.99	1.00	0.85	0.98	0.98	0.98
	crste	1.00	0.79	0.87	0.58	0.75	0.67	0.67	0.58	0.70	0.74
2013	vrste	1.00	1.00	0.89	0.67	0.78	0.69	0.80	0.65	0.71	0.74
	scale	1.00	0.79	0.98	0.86	0.96	0.96	0.84	0.89	0.99	0.99
	crste	1.00	0.77	0.85	0.57	0.71	0.65	0.64	0.63	0.69	0.72
2014	vrste	1.00	1.00	0.89	0.67	0.76	0.68	0.79	0.66	0.69	0.73
	scale	1.00	0.77	0.96	0.85	0.94	0.95	0.81	0.96	1.00	1.00
	crste	1.00	0.77	0.80	0.54	0.67	0.61	0.61	0.58	0.66	0.68
2015	vrste	1.00	1.00	0.88	0.64	0.76	0.65	0.77	0.64	0.69	0.70
	scale	1.00	0.77	0.91	0.84	0.88	0.93	0.79	0.89	0.96	0.97
	crste	1.00	0.83	0.96	0.66	0.81	0.77	0.67	0.60	0.73	0.75
2016	vrste	1.00	1.00	0.98	0.72	0.81	0.77	0.78	0.63	0.74	0.77
	scale	1.00	0.83	0.98	0.92	1.00	1.00	0.85	0.96	0.99	0.98
	crste	1.00	0.80	0.82	0.54	0.64	0.58	0.54	0.50	0.58	0.59
2017	vrste	1.00	1.00	0.94	0.65	0.77	0.63	0.76	0.67	0.68	0.68
	scale	1.00	0.80	0.88	0.83	0.83	0.92	0.72	0.74	0.85	0.86
	mean	1.00	0.82	0.90	0.61	0.76	0.71	0.65	0.72	0.68	0.71

		Guang dong	Guangxi	Hainan	Chong qing	Sichuan	Guizhou	Yunnan	Shanxi	Gansu	Qinghai	Ningxia	Xinjiang	China
	crste	1.00	0.66	0.81	0.52	0.55	0.57	0.51	0.57	0.61	0.48	0.46	0.60	0.62
2007	vrste	1.00	0.67	1.00	0.53	0.55	0.62	0.53	0.58	0.67	1.00	0.96	0.65	1.00
	scale	1.00	0.98	0.81	0.99	0.99	0.92	0.96	0.99	0.91	0.48	0.48	0.92	0.62
	crste	1.00	0.68	0.90	0.56	0.57	0.61	0.53	0.63	0.62	0.55	0.55	0.63	0.65
2008	vrste	1.00	0.69	1.00	0.57	0.57	0.66	0.55	0.64	0.68	1.00	1.00	0.69	1.00
	scale	1.00	0.98	0.90	0.99	0.99	0.92	0.96	0.99	0.91	0.55	0.55	0.92	0.65
	crste	1.00	0.68	0.81	0.57	0.59	0.62	0.54	0.62	0.60	0.53	0.55	0.59	0.66
2009	vrste	1.00	0.69	1.00	0.57	0.60	0.67	0.56	0.63	0.66	1.00	0.90	0.64	1.00
	scale	1.00	0.98	0.81	0.99	0.99	0.92	0.97	0.99	0.91	0.53	0.62	0.92	0.66
	crste	1.00	0.68	0.86	0.57	0.59	0.60	0.52	0.64	0.60	0.57	0.58	0.64	0.66
2010	vrste	1.00	0.69	1.00	0.58	0.59	0.65	0.53	0.64	0.65	1.00	0.99	0.69	1.00
	scale	1.00	0.99	0.86	0.99	0.99	0.93	0.97	0.99	0.92	0.57	0.59	0.93	0.66
	crste	1.00	0.69	0.81	0.61	0.62	0.64	0.56	0.65	0.61	0.59	0.61	0.67	0.68
2011	vrste	1.00	0.71	1.00	0.62	0.62	0.69	0.58	0.66	0.67	1.00	0.90	0.72	1.00
	scale	1.00	0.99	0.81	0.98	0.99	0.92	0.97	0.99	0.92	0.59	0.67	0.93	0.68
	crste	1.00	0.68	0.80	0.61	0.63	0.67	0.59	0.66	0.61	0.59	0.60	0.68	0.68
2012	vrste	1.00	0.69	1.00	0.62	0.63	0.73	0.61	0.66	0.67	1.00	0.89	0.73	1.00
	scale	1.00	0.99	0.80	0.98	0.99	0.92	0.96	0.99	0.91	0.59	0.68	0.93	0.68
	crste	1.00	0.64	0.75	0.61	0.61	0.65	0.58	0.62	0.56	0.57	0.57	0.64	0.67
2013	vrste	1.00	0.65	1.00	0.62	0.61	0.72	0.61	0.63	0.63	1.00	0.89	0.69	1.00
	scale	1.00	0.99	0.75	0.99	1.00	0.91	0.95	0.99	0.89	0.57	0.64	0.93	0.67
	crste	1.00	0.61	0.71	0.60	0.59	0.63	0.54	0.59	0.51	0.52	0.51	0.60	0.64
2014	vrste	1.00	0.61	1.00	0.60	0.60	0.70	0.57	0.60	0.58	1.00	0.86	0.64	1.00
	scale	1.00	0.99	0.71	0.99	0.99	0.90	0.95	0.99	0.88	0.52	0.60	0.94	0.64
	crste	1.00	0.57	0.65	0.58	0.57	0.61	0.53	0.53	0.44	0.46	0.46	0.51	0.61
2015	vrste	1.00	0.58	1.00	0.59	0.59	0.65	0.54	0.55	0.49	1.00	0.87	0.53	1.00
	scale	1.00	0.99	0.65	0.99	0.96	0.93	0.98	0.96	0.91	0.46	0.54	0.96	0.61
	crste	1.00	0.69	0.81	0.61	0.62	0.64	0.56	0.65	0.61	0.59	0.61	0.67	0.68
2016	vrste	1.00	0.71	1.00	0.62	0.62	0.69	0.58	0.66	0.67	1.00	0.90	0.72	1.00
	scale	1.00	0.99	0.81	0.98	0.99	0.92	0.97	0.99	0.92	0.59	0.67	0.93	0.68
	crste	0.99	0.45	0.56	0.53	0.53	0.50	0.46	0.47	0.35	0.36	0.42	0.42	0.55
2017	vrste	1.00	0.51	1.00	0.58	0.61	0.51	0.47	0.54	0.38	1.00	0.88	0.43	1.00
	scale	0.99	0.88	0.56	0.91	0.88	0.99	0.99	0.87	0.93	0.36	0.48	0.98	0.55
	mean	1.00	0.64	0.77	0.58	0.59	0.61	0.54	0.60	0.56	0.53	0.54	0.60	0.64

We can divide the regions into five echelons according to the comprehensive energy utilization efficiency. The first echelon achieves DEA efficiency, including Beijing, Shanghai and Guangdong. The second echelon is the weak effective echelon, including Tianjin, Zhejiang and Jiangsu, which can strive to move closer to the first echelon. The third echelon includes Hainan, Fujian, Hunan, Jiangxi and Hubei. The comprehensive efficiency of this level is lower than 0.8 but higher than the national overall efficiency, which has promoted the national energy utilization efficiency for a long time. The energy efficiency of the fourth echelon is about the national overall efficiency, including Heilongjiang, Henan, Shandong and Guangxi. These provinces can improve energy efficiency through some structural adjustment. The fifth echelon includes 14 provinces, namely Inner Mongolia, Guizhou, Anhui, Xinjiang, Sichuan, Shanxi, Liaoning, Chongqing, Hebei, Gansu, Jilin, Ningxia, Yunnan and Qinghai. Their comprehensive efficiency has been lower than that of the whole country for a long time. To improve the overall energy utilization efficiency, the fifth echelon is the main target.

From the perspective of average comprehensive efficiency (Figure 3), the general characteristics of energy utilization comprehensive efficiency of 31 regions during 2007–2017 are as follows: the eastern region is larger than the central region, and the central region is larger than the western region. The reasons for the differences may come from three aspects: Firstly, from the perspective of capital stock and labor force, the capital investment in the eastern region is much higher than that in the central and western regions due to the differences in regional economic development level. Secondly, the central and western regions are located in the interior of China, and the development environment is relatively closed, which leads to the low level of input allocation of relevant factors and technology development. Finally, from the perspective of output, the large difference in output among the eastern, central and western regions directly affects the utilization efficiency of input factors.

Figure 3. Distribution of average comprehensive efficiency in China during 2007–2017.

2.3.2. Comparison of Energy Efficiency in Four Economic Zones

According to the standard of National Bureau of statistics, 30 regions are divided into four major economic zones, including eastern, central, western and northeastern regions. The eastern region includes 10 provinces: Beijing, Shanghai, Tianjin, Hebei, Jiangsu, Zhejiang, Fujian, Shandong, Guangdong and Hainan. The central region includes 6 provinces: Anhui, Jiangxi, Henan, Hubei, Shanxi and Hunan. The western region includes 11 provinces: Inner Mongolia, Guangxi, Sichuan, Chongqing, Guizhou, Yunnan, Shaanxi, Gansu, Qinghai, Ningxia and Xinjiang. The Northeast region includes 3 provinces: Liaoning, Jilin and Heilongjiang. The energy efficiency of the four economic zones during 2007–2017 is shown in Table 2.

Table 2. Comprehensive efficiency in four major economic zones during 2007–2017.

	Eastern	Central	Western	Northeastern
2007	1	0.804	0.735	0.817
2008	1	0.831	0.778	0.868
2009	1	0.819	0.78	0.869
2010	1	0.829	0.78	0.885
2011	1	0.849	0.814	0.922
2012	1	0.84	0.822	0.924
2013	1	0.812	0.797	0.889
2014	1	0.796	0.774	0.848
2015	1	0.78	0.745	0.823
2016	1	0.783	0.718	0.681
2017	1	0.791	0.693	0.599

It can be seen from Table 2 that the average energy utilization efficiency in the eastern region is the highest, which has always been in the effective frontier. The comprehensive efficiency in the central region is basically above 0.8, which is weak effective. The efficiency in the western region is relatively weak, hovering around 0.7–0.8 for a long time, and has a downward trend. The efficiency in the northeast region has declined year by year in recent five years, and the efficiency in 2016 and 2017 was even lower than 0.7. From the perspective of scale efficiency and pure technical efficiency (Figure 4), the central region needs to improve scale efficiency and pure technical efficiency, and scale efficiency should be given more attention. The western region also needs to improve both, but the focus is on improving pure technological efficiency. The northeast region is obviously lacking in scale efficiency, but the pure technical efficiency is always in a good state.

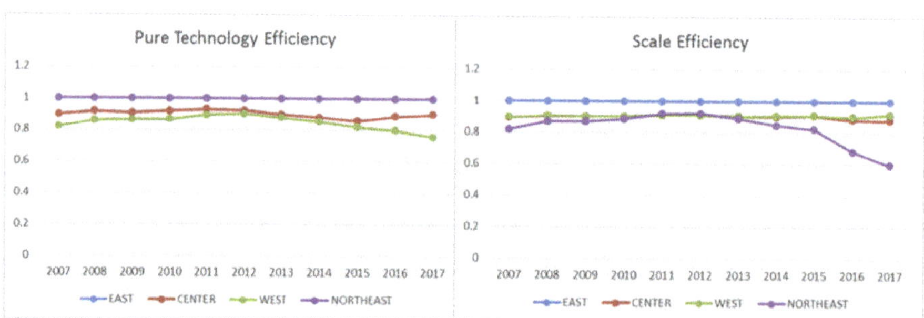

Figure 4. Pure technology efficiency and scale efficiency in four major economic zones.

The energy utilization efficiency in the four economic zones in China is basically in line with the actual economic situation. Compared with the inland areas, the eastern coastal areas have developed earlier and have great advantages in economic output, talent reserve and resource allocation. The economic development of the western region is relatively backward, and scientific and technological means should be adopted to improve energy utilization efficiency. There is still a big gap between Northeast China and Eastern China, and we should improve energy efficiency in terms of scale effect. In addition, there are significant improvements in energy efficiency in different regions compared to inter-provincial energy efficiency. That is to say, the overall energy efficiency difference of China's sub regions is gradually narrowing, but the energy efficiency within regions is extremely unbalanced, resulting in the huge energy efficiency differences among provinces.

3. The Relationship between Openness, Marketization, Urbanization and Energy Utilization Efficiency

3.1. The Definition of Variables

Due to the large gap in the resource status and economic development level of different regions, the time series model would ignore the individual differences among regions, while the cross-sectional data model cannot reflect the dynamic trend of economic change. In order to overcome these two shortcomings, this paper uses panel data model to analyze the impact of openness, marketization and urbanization on energy efficiency. The data are from China Statistical Yearbook and CEInet statistical database. The definition of variables is showed as follows:

(1) **Energy utilization efficiency (EE)**: The result of the DEA model in the previous section is energy efficiency.
(2) **Openness (OP)**: The proportion of total imports and exports in GDP is used to measure the degree of economic openness.
(3) **Marketization (MK)**: The marketization level is measured by (1- local fiscal expenditure/GDP). Generally speaking, the larger the proportion of government fiscal expenditure, the more government intervention, the lower the marketization degree.
(4) **Urbanization (UB)**: The urbanization level is measured by the proportion of urban population in the total population of regions.
(5) **Industrial structure (IS)**: It is measured by the proportion of added value of the tertiary industry in GDP.

3.2. Unit Root Test and Cointegration Test of Panel Data

In order to avoid spurious regression and guarantee the validity of estimation results, it is necessary to test the stability of panel data by a unit root test before regression. If the results of the unit root test have the same order among variables, the co-integration test can be used to study the long-term equilibrium relationship. If there is a long-term stable equilibrium relationship among variables, the regression residual of the equation is stable. On this basis, regression analysis will be more accurate.

There are many unit root test methods, including LLC, IPS, Breitung, Fisher-ADF, Fisher-PP test and so on. In this paper, we mainly use the common root test LLC and the different root tests Fisher-ADF and Fisher-PP. If they all reject the null hypothesis, the sequence is considered to be stationary. Otherwise, it is unstable. Then, the differential test is continued until it is stable. The unit root test result of panel data is shown in Table 3.

Table 3. The result of unit root test.

Methods / Variables	EE	ΔEE	OP	ΔOP	MK	ΔMK	UB	ΔUB	IS	ΔIS
LLC Test	−3.8587 **	−17.175 **	−12.010 **	−26.299 **	−7.4667 **	−13.495 **	−4.3202 **	−17.758 **	−3.7790 **	−21.562 **
Fisher-ADF Test	28.1527	132.902 **	116.860 **	206.809 **	68.4116	140.244 **	56.4507	150.582 **	41.6331	155.626 **
Fisher-PP Test	38.1522	150.277 **	124.424 **	213.927 **	77.1731	293.208 **	78.6062 *	208.198 **	62.1979	158.717 **

Note: **, * indicate that the variable is significant at the level of 1% and 5% respectively, the same below.

The results in Table 4 show that all variables are I (1) process. On this basis, the co-integration test is carried out to test whether there is a long-term stable equilibrium relationship between nonstationary sequences. The Pedroni method is used in this paper and the result is shown in Table 4.

Table 4. The result of co-integration test.

	Statistics	Prob.	Results
Panel v-Statistic	−2.319973	0.9898	Accept
Panel rho-Statistic	4.184402	1.0000	Accept
Panel PP-Statistic	−6.727107	0.0000	Reject
Panel ADF-Statistic	−4.357801	0.0000	Reject
Group rho-Statistic	6.557975	1.0000	Accept
Group PP-Statistic	−12.49921	0.0000	Reject
Group ADF-Statistic	−3.656433	0.0001	Reject

It can be seen from Table 4 that Panel v-Statistic, Panel rho-Statistic and Group rho-Statistic fail to pass the test at the significance level of 5%, indicating that they accept the null hypothesis and there is no co-integration relationship. However, the rest of the statistical data have passed the significance test. Panel ADF-Statistic and Group ADF-Statistic are more suitable for small sample data in the Pedroni test. Since the data used in this paper is relatively small (T < 20) and the above two statistics have passed the significance test, it can be concluded that there is a long-term stable equilibrium relationship among variables.

3.3. Estimation and Analysis of Panel Data Model

We take energy efficiency as a dependent variable, and openness, marketization, urbanization and industrial structure as independent variables to construct a panel data regression model. In this paper, the F test and Hausman test are used to determine the form of the model to avoid model setting errors and improve the effectiveness of parameter estimation.

From the results of F test in Table 5, the null hypothesis should be rejected. In other words, compared with a pooled regression model, we should establish an entity fixed effect model. Similarly, from the results of the Hausman test, an entity fixed effect model should be established compared with an entity random effect model. The model is as follows:

$$EE_{it} = C + \beta_1 OP_{it} + \beta_2 MK_{it} + \beta_3 UB_{it} + \beta_4 IS_{it} + D_1 + D_2 + \cdots D_{30} + u_{it} \quad (2)$$

Table 5. Results of F test and Hausman test.

	Statistics	Prob.
F test	24.573998	0.0000
Hausman test	180.561486	0.0000

The dummy $D_1, D_2, \cdots D_{30}$ variable is defined as:

$$D_i = \begin{cases} 1, (If\ it's\ the\ a - th\ individual) \\ 0, (others) \end{cases} \quad i = 1, 2, 3, \cdots 30;\ t = 2007, 2008, 2009, \cdots 2017$$

In this model, i represents province i and t stands for year t. EE, OP, MK, UB, IS, respectively represent the energy utilization efficiency, openness, marketization, urbanization and industrial structure variable. C is the public intercept; $D_1, D_2, \cdots D_{30}$ stands for individual difference; $\beta_1, \beta_2, \beta_3, \beta_4$ are corresponding coefficients; u_{it} is the random disturbance term.

To avoid heteroscedasticity and autocorrelation, the cross-section weights estimation method is adopted to estimate the entity fixed effects model of panel data. The results are shown in Table 6.

Table 6. Panel data model estimation results 1.

Variable	Coefficient	t-Statistic	Prob.
C	144.8916 **	16.66261	0.0000
OP	−0.182832 **	−10.49266	0.0000
MK	−0.151994 *	−1.978842	0.0488
UB	−0.26349 **	−3.573559	0.0004
IS	−1.071729 **	−16.27322	0.0000
Adj.R^2	0.971921	F-Statistic	310.4761
D-W	0.648526	Prob.(F-Statistic)	0.0000

** 5%, * 10%.

From Table 6, R^2 of this model is 0.9719 and the value of F-statistic is 310.4761. The regression equation is overall significant and its goodness of fit is very high. However, the explanatory variables OP, UB fail to pass the test at the significance level of 5%. The value of D-W statistic is 0.6540, which indicates that there is serious positive autocorrelation. To eliminate the autocorrelation, AR (1) is added to this model. The estimation results are shown in Table 7.

Table 7. Panel data model estimation results 2.

Variable	Coefficient	t-Statistic	Prob.
C	156.3528 **	15.47954	0.0000
OP	−0.04625 **	−4.768227	0.0000
MK	0.165401 *	1.757011	0.0401
UB	−1.268613 **	−9.512252	0.0000
IS	−0.649303 **	−10.03062	0.0000
AR(1)	0.686154 **	26.37959	0.0000
Adj.R^2	0.991604	F Statistic	922.7332
D-W	1.737583	Prob.(F-Statistic)	0.0000

** 5%, * 10%.

From Table 7, R^2 of this model is 0.9916. Its goodness of fit is high and the value of F-statistic is 922.7332. The regression equation is overall significant. All explanatory variables pass the test at the significance level of 5%, and the value of D-W statistic is 1.7375, indicating that autocorrelation has been basically eliminated. Therefore, the final model is as follows:

$$\begin{cases} EE_{it} = 156.3528 - 0.0463OP_{it} + 0.1654MK_{it} - 1.2686UB_{it} - 0.6493IS_{it} \\ \quad +97.0395D_1 + 66.6155D_2 + \cdots - 16.4617D_{29} - 24.9672D_{30} + u_{it} \\ u_{it} = 0.6862u_{it-1} + e_{it} \end{cases} \quad (3)$$

$$D_i = \begin{cases} 1, (If\ it's\ the\ a - th\ individual) \\ 0, (others) \end{cases}$$

$$i = 1, 2, 3, \cdots 30;\ t = 2007, 2008, 2009, \cdots 2017$$

According to the results of regression estimation, openness, marketization, urbanization and industrial structure are important factors affecting regional energy utilization efficiency. Openness has a significant negative impact on energy utilization efficiency, and its regression coefficient is −0.0463, which indicates that if the proportion of total imports and exports in GDP increases by 1%, China's energy efficiency will decrease by 0.0463%. Opening up is a double-edged sword. On the one hand, it can attract foreign advanced technology and experience. It can also produce technology spillovers, promote the transformation of the economic development mode, and improve the environment and efficiency. On the other hand, some high-polluting enterprises will flow into the country with the improvement of openness degree. They can cause serious environmental pollution and reduce energy efficiency. Generally speaking, the environmental pollution caused by economic opening partially offsets the technology spillover and affects the energy efficiency.

Marketization has a positive impact on energy efficiency. The regression coefficient of marketization is 0.1654, which passes the test at the significance level of 5%. If the proportion of local fiscal expenditure in GDP decreases by 1%, energy utilization efficiency in China will increase by 0.1654%. This shows that market-oriented reform is beneficial to the improvement of China's energy efficiency to some extent and promotes enterprises to increase independent innovation. However, the government should also play its role of macro-control and supervision to make up for the failure of the market in environmental governance and promote sustainable economic development.

Urbanization has the greatest negative impact on energy utilization efficiency, and its regression coefficient is −1.2686, indicating that every 1% increase in the urbanization rate will cause a 1.2686% decrease in energy utilization efficiency. It indicates that with the improvement of urbanization level, China's energy use efficiency has not improved, but decreased. The reason may be that the employment, infrastructure construction, transportation, capital and technology pressure brought by population aggregation make the urbanization quality of China's provinces not high. Moreover, many foreign-funded enterprises focus on the cheap labor force and broad market provided by China's urbanization, which is not conducive to the urbanization level to promote technological progress and improve energy utilization efficiency through the spillover effect of foreign-funded technologies.

Industrial structure has a significant negative influence on energy efficiency at the level of 1%. Its coefficient is −0.6493, which indicates that if the proportion of the added value of the tertiary industry in GDP increases by 1%, China's energy efficiency will decrease by 0.6493%. This is not consistent with expected results and conclusions of previous scholars. Why do we get this opposite conclusion? Although the proportion of the tertiary industry in China's industrial structure is increasing year by year, the proportion of the secondary industry is still high, with an average of 40.5%. Besides, energy-intensive industries occupy a large proportion in the secondary industry of China, which offset the improvement of energy utilization efficiency by the tertiary industry. Therefore, we conclude that the impact of industrial structure variables is negative.

In order to further investigate the influence of various factors on different regions, this paper not only carries out a regression analysis on the national panel data, but also makes a corresponding analysis on each region. The regression results of four major economic zones are shown in Table 8.

Table 8. Estimation results of four major economic zones.

Explanatory Variable	China	East	Central	West	Northeast
OP	−0.04625 **	−0.045487 *	−0.003827	0.101132	−0.347372
MK	0.165401 *	−0.015305	0.024308	0.117862 **	−1.372778 **
UB	−1.268613 **	−1.071398 **	−0.717434 **	−1.315832 **	−0.661520
IS	−0.649303 **	−0.797692 **	−0.894773 **	−0.557170 **	−1.495912 **
Adj. R^2	0.991604	0.976768	0.925243	0.924013	0.807909
D-W	1.737583	1.849916	2.090744	2.196564	1.925070

** 5%, * 10%.

As can be seen from Table 8, from a national perspective, openness, marketization, urbanization and industrial structure all have significant impacts on energy efficiency. However, from a regional perspective, their impacts are different and not entirely significant. The energy efficiency of the eastern region is the highest. Openness, urbanization and industrial structure all have a negative impact on its efficiency, while the impact of marketization on efficiency is not significant. Compared with other regions, the economic development level and opening-up degree are relatively high in the east. Meanwhile, by introducing and learning foreign advanced technology and experience, some enterprises with high pollution, high energy consumption and high emissions have flowed into China, which have a negative influence on the environment. For the eastern region, we should not blindly introduce foreign investment, but transfer from "quantity" to "quality". We should build a green and healthy urban environment while pursuing the speed of urbanization and improving infrastructure construction.

In western regions, urbanization and industrial structure have significant negative effects on energy efficiency, while marketization has a significant positive effect. However, the impact of openness is not significant. The high proportion of the secondary industry and energy-intensive industries in the western region offsets the improvement of energy utilization efficiency by the tertiary industry. Therefore, the government should strengthen its support and influence on the western region and promote the adjustment of economic structure and the transformation of the development mode. Meanwhile, it is of great importance to speed up the process of marketization and improve the quality of urbanization.

The impact of urbanization and industrial structure on energy efficiency in Central China is negative. The central region is an inland area, and there is a big gap between the central region and the eastern coastal area in terms of introducing foreign capital. Therefore, we should improve the economic openness degree and learn from foreign advanced technology and management experience. The urbanization process should also be controlled to avoid blind urban agglomeration. Otherwise, it will hinder the optimal allocation of resources and lead to the reduction of efficiency.

Marketization and industrial structure have a significant negative impact on the energy efficiency in Northeast China, with coefficients as high as -1.3728 and -1.4959, respectively, while other variables have no significant impact. As an old industrial base, the proportion of secondary industry in the northeast has been relatively high. There are a lot of energy-intensive industries, which offset the improvement of energy efficiency by tertiary industry and also make the influence of industrial structure negative. In order to improve the energy efficiency in Northeast China, we should not only pay attention to the adjustment of industrial structure, but also accelerate the transformation of the economic development mode and curb the excessive growth of energy-intensive industries.

4. Conclusions and Suggestions

Based on the analysis results, the following conclusions can be drawn:

(1) With the development of the economy and the progress of science and technology, energy utilization efficiency in China has not improved. There is a big gap in energy efficiency among the four major economic zones in China. How to improve energy utilization efficiency and find reasonable ways of energy conservation and emission reduction is extremely urgent.

(2) From the national perspective, the changes in economic structure, including openness, marketization and urbanization, are all important factors affecting energy utilization efficiency. From the regional perspective, the influence of openness, marketization, urbanization and industrial structure are different and not entirely significant. Therefore, regional differences must be fully taken into account when taking measures to improve energy efficiency.

Based on the above conclusions, we put forward the following suggestions for improving energy utilization efficiency in China:

(1) We should continue to take accelerating the transformation of the mode of economic development as a long-term and arduous task. China must strive to break the extensive growth mode of high input and low output, pay more attention to the quality of economic development, energy conservation and emission reduction, and realize the effective utilization of resources. In addition, energy conservation policies and objectives must take into account regional differences.

(2) The four major economic zones should encourage the rational flow of capital, technology, talent and resources, so as to promote the optimal allocation of resources, mutual exchange and cooperation. In order to improve the overall energy efficiency

of our country, the high-efficiency regions should ensure the stable improvement of efficiency and guide the low-efficiency areas to narrow the regional gap.

Author Contributions: W.P. contributed to study design. G.H., W.-L.P. and C.H. collected and analyzed the data. G.H., C.H. and W.-Q.D. interpreted results. W.P. wrote the manuscript. Revision of the manuscript and the final approval of the version to be published: all authors. All authors have read and agreed to the published version of the manuscript.

Funding: This research received no external funding.

Data Availability Statement: All data have been included in this paper.

Acknowledgments: This study is supported by the National Natural Science Foundation of China (NSFC) (Grant no. 71871169, U1933120). The funders had no role in study design, data collection and analysis, decision to publish or preparation of the manuscript.

Conflicts of Interest: The authors declare no conflict of interest.

References

1. Patterson, M.G. What is energy efficiency?: Concepts, indicators and methodological issues. *Energy Policy* **1996**, *24*, 377–390. [CrossRef]
2. Halkos, G.; Petrou, K.N. Assessing 28 EU member states' environmental efficiency in national waste generation with DEA. *J. Clean. Prod.* **2019**, *208*, 509–521. [CrossRef]
3. Yang, Z.; Wei, X. The measurement and influences of China's urban total factor energy efficiency under environmental pollution: Based on the game cross-efficiency DEA. *J. Clean Prod.* **2019**, *209*, 439–450. [CrossRef]
4. Huo, T.; Tang, M.; Cai, W.; Ren, H.; Liu, B.; Hu, X. Provincial total-factor energy efficiency considering floor space under construction: An empirical analysis of China's construction industry. *J. Clean. Prod.* **2020**, *244*, 118749. [CrossRef]
5. Cheng, Z.; Liu, J.; Li, L.; Gu, X. Research on meta-frontier total-factor energy efficiency and its spatial convergence in Chinese provinces. *Energ Econ.* **2020**, *86*, 104702. [CrossRef]
6. Geng, Z.; Zeng, R.; Han, Y.; Zhong, Y.; Fu, H. Energy efficiency evaluation and energy saving based on DEA integrated affinity propagation clustering: Case study of complex petrochemical industries. *Energy* **2019**, *179*, 863–875. [CrossRef]
7. Iftikhar, Y.; Wang, Z.; Zhang, B.; Wang, B. Energy and CO_2 emissions efficiency of major economies: A network DEA approach. *Energy* **2018**, *147*, 197–207. [CrossRef]
8. Fernández, D.; Pozo, C.; Folgado, R.; Jiménez, L.; Guillén-Gosálbez, G. Productivity and energy efficiency assessment of existing industrial gases facilities via data envelopment analysis and the Malmquist index. *Appl. Energy* **2018**, *212*, 1563–1577. [CrossRef]
9. Wen, Q.; Hong, J.; Liu, G.; Xu, P.; Tang, M.; Li, Z. Regional efficiency disparities in China's construction sector: A combination of multiregional input–output and data envelopment analyses. *Appl. Energy* **2020**, *257*, 113964. [CrossRef]
10. Zhu, B.; Zhang, M.; Zhou, Y.; Wang, P.; Sheng, J.; He, K.; Wei, Y.M.; Xie, R. Exploring the effect of industrial structure adjustment on interprovincial green development efficiency in China: A novel integrated approach. *Energy Policy.* **2019**, *134*, 110946. [CrossRef]
11. Wang, F.; Sun, X.; Reiner, D.M.; Wu, M. Changing trends of the elasticity of China's carbon emission intensity to industry structure and energy efficiency. *Energy Econ.* **2020**, *86*, 104679. [CrossRef]
12. Xiong, S.; Ma, X.; Ji, J. The impact of industrial structure efficiency on provincial industrial energy efficiency in China. *J. Clean Prod.* **2019**, *215*, 952–962. [CrossRef]
13. Zhao, H.; Lin, B. Impact of foreign trade on energy efficiency in China's textile industry. *J. Clean. Prod.* **2020**, *245*, 118878. [CrossRef]
14. Koengkan, M. The positive impact of trade openness on consumption of energy: Fresh evidence from Andean community countries. *Energy* **2018**, *158*, 936–943. [CrossRef]
15. Peng, L.; Zhang, Y.; Wang, Y.; Zeng, X.; Peng, N.; Yu, A. Energy efficiency and influencing factor analysis in the overall Chinese textile industry. *Energy* **2015**, *93*, 1222–1229. [CrossRef]
16. Wang, J.; Shi, Y.; Zhang, J. Energy efficiency and influencing factors analysis on Beijing industrial sectors. *J. Clean Prod.* **2017**, *167*, 653–664. [CrossRef]
17. Zhao, H.; Guo, S.; Zhao, H. Provincial energy efficiency of China quantified by three-stage data envelopment analysis. *Energy* **2019**, *166*, 96–107. [CrossRef]
18. Morfeldt, J.; Silveira, S. Capturing energy efficiency in European iron and steel production—comparing specific energy consumption and Malmquist productivity index. *Energy Effic.* **2014**, *7*, 955–972. [CrossRef]
19. Li, K.; Fang, L.; He, L. How urbanization affects China's energy efficiency: A spatial econometric analysis. *J. Clean. Prod.* **2018**, *200*, 1130–1141. [CrossRef]
20. Charnes, A.; Cooper, W.W.; Rhodes, E. Measuring the efficiency of decision making units. *Eur. J. Oper. Res.* **1978**, *2*, 429–444. [CrossRef]

Article

Cybersecurity Policy and the Legislative Context of the Water and Wastewater Sector in South Africa

Masike Malatji [1], Annlizé L. Marnewick [1,*] and Suné von Solms [2]

[1] Postgraduate School of Engineering Management, University of Johannesburg, Auckland Park, PO Box 524 Johannesburg, South Africa; masikem@gmail.com
[2] Department of Electrical & Electronic Engineering Science, University of Johannesburg, PO Box 524 Johannesburg, South Africa; svonsolms@uj.ac.za
* Correspondence: amarnewick@uj.ac.za

Abstract: The water and wastewater sector is an important lifeline upon which other economic sectors depend. Securing the sector's critical infrastructure is therefore important for any country's economy. Like many other nations, South Africa has an overarching national cybersecurity strategy aimed at addressing cyber terrorism, cybercriminal activities, cyber vandalism, and cyber sabotage. The aim of this study is to contextualise the water and wastewater sector's cybersecurity responsibilities within the national cybersecurity legislative and policy environment. This is achieved by conducting a detailed analysis of the international, national and sector cybersecurity stakeholders; legislation and policies; and challenges pertaining to the protection of the water and wastewater sector. The study found some concerning challenges and improvement gaps regarding the complex manner in which the national government is implementing the cybersecurity strategy. The study also found that, along with the National Cybersecurity Policy Framework (the national cybersecurity strategy of South Africa), the Electronic Communications and Transactions Act, Critical Infrastructure Protection Act, and other supporting legislation and policies make provision for the water and wastewater sector's computer security incidents response team to be established without the need to propose any new laws or amend existing ones. This is conducive for the immediate development of the sector-specific cybersecurity governance framework and resilience strategy to protect the water and wastewater assets.

Keywords: cybersecurity; cybercrime; legislation; policy; systems thinking; water

Citation: Malatji, M.; Marnewick, A.L.; von Solms, S. Cybersecurity Policy and the Legislative Context of the Water and Wastewater Sector in South Africa. *Sustainability* **2021**, *13*, 291. https://doi.org/10.3390/su13010291

Received: 11 November 2020
Accepted: 23 December 2020
Published: 30 December 2020

Publisher's Note: MDPI stays neutral with regard to jurisdictional claims in published maps and institutional affiliations.

1. Introduction

Goal 16 of the United Nations' (UN) 17 sustainable development goals is intended to "promote peaceful and inclusive societies for sustainable development, provide access to justice for all and build effective, accountable and inclusive institutions at all levels" [1]. But peace, justice and strong institutions [1] require strengthening coordination among various international and domestic stakeholders. Critical infrastructure protection also requires the strengthening of coordination among international and domestic stakeholders. The United States of America (USA) defines critical infrastructure according to the 2013 Presidential Policy Directive No. 21, as "systems and assets, whether physical or virtual, so vital to the United States that the incapacity or destruction of such systems and assets would have a debilitating impact on security, national economic security, national public health or safety, or any combination of those matters." [2] (p. 37). The study has adopted this as the baseline definition of a critical infrastructure.

An example of a water-specific critical infrastructure is the Latvian water supply and sewerage enterprises association [3] which oversees 27 member organisations [4]. In Austria, there are approximately 5500 water utilities, 1900 community-based utilities, 165 water supply associations and 3400 water supply cooperatives [5]. Having a regularly updated inventory list of such critical infrastructures is a good practice [6]. However, an effective cyberlegislation is not only vital for identifying and classifying but maintaining a country's infrastructure and protecting its citizens [6,7].

In many countries, the water and wastewater supply systems are classified as critical infrastructure as they are vital to national public health and economic security. Thus, prolonged interruptions of such critical infrastructures would naturally result in deteriorating public health and economic losses [5]. It is therefore crucial to understand the cybersecurity policy trends and discussions [7] to ensure proper coordination of cybersecurity activities in a country. This paper explores South Africa's water and wastewater sector cybersecurity responsibilities within the national and international policy context. This highlights how well-defined policy regulations in any country could ensure coordination of stakeholder roles and responsibilities for carrying out water-specific critical infrastructure cybersecurity activities. Thus, failure to define and implement effective cyberlegislation and policies could have devastating impact on the protection of water and wastewater critical infrastructure.

In South Africa, the government gazetted the National Cybersecurity Policy Framework (NCPF) in 2015, which aimed at addressing cyber terrorism, cybercriminal activities, cyber vandalism, and cyber sabotage [8,9]. As the overarching national cybersecurity strategy of South Africa [9], the NCPF provides a governance process and guidelines to respond to cybersecurity threats and attacks against the country [8,9]. In the cybersecurity domain, policies outline the objectives and limitations of a strategy [10] to provide for measures to be put in place for the protection, safeguarding, and resilience of assets [11]. Thus, adopting the most recent cybersecurity technologies is only effective when deployed within the guidelines of a clearly defined and enforceable policy [10]. Since the adoption of the NCPF, South Africa has been actively conducting cybersecurity assessments, audits, and readiness exercises in different public sector entities as part of the implementation of the cybersecurity strategy. Water and wastewater is one such sector that needs to conduct its own cybersecurity assessments, audits, and readiness exercises. Failure to conduct these periodically could increase the risk and intensify severity of a cyberattack to critical water infrastructure [12].

For example, an attacker may use the cyber kill chain—reconnaissance, weaponisation, delivery, exploitation, installation, command and control, and action on their objectives—to gain entry into the victim's environment through the corporate information technology (IT) domain and then move laterally to the operational technology (OT) domain to launch attacks on critical infrastructure [13]. OT is a collective term for industrial control systems (ICSs), supervisory control and data acquisition (SCADA) systems, and other industrial monitoring and control processes [14,15]. ICSs and SCADA systems are essentially the backbone of critical infrastructures worldwide, including water supply systems, electricity grids, and transportation and telecommunication networks [16,17]. A well-documented cyberattack of a water supply system which took three months to detect occurred at the Maroochy water treatment plant in Australia [18]. This cyberattack took place in 2000, when SCADA systems began experiencing loss of communication, false alarms, and loss of pump controllability due to altered configurations [12,13,19]. This resulted in nearly 1 million litres of raw sewage spilling into rivers, parks, and residential areas, causing damage to the environment and costing society a lot of money [14,16,20,21].

The cyberattack example above demonstrates that cybersecurity can significantly affect sustainability. All three pillars of sustainability—social, environmental or ecological, and economic [19]—were impacted. The social pillar was impacted as a result of the raw sewage spillage in residential areas, including the grounds of a hotel [20]. The death of marine life and unbearable stench, as reported by the Australian Environmental Protection Agency [16], shows the extent to which the environmental pillar was affected. Lastly, all these damages cost the Maroochy Shire Council and the state of Queensland money to clean up and rehabilitate the environment. Thus, the economic pillar of sustainability was also greatly impacted upon. It is also clear from this incident that the sustainability pillars can also be viewed as three distinct and yet interacting systems [21]. That is, if one system/pillar is compromised, the other two will be equivalently affected in an attempt to return to the natural state of equilibrium [22,23].

In light of this, the paper aims to contextualise the water and wastewater sector's cybersecurity responsibilities within the national cybersecurity legislative and policy environment of South Africa. This will determine if and whether there is a need to propose any new legislation and/or policies, or amend existing ones, to address the cybersecurity requirements of the sector. A systems thinking method is adopted to achieve the study's aim by examining the interrelationships between the water and wastewater sector and national cybersecurity legislative and policy environments as one system rather than independent and unrelated elements.

This introductory section provides the background and context of the study problem. The rest of the paper is structured as follows: Section 2 outlines the international, national (South Africa), and sector (South African water and wastewater sector) cybersecurity policy and legislative environments; Section 3 describes the systems thinking research methodology adopted in the paper to contextualise the water and wastewater sector's cybersecurity responsibilities within the South African cybersecurity legislative and policy environment; Section 4 presents the results; and Section 5 discusses the findings. The policy recommendations of the study are outlined in Section 6 and the conclusion presented in Section 7.

2. Cybersecurity Policy and Legislative Environment

A cybersecurity policy helps to chart a course of action for ensuring security of cyberspace by defining collective and individual regulatory, legal, technical, behavioural, organisational, and international responsibilities in pursuit of cybersecurity [24,25]. Cybersecurity is therefore a shared responsibility for national governments, economic sectors, and organisations and/or individual digital device end-users [26]. The shared cyber defence responsibilities are usually coordinated by nation states to develop capabilities to achieve cyber resilience, reduce cybercrime, and secure critical national infrastructure while developing industrial and technological resources for cybersecurity [27]. In this section, the researchers reviewed the international, national, and sector (water and wastewater) cybersecurity literature to identify the stakeholders involved and existing policy and legal environment.

2.1. International System

In the digital era, cybersecurity is of paramount importance for economic competitiveness and continuity of trade for organisations of all types and sizes. As the United Nations Economic Commission for Europe (UNECE) [28,29] asserts, cyberthreats cut across any social and economic activities nationally, regionally, and internationally. It is therefore prudent to explore available international cybersecurity cooperation mechanisms for the protection of critical infrastructure, including water and wastewater critical infrastructure. Of particular focus in this section are the key international cybersecurity stakeholders involved, applicable laws, and the challenges encountered when implementing cybersecurity practices.

2.1.1. International Cybersecurity Stakeholders

In the protection of critical water-related infrastructure cybersecurity webinar held on 18 November 2020 by the World Meteorological Organisation [30], it was indicated by one of the UNECE speakers that work encouraging common regulatory frameworks in specific sectors with critical impact on sustainable development is under way at the UN. This includes a report on the sectoral initiative on cybersecurity by the UNECE [28], albeit not one specifically focused on the water-related infrastructure sector. This makes the UN one of the important international cybersecurity cooperation stakeholders. In addition, some of the regional and other international stakeholders relevant to South Africa's cybersecurity endeavours were reviewed in Appendix A and are as follows:

- African Union

- African Network Information Centre
- Council of Europe
- Forum of Incident Response and Security Teams (FIRST)
- International Criminal Police Organisation (Interpol)
- International Telecommunication Union
- Southern African Development Community
- United Nations

The African Network Information Centre is missing in Appendix A and is regarded by Dlamini [31] as a relevant stakeholder on the African continent regarding security of cyberspace. The next section explores some of the available treaties and conventions governing international cybersecurity cooperation and the interrelationships between the stakeholders mentioned above.

2.1.2. International Cybersecurity Laws

The 2001 Budapest Convention, which is the Convention on international cybercrime by member states of the Council of Europe and other non-member states [32], is the first international cooperation mechanism on issues relating to cybersecurity and cybercrime [33]. It attempts to provide signatory states with a common international policy to fight harmoniously against cybercriminals [34]. Of the 47 member states of the Council of Europe, only one—the Russian Federation—has not signed [35], citing infringement of its (internet) sovereignty [36]. Ireland and Sweden are the only two member states that have signed but never ratified [35].

There are several non-member states that have not signed and/or ratified the Budapest Convention. These include countries such as Brazil, Nigeria, and New Zealand. In the Brazil-Russia-India-China-South Africa (BRICS) bloc, only South Africa has signed the Convention but has never ratified [37,38]. Thus, the total number of signatures not followed by ratifications stands at three—South Africa, Ireland, and Sweden—as of 10 November 2020. In addition, the total number of ratifications now stands at 65 [35]. Since accession to the Convention is by invitation only for non-member states such as those in the BRICS bloc, no truly binding international cybersecurity and cybercrimes agreement is currently in place [33]. On the African continent however, the African Union (AU) adopted the AU Convention—Convention on Cyber Security and Personal Data Protection in June 2014 [36,38,39]. According to Coleman [39], the AU Convention provides a framework for personal data protection which member countries may transpose into their domestic legislation but requires at least 15 countries to be ratified and take effect. At the time of writing, the AU Convention had been signed by 14 member countries out of 55, and ratified by 8 [40]. South African has not yet signed the AU Convention.

There has since been other efforts for international cooperation regarding cybersecurity and cybercrimes, such as the UN General Assembly resolution 70/237 adopted on 23 December 2015 [41]; the world summit on the information society's (WSIS) Geneva Plan of Action [42]; Global Cybersecurity Agenda by the International Telecommunication Union [33]; the Open-Ended Working Group based on UN General Assembly resolution 73/27 [43]; and the Group of Governmental Experts (GGE) based on UN General Assembly resolution 73/266 [44]. South Africa is a member of the GGE and, along with 24 other member states, is expected to submit a final report to the UN General Assembly in 2021 [44]. In summary, some of the most pertinent international cybersecurity laws are as follows:

- The Budapest Convention
- The ITU Global Cybersecurity Agenda
- UN General Assembly resolution 70/237
- UN General Assembly resolution 73/27
- UN General Assembly resolution 73/266
- WSIS Geneva Plan of Action

Apart from the Budapest Convention of 2001, none of these international cooperation measures are binding as yet. This leaves the Budapest Convention on international cybercrime as the only treaty that is binding to its member states. Clough [33] (p. 725), however, cautions that the Convention is only effective when all member states have capacity in place to enact "domestic legislation across the spectrum of substantive and procedural laws and to put in place mechanisms for international cooperation." Some of the international cybersecurity implementation gaps and challenges in the water and wastewater sector are explored in the next section.

2.1.3. International Water-Specific Cybersecurity Challenges

It was mentioned earlier that ICSs are essentially the backbone of critical infrastructures worldwide, including of the water and wastewater critical infrastructure. The introduction of cyber connectivity into ICS environments has increased the vulnerability of all types of critical infrastructures to cyberattacks [3,45–47]. Recently, the USA's cybersecurity and infrastructure security agency (CISA) [48] has reported compromises on critical infrastructures, government agencies, and private sector organisations through a third-party contractor network management tool called SolarWinds Orion platform. According to CISA [48], this advanced persistent threat (APT) [49] began approximately in March 2020, with evidence suggesting that there are additional initial access vectors other than the SolarWinds Orion platform. APTs are cyberattacks carried out repeatedly over an extended period of time by actors with significant resources and sophisticated levels of expertise [20].

The Australian and USA critical infrastructure cyberattacks point to supply chain compromises [11,25,50,51]. Some of the challenges of implementing cybersecurity safeguards on critical infrastructures, including the water and wastewater critical infrastructure, are summarised in Table 1.

Table 1. International water-related cybersecurity implementation challenges.

Challenge	Description	Source
Supply chain compromises	Third-party contractors and vendors are used as access vectors to the intended victim's computer networks.	[12,48,52]
Increased cyber connectivity	Introduction of internet communication protocols to industrial control systems (ICSs) exposes them to security risks through the IT domain.	[12,13,53]
False sense of security by obscurity	Older supervisory control and data acquisition (SCADA) systems were isolated from corporate IT networks. With increasing cyber connectivity, they become difficult to secure due to design for safety and performance.	[53,54]
Network misconfigurations	Vulnerable computer network as a result of the misconfiguration of the firewall and related tools.	[45,55,56]
No media protection enforcement	Data theft due to a lack of removable media policy enforcement.	[57]
Unsecured remote access	Remote access to ICSs through untrusted devices, usually by third-party contractors and vendors increases cyber risk.	[53,58]
Undocumented policies and procedures	Undocumented cybersecurity policies and procedures make enforcement and compliance difficult. This inevitably increases organisational cyber risk.	[20,56]
Untrained personnel	Training and awareness of staff achieves significant cybersecurity improvements. The opposite also applies.	[20,59,60]

The above-mentioned challenges of implementing water-related and other critical infrastructure cybersecurity safeguards are mostly at an organisational level [61]. However, government policy and legislation and international cooperation on fighting cybercrime can help deter the would-be attackers in various ways. For example, they can regulate and help improve the information flows, enable collaborative interrelationships, highlight best practices for different sectors, track and monitor emerging cybersecurity technologies, and increase cyber risk awareness and training among citizens [26]. South Africa's national cybersecurity legislation and government policies are reviewed in this regard.

2.2. National System

To develop an effective cybersecurity strategy for the water and wastewater sector, it is prudent to first understand policy discussions at the national level [7]. On 23 March 2012, the NCPF was adopted by the South African Cabinet [36,62–64] and gazetted by the Minister of State Security on 23 September 2015 [65]. As the national cybersecurity strategy, the NCPF has six key objectives that can be summarised as "centralise coordination of cybersecurity activities, by facilitating the establishment of relevant structures, policy frameworks and strategies in support of cybersecurity in order to combat cybercrime, address national security imperatives and to enhance the information society and knowledge-based economy" [65] (p. 15). The NCPF's supporting legislation and policies were reviewed to determine where and how the water and wastewater sector fits in, if at all.

A review of the NCPF has since been done by various other researchers over the years, as detailed in Appendix A of this paper. Appendix A could have excluded all work published prior to September 2015, which was when the NCPF was officially gazetted. This is because, as discussed in later sections, some of the conclusions drawn from such work might currently be invalid or partially valid due to subsequent insertions, substitutions, and/or repeals of some pieces of legislation supporting the NCPF, notwithstanding the mergers and renaming of some government departments. However, it was decided that the essence of the content of some of the previous research work—such as stakeholders involved, coordination structure, and perceived gaps and challenges—remained relevant. Appendix A therefore includes the NCPF review work from 2013 onwards, that is, the period after which the South African Cabinet adopted the NCPF in 2012.

2.2.1. National Cybersecurity Stakeholders

Review work of the national cybersecurity stakeholders was conducted in Appendix A. Stakeholders that are mentioned multiple times in Appendix A are listed once below as either domestic or foreign. All other stakeholders are listed below without exception. It should thus be noted that not all of these are necessarily key stakeholders to the implementation of the national cybersecurity strategy. The domestic stakeholders relevant to the national cybersecurity endeavours as reviewed in Appendix A are as follows:

- State Security Agency (SSA)
    - Electronic Communications Security—Cyber Security Incidents Response Team (ECS-CSIRT)
    - Cybersecurity Centre
- Department of Communications and Digital Technologies (DCDT)
    - National Cybersecurity Hub
    - Cyber Inspectorate
    - National Cybersecurity Advisory Council
- Department of Defence (DoD)
    - Cyber Command
    - Centre Headquarters
- South African Police Service (SAPS)
    - Cyber Crime Centre
- Department of Justice and Constitutional Development
    - National Prosecuting Authority
- Department of Trade, Industry and Competition
- Department of Public Service and Administration
- Department of International Relations and Cooperation
- Department of Science and Innovation
- Public sector Cyber Security Incidents Response Teams (CSIRTs)
- Industry CSIRTs
- State Information Technology Agency

- South African Revenue Service

The key national and domestic stakeholders as defined in the NCPF can be represented, as shown in Figure 1. As shown in Figure 1 and delineated in the NCPF, the key organs of state that play a critical role in the implementation of the cybersecurity strategy [65] are dominated by the Justice, Crime Prevention and Security (JCPS) cluster [66]. According to the Government of South Africa [67], the JCPS cluster is made up of the Presidency, the Ministry of Defence and Military Veterans, the Ministry of State Security, the Ministry of Justice and Correctional Services, the Ministry of Police, the Ministry of Home Affairs, the Ministry of International Relations and Cooperation, the Ministry of Finance, the Ministry of Small Business Development, the Ministry in the Presidency for Women, Youth and Persons with Disabilities, and the Ministry of Social Development. In Figure 1, the bidirectional arrows are not reporting lines. They represent information flow within and outside the national cybersecurity system.

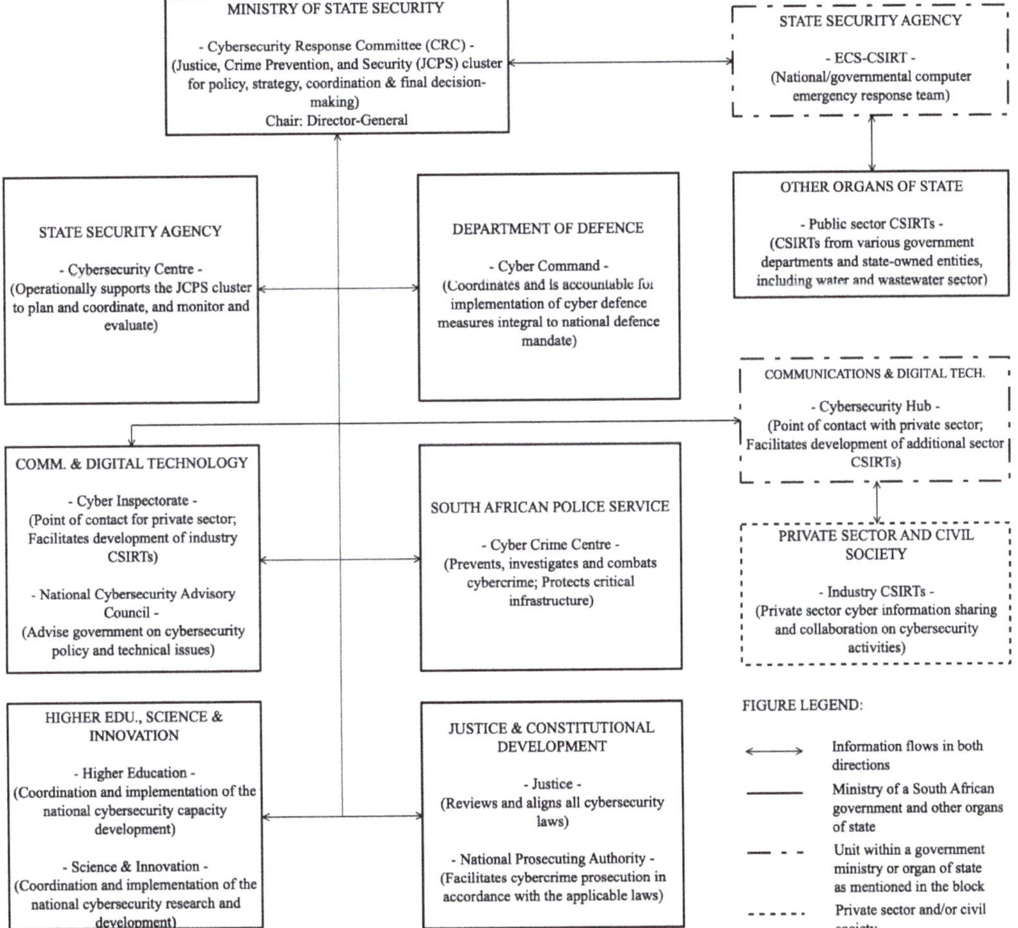

Figure 1. National cybersecurity governance structure in South Africa.

All other organs of state, including but not limited to those listed above, are required to align their cybersecurity and Information and Communications Technology (ICT) policies and practices with the NCPF [65]. Effectively, Figure 1 shows the cybersecurity coordination and management structure in South Africa. The coordination is performed by the JCPS Cybersecurity Response Committee (CRC) [67] that is operationally supported by the Cybersecurity Centre in the SSA [65]. This inter-ministerial coordination is managed and facilitated through various pieces of legislation and government policies.

2.2.2. National Cybersecurity Legislation and Policies

Review work of legislation and government policies used for the implementation of the national cybersecurity strategy was conducted in Appendix A. Similarly, pieces of legislation and policies that are mentioned multiple times in Appendix A are listed once below. All other pieces of legislation and policy are listed below without exception. It is therefore acknowledged that not all of these are necessarily key cybersecurity legislation and policies for the implementation of the national cybersecurity strategy. It is also acknowledged that not all cybersecurity-relevant legislation and policies are reflected in Appendix A. For example, as mentioned in the NCPF [65], the Electronic Communications Security Proprietary (Pty) Limited (Ltd) Act 68 of 2002 was not reflected in the review work in Appendix A. Nonetheless, the legislation and policies relevant to the national cybersecurity endeavours as reviewed in Appendix A are as follows:

- Constitution of the Republic of South Africa of 1996
- Broadband Infraco Act 33 of 2007
- Companies Act 71 of 2008
- Consumer Protection Act 68 of 2008
- Competition Act 89 of 1998
- Copyright Act 98 of 1978
- Corporate Governance of Information and Communications Technology Framework
- Critical Infrastructure Protection Act (CIPA) 8 of 2019
- Cryptography regulations
- Cybercrimes Bill of 2019 (waiting for assent by the President)
- Cyber Warfare Strategy
- Defence Review
- Designs Act 195 of 1993
- E-government strategy and roadmap (national)
- E-government strategy for each province
- Electronic Communications and Transactions Act 25 of 2002 (ECT Act)
- Electronic Communications Act 36 of 2005
- Films and Publications Act 65 of 1996
- Financial Intelligence Centre Act 38 of 2001
- Independent Communications Authority of South Africa Act 13 of 2000
- Inter-Governmental Relations Framework of 2005
- King IV Report on Corporate Governance
- National Archives and Record Service of South Africa Act 43 of 1996
- National Development Plan
- National Cybersecurity Policy Framework
- National Prosecutions Act 32 of 1998
- Prevention of Organized Crime Act 38 of 1999
- Promotion of Access to Information Act (PAIA) 25 of 2002
- Protection of Constitutional Democracy against Terrorism and Related Activities Act 33 of 2004
- Protection of Personal Information (POPI) Act 4 of 2013
- Protection of State Information Bill
- Protection from Harassment Act 17 of 2011
- Promotion of Equality and Prevention of Unfair Discrimination Act 4 of 2000
- Public Service Act: Regulation

- Regulation of Interception of Communications and Provision of Communication-related Information Act 70 of 2002 (or RICA)
- State Information Technology Agency Act 88 of 1998
- Trade Marks Act 194 of 1993

Achievement of the six key objectives of South Africa's national cybersecurity strategy is therefore distributed among 37, and probably more, different pieces of legislation and government policies [37,38]. This is the legal framework for national cybersecurity governance and resilience in South Africa. Harmonising and aligning these [37] could make the currently complex coordination and management of the national cybersecurity endeavours [38] a bit easier. In addition to the Constitution [68], it would appear from Appendix A that seven pieces of legislation and government policies in particular are key to the implementation of the national cybersecurity strategy as they are repeatedly mentioned. These are shown in Figure 2 [65,69–74].

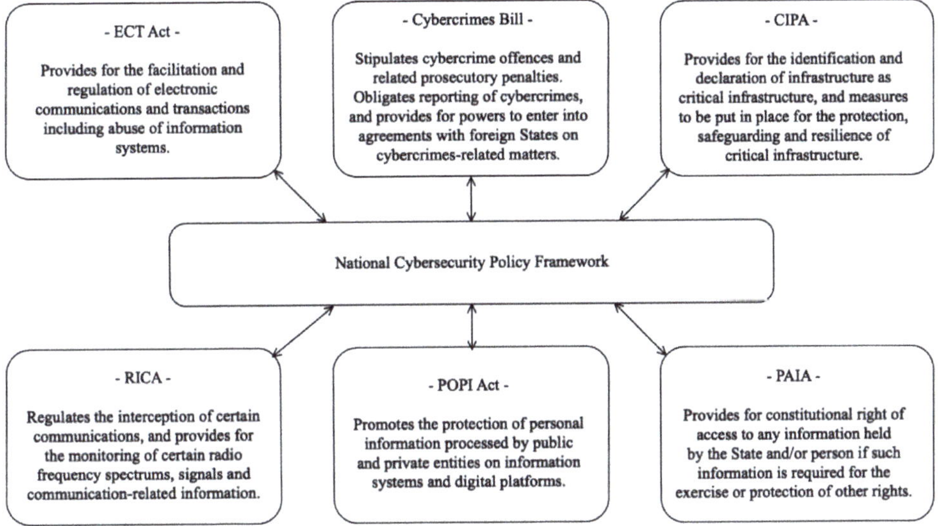

Figure 2. Key national cybersecurity policy and legislation in South Africa.

Review of the six individual pieces of legislation and one policy in Figure 2 revealed that some older laws—those enacted prior to the democratic dispensation in 1994—have since been repealed while others have been amended to respond to changing needs and to align with the country's constitution. It is worth highlighting a few of these in Table 2 as they relate to cybersecurity and cybercrimes in South Africa.

There are many other repeals and amendments but those are beyond the scope of the study. However, as one of the key cybersecurity laws in South Africa, it is imperative to highlight that, as shown in Table 2, sections 85 to 88 (cybercrime offences) of the ECT Act [73] have since been repealed and substituted by sections 2 to 12 of the newly approved Cybercrimes Bill [69]. Moreover, section 89 (cybercrime penalties) of the ECT Act has also been amended as outlined in section 58 of the Cybercrimes Bill. A review of the NCPF also revealed a few implementation gaps and challenges.

2.2.3. National Cybersecurity Challenges

The review work in Appendix A revealed that, apart from the fact that the current coordination and management of the national cybersecurity strategy of South Africa is complex and should be simplified [37,38], a few challenges were identified. Although

Appendix A revealed more than ten gaps and challenges, these can be aggregated into the ten described in Table 3.

Table 2. National cybersecurity legislation amendments and repeals.

Legislation	Current Status
Computer Evidence Act 57 of 1983	Repealed by the ECT Act 25 of 2002.
Copyright Act 98 of 1978	Amended after 1994.
Critical Infrastructure Bill of 2017	Signed into law on 28 November 2019, and it is now the Critical Infrastructure Protection Act 8 of 2019 (Critical Infrastructure Act).
Cybercrimes and Cybersecurity Bill of 2017	Revised and approved as the Cybercrimes Bill by the National Council of Provinces on 1 July 2020.
Monitoring and Prohibition Act 127 of 1992	Repealed by RICA.
National Key Points Act 102 of 1980	Repealed by CIPA.
Sections 85 to 88 of the ECT Act	Repealed and substituted by sections 2 to 12 of the newly approved Cybercrimes Bill.
Section 89 of the ECT Act	Amended as outlined in section 58 of the Cybercrimes Bill.

Some of the challenges in Table 3 are similar to those experienced in other countries, for example, the limited collaboration and information sharing among various sectors and inadequate cybersecurity skills in Turkey [75]. Identifying and classifying critical infrastructure and updating the inventory on a regular basis is a challenge [6]. This is highlighted by White [2] in regards to the USA's Department of Homeland Security's need to develop guidelines to classify critical infrastructure sectors. In the case of Turkey, what [75] found was that if a sector is predominantly managed by private entities, the general cybersecurity posture tends to be more mature, and vice versa. In the case of the USA, however, the Department of Homeland Security is not a private entity. Perhaps cybersecurity issues are not that straightforward as stakeholder roles and responsibilities are often not as obvious, and moreover, the required security levels are also difficult to define [76]. The complex nature of the current coordination and management of the national cybersecurity strategy [37,38] may not be unique to South Africa after all. It is, however, important to understand how the cybersecurity gaps and challenges in Table 3 impact the water and wastewater sector's cybersecurity responsibilities. In this regard, the water and wastewater legal context was reviewed to determine whether and how it addresses protection of the sector's critical cyber infrastructure.

2.3. Sector System

The Constitution of South Africa and specifically the Bill of Rights enshrines the basic human right to have access to adequate drinking water in section 27(1)(b), an environment that is not harmful to human health or well-being in section 24(a), and a healthy and safe environment in section 152(1)(d) [68]. These constitutional rights mandate the state in section 27(2) of the Constitution [68], through the Department of Water and Wastewater (DWS), to ensure that the water resources of the country are sustainably consumed and managed as well as protected [77].

Table 3. National cybersecurity challenges.

Challenge	Description
Poor public-private partnerships track record	There is generally a poor track record of inter-ministerial coordination of government projects. It becomes even complex when stakeholders from industry, civil society, and special interest groups are involved.
Insufficient technical cybersecurity skills and user awareness education in South Africa	Development of technical cybersecurity skills must be prioritised by government. Public user education and awareness are pertinent aspects to preventing spoofing and phishing related cybercrimes in the country.
Independent and uncoordinated cybersecurity awareness initiatives	Currently, disparate and uncoordinated cybersecurity awareness training initiatives do exist. An integrated and coordinated approach to educating the public digital user about the dangers of cyberspace would be more effective.
Missing sector CSIRTs	With the exception of the banking sector which has the South African Banking Risk Information Centre (SABRIC), missing sector CSIRTs refers to the absence of CSIRTs in major sectors of the country, for example, in the mining, aviation, and agricultural sectors. These would be effective in sector information sharing and national coordination of cybersecurity incident responses.
Requirement for the establishment of new and dedicated cybersecurity institutions	The most critical cyber threats in South Africa are to the national critical infrastructure, intelligence agencies, and military. While the military and intelligence agencies are to some degree equipped to tackle cybersecurity, the provincial and local governments as well as the private sector operate and manage the vast majority of the national critical infrastructure. These entities must also be equipped to effectively protect the national critical infrastructure in a coordinated manner. This warrants the establishment of new and dedicated cybersecurity institutions.
Implementation of critical infrastructure protection still in abeyance	Protection of critical infrastructure is key in advanced cybersecurity strategies and must include strategies for cyber resilience and crisis management. Regulations are yet to be promulgated to implement the Critical Infrastructure Act.
Outstanding commitment to existing security conventions	There are no visible commitments to existing conventions such as the Budapest and African Union Convention on Cyber Security and Personal Data Protection. This would help in international collaboration on fighting cybercrimes, capacity building, and information sharing.
Lack of capacity and capability by law enforcement agencies	There is a huge gap between enacted laws and practical enforcement capability on the ground in most emerging and developing countries such as South Africa. This speaks to the point regarding the development of technical cybersecurity skills and user education and awareness.
Missing Cyber Inspectorate unit	A Cyber Inspectorate unit with powers to inspect, search, and seize cyber content in pursuit of unlawful digital acts was never established as clearly delineated in the ECT Act enacted in 2002. This is exacerbated by a poor track record of inter-ministerial coordination of complex government programmes.
International cooperation	South Africa is a non-member state signatory to the Council of Europe's international Convention on cybercrime—the Budapest Convention. However, a clear commitment to the Convention is lacking as it is yet to be ratified since its signing on 23 November 2001.

2.3.1. Water Stakeholders

Two water and sanitation strategic documents were reviewed to identify the stakeholders legally mandated to provide water and wastewater services in South Africa. These are the national water and sanitation master plan [78] and the latest Department of Water and Sanitation (DWS) annual report [77]. In these two documents, the key water and wastewater stakeholders from the public sector and their roles and responsibilities are clearly defined. The following are the identified key stakeholders in the water and wastewater sector of South Africa [77,78]:

- Parliament Portfolio Committee
- National Department of Water and Wastewater
- Regional Department of Water and Wastewater
- Provincial governments
- Local governments (municipalities as water service authorities, or water service providers through subcontractors)
- Water boards/regional water utilities
- Catchment management agencies
- Water-user associations
- Water Research Commission
- Trans-Caledon Tunnel Authority
- Water Tribunal
- Water trading entity

Note that the water boards/regional water utilities, catchment management agencies, water service authorities, water service providers and water-user associations are stakeholder categories that represent many water organisational entities. For example, the water service providers category includes both the public and private sector entities. Thus, the stakeholder categories above are representative of all the key stakeholders in the water and wastewater sector of South Africa. In addition to the stakeholders, the appropriate water legal framework is required for ensuring that the water resources of the country are sustainably consumed, managed, and protected.

2.3.2. Water Legislation and Policies

Sources from [79–82] were reviewed to identify legislation and policies governing the water and wastewater sector of South Africa. Similar pieces of legislation and government policies in the sources were listed once below. All other pieces of legislation and policies are listed without exception below:

- Constitution of the Republic of South Africa of 1996—Chapter 2, sections 10, 24(a), 27(1)(b), 27(2), and 152(1)(d); Chapter 6, section 139(1); Chapter 7, section 154(1); Schedule 4, Part B
- Housing Act 107 of 1997
- National Water Act 36 of 1998
- Water Services Act 108 of 1997
- Water Research Act 34 of 1971
- National Environmental Management Act 107 of 1998
- Local Government: Municipal Structures Act 117 of 1998
- Local Government: Municipal Systems Act 32 of 2000
- Strategic Framework on Water Services of 2003
- Chapter 4 of the National Development Plan
- National Water Policy Review of 2013
- National Wastewater Policy of 2016
- Water and Wastewater Climate Change Policy of 2017
- National Water Resources Strategy, Second Edition, of 2013
- White Paper on Basic Household Wastewater of 2001
- White Paper on National Water Policy for South Africa of 1997

- White Paper on Water Supply and Wastewater of 1994
- National Water and Wastewater Master Plan of 2019

The words "secure", "security" and "protection" were searched in each of the pieces of legislation and policies above. The idea was to determine if and whether provisions for cyber critical infrastructure protection are made. The review revealed water cybersecurity gaps and challenges as discussed in the next section.

2.3.3. Water Cyber Critical Infrastructure Protection Challenges

A review of the legislation and policies identified in the previous section revealed that their purposes are essentially about providing for an integrated water resources management agenda [83]; a technique for planning, monitoring, and managing water resources in a coordinated manner. The legislation and policies contain nothing relating to the protection of critical cyber and physical infrastructure as described in Table 4.

Table 4. Water cyber critical infrastructure protection challenges.

Challenge	Description
National Water Act provides for protection of raw water	This does not refer to the protection of raw water cyber critical infrastructure. Instead, it refers to the planning, monitoring and managing of water resources in a coordinated manner.
The Strategic Framework on Water Services of 2003 provides for protection of water assets	This does not refer to the cyber protection of water assets. Instead, it refers to the repair, maintenance, and rehabilitation of water systems.

Table 4 indicates that the closest reference to some kind of protection is in the National Water Act, which in addition to the protection of raw water in South Africa, provides for the governance of raw water, including the development, consumption, management, and control of aquatic ecosystems [78]. The Strategic Framework on Water Services of 2003 also mentions protection of water assets albeit as it pertains to the repair, maintenance, and rehabilitation of water systems. Therefore, no provision for critical cyber and physical infrastructure protection is made in all the water and wastewater legislation and policies. A review of the existing international, national, and sector (water and wastewater) cybersecurity legislative and policy environments has been conducted in this section. The review identified the national and water and wastewater sector cybersecurity gaps and challenges. What is not clear thus far is how the water and wastewater sector interrelates with the national cybersecurity legislative and policy environment.

2.4. *Systems Interrelationships*

The previous sections discussed three interdependent cybersecurity systems, each with its own unique purpose. These were the international, national, and sector cybersecurity systems. The interdependent relationships between these dynamic systems as well as how they can interoperate effectively is illustrated in Figure 3 as derived from [26].

The arrows in Figure 3 represent cybersecurity information flow within and between the three interdependent systems. Clough [33] indicated that nation states should put in place domestic legislation that is conducive for international cooperation such as the Budapest Convention. Coleman [39] concurs with this and argues that collaborations such as the AU Convention on Cyber Security and Personal Data Protection provide a legal template that could be aligned with but also customised according to domestic legislation and policy requirements. This indicates that the dynamic relationships within and between the three systems are governed by legislation and government policy. While the international and national systems in Figure 3 have clear cybersecurity-related policies and/or legislation, no cybersecurity-related legislation and/or government policy is defined specifically for the water and wastewater sector. By utilising the systems thinking approach, the interrelationships between the water and wastewater sector (sector system in Figure 3) and national cybersecurity legislative and policy environment (national system in Figure 3) were examined further. The research methodology on how to achieve this is described in the next section.

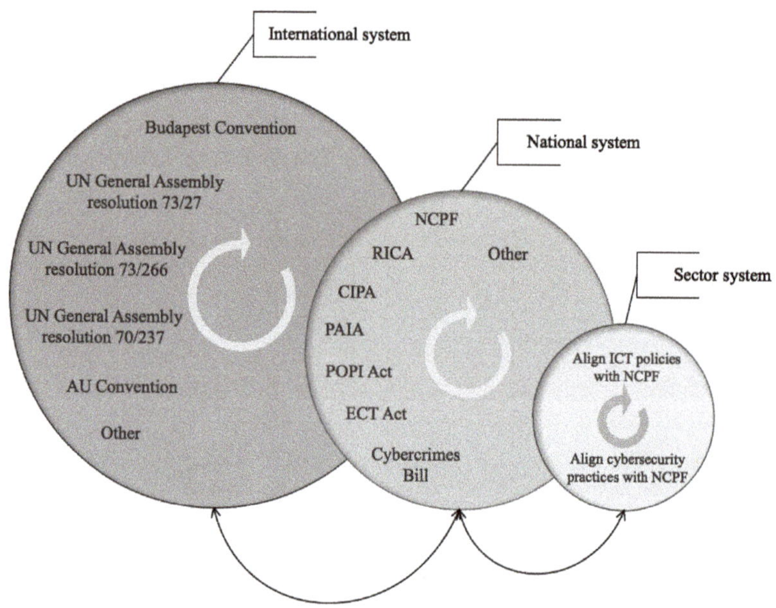

Figure 3. Cybersecurity systems dynamic interrelationships.

3. Materials and Methods

The systems thinking approach [84,85] is employed to achieve the research aim of this study. The approach is deemed suitable as it helps examine dynamic patterns and events by holistically focusing on the interrelationships between a system's parts rather than seeing the constituent parts as static, standalone, and unrelated elements [84,85]. It is an analysis tool to identify and understand how the parts interconnect within the entire system [86]. This is especially useful when considering the complex nature of government policy and the different parties involved in effecting legislation. In this study, a system is perceived as a group of interdependent elements assembled to create an emergent character or behaviour of the group as a whole [22,23,87,88]. As shown in Figure 4, the national cybersecurity strategy of South Africa is considered a system in this study, and its underlying structure comprises three main parts: (i) Function; (ii) Elements; and (iii) Interconnections.

Firstly, the stated function of a system is its purpose, which sets out how that system is expected to behave [87]. Altering the function of a system has the greatest impact on the entire system and may render it unrecognisable [84]. Secondly, the elements of a system are the most visible and are the actors in the system [87]. It is however acknowledged that some elements can be more important than others [84]. Changing system elements has the least impact on a system [84], provided that the function of the system remain unaltered [87]. Thirdly, interconnections are oftentimes harder to see but more critical in the system than elements [84,87]. They are the signals that enable one element of a system to respond to other elements through action or decision points [84]. Oftentimes, interconnections are not physical flows [84,87], but rather the flow of influences, energy, or information inside and outside the system as it strives towards a state of equilibrium [22,23]. The interconnections of a system's elements are configured in such a way as to generate their own characteristic or emergent behaviour, which may start to differ from the espoused or defined purpose [22,84,87]—which is why systems are firm and very difficult to change [89].

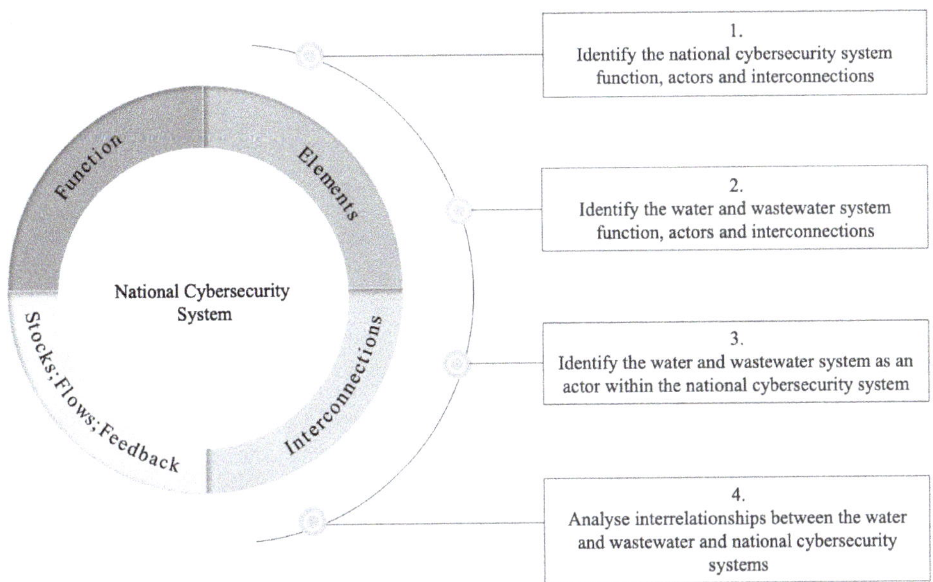

Figure 4. Systems thinking approach.

In addition to system elements/actors, interconnections and function, three more parts make up a system [84]: (i) Stocks, which are the snapshots or historical views of a system, showing the changing flows in the system; (ii) Flows, which are the inflow and outflow activities of a system impacting the levels of stock; and (iii) Feedback loops, which occur when a change—reinforcing or balancing loop [85]—in stock levels leads to additional positive or negative changes [84,87,89,90]. However, these did not form the central aim of the study. To closely examine the interrelationships between the water and wastewater sector and national cybersecurity legislative and policy environment, the four steps in Figure 4 are sequentially operationalised.

Ultimately, the goal of a systems thinking approach is leverage—identifying where changes and concomitant actions in the underlying structure of a system can result in significant and lasting improvements [86]. In the next section, a review of the national and sector cybersecurity literature is conducted to identify the underlying structure of the national cybersecurity system. This should shed light on the key stakeholders and government policies and legislation required to realise significant and lasting improvements to national and, more specifically, water and wastewater sector, cybersecurity endeavours.

4. Results

In this study, South Africa's water and wastewater sector and the national cybersecurity legislative and policy environment were analysed. The analysis was conducted to contextualise the water and wastewater sector's cybersecurity responsibilities within the national cybersecurity legislative and policy environment and determine whether there is a need to propose any new legislation and/or policies, or amend existing ones, to address cybersecurity requirements of the sector. The findings are summarised in Table 5.

In Table 5, the "international cybersecurity system" means the international laws and stakeholders on fighting cybercrime, and the "national cybersecurity system" means the South African cybersecurity legislative and policy environment inclusive of key stakeholders. Similarly, the "water and wastewater sector as a system" means the water and wastewater legislative and policy environment inclusive of the sector's key stakeholders, and the "water and wastewater sector as a stakeholder" means the sector as one of the key stakeholders within the national cybersecurity system. The findings in Table 5 are discussed in the next four sections.

Table 5. Summary of study findings.

	Cybersecurity Purpose (System Function)	Cybersecurity Stakeholders (System Elements/Actors)	Cybersecurity Legislation and Policies (System Interconnections)
International cybersecurity system	Defined	Partially defined	Partially defined
National cybersecurity system	Defined	Defined	Defined
Water and wastewater sector as a system	Not defined	Not defined	Not defined
Water and wastewater sector as a stakeholder	Defined	Defined	Defined

4.1. Identify the National Cybersecurity System Function, Actors and Interconnections

The purpose of this analysis exercise was to identify key national cybersecurity stakeholders (actors) responsible for the implementation of the six key objectives of the national cybersecurity (function), as well as to identify legislation and policies (interconnections) governing the interrelationships among stakeholders. The function of the national cybersecurity strategy has already been defined in Section 2.2 as to "centralise coordination of cybersecurity activities, by facilitating the establishment of relevant structures, policy frameworks and strategies in support of cybersecurity in order to combat cybercrime, address national security imperatives and to enhance the information society and knowledge-based economy" [65] (p. 15). On the one hand, the national cybersecurity strategy function is implemented by domestic stakeholders such as the SSA, SAPS, and DCDT supported by foreign stakeholders such as the African Union, Interpol, and FIRST. The national cybersecurity stakeholders are the defined actors or elements of the national cybersecurity system.

On the other hand, six key pieces of legislation—such as the ECT Act, Cybercrimes Bill, and POPI Act—and one policy, the NCPF, were found to determine the interrelationships among the stakeholders in the national cybersecurity system. These are the interconnections of the national cybersecurity legislative and policy environment. As argued by Sutherland [38] and Detecon [37], the current coordination and management of the national cybersecurity programme is complex. To demonstrate how complex the current implementation of the national cybersecurity strategy is, a few gaps and challenges were identified in the national cybersecurity legislation and policy environment. These are summarised as follows:

- Subsections 16.4(b) and 16.4(c) of the NCPF mandate the DCDT to establish the National Cybersecurity Advisory Council and Cybersecurity Hub, which in turn is tasked to encourage and facilitate the establishment of industry CSIRTs, whereas Chapter 12 of the ECT Act mandates the same government department to establish a Cyber Inspectorate unit and appoint cyber inspectors. Firstly, no Cyber Inspectorate unit has ever been established and no cyber inspectors were ever appointed to date. Secondly, except for the banking industry, which has SABRIC, there are few other industry CSIRTs, even those are not actively coordinated for information sharing and incidents recording in a national database. Lastly, the National Cybersecurity Advisory Council is non-existent or at least its activities, if any, are not visible.
- The NCPF recognises and encourages cybersecurity education for technical skills development, user awareness campaigns, and research and development in Section 2.7 of the policy. However, there are no visible and coordinated nation-wide activities to address insufficient technical cybersecurity skills and user awareness campaigns in the country.

- The CIPA provides for infrastructure resilience, albeit without explicitly stating whether this includes cyber resiliency. Moreover, the SAPS is yet to develop regulations to implement the Act.
- Despite the existence of the different pieces of cybersecurity-related legislation and policies, there seems to be a lack of capacity and capability by law enforcement agencies in fighting cybercrimes in South Africa.

4.2. Identify the Water and Wastewater System Function, Actors and Interconnections

The purpose of this analysis exercise was to identify all the important stakeholders (actors) for the provision of quality water and wastewater services as well as cyber protection of the water infrastructure (function), which legislation and policies (interconnections) are responsible for the functions, and whether these delineate cybersecurity-related roles and responsibilities. On the one hand, the key stakeholders, such as the DWS, water boards and Trans-Caledon Tunnel Authority responsible for the provision of quality water and wastewater services, were identified in Section 2.3.1. On the other hand, pieces of legislation, such as the National Water Act, Water Services Act and Water Research Act, and policy, such as the National Water and Wastewater Master Plan, were identified in Section 2.3.2. These determine the interrelationships among the stakeholders in the water and wastewater sector for the provision of quality water and wastewater services. However, further analysis revealed that no cybersecurity-related roles and responsibilities are defined in the water and wastewater sector legislation and policies. This means that the water and wastewater sector is what SEBoK Editorial Board [88] refers to as an independent system (see sector system in Figure 3) comprised of its own components configured in such a way as to achieve its unique purpose within the national system.

4.3. Identify the Water and Wastewater System as an Actor in the National Cybersecurity System

The purpose of this analysis exercise was to identify which of the national cybersecurity stakeholders represent the water and wastewater sector. Analysis revealed that the *Public sector CSIRTs* in the 'OTHER ORGANS OF STATE' block in Figure 5 represents the water and wastewater sector as an actor or stakeholder within the bigger national cybersecurity system. Moreover, all national, provincial, and local government departments as well as state-owned entities are also represented by the public sector CSIRTs. As shown in Figure 5, the public sector CSIRTs have a direct interconnected relationship with the ECS-CSIRT located in the SSA.

According to Sutherland [38], the ECS-CSIRT is actually Electronic Communications Security (Pty) Ltd. or COMSEC Pty Ltd., a private enterprise established in 2002 and mandated by the SSA to ensure protection of critical electronic communications. Like many other public sector and industry CSIRTs, the water and wastewater sector CSIRT is yet to be established. Since no cybersecurity-related roles and responsibilities are defined in the water and wastewater legislative and policy environment, only one option is left: the national cybersecurity legislative and policy environment. To determine whether and how the existing national cybersecurity legislative and policy environment delineates the water and wastewater cybersecurity responsibilities, the interconnected relationships between the two systems were analysed.

4.4. Analyse Interrelations between the Water and Wastewater and National Cybersecurity Systems

The purpose of this analysis exercise was to determine if and whether the existing national cybersecurity legislation and government policies delineate water and wastewater cybersecurity role and responsibilities. It was found that the water and wastewater legislation and policies give no provision for the sector's critical cyber and physical infrastructure protection. Instead, analysis revealed that the cybersecurity roles and responsibilities to provide for the sector's critical cyber and physical infrastructure protection, and indeed those of other sectors, are drawn mainly from the NCPF [65], Cybercrimes Bill [69], CIPA [70], POPI Act [71], RICA [72], ECT Act [73], and PAIA [74]. For example, the NCPF states that the SSA shall, among other things, be required to "initiate and lead a process" [65] (p. 27) for the establishment of public sector CSIRTs while the Cybersecurity Hub at the DCDT should do the same with private sector CSIRTs and civil society stakeholders [65] (p. 18).

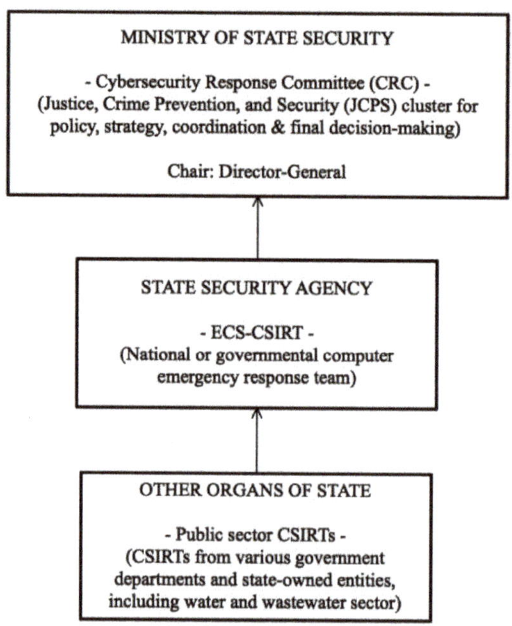

Figure 5. Water and wastewater system as an actor within the national cybersecurity system.

It has already been established in the previous section that the water and wastewater sector is represented by the public sector CSIRTs block in the national cybersecurity governance structure. The cybersecurity roles and responsibilities of sector CSIRTs are delineated in Section 6.3.6 of the NCPF and require, among others, that sector CSIRTs "establish national security standards and best practices for the sector in consultation with the Cybersecurity Centre (located in the Ministry of State Security) and the JCPS CRC, which are consistent with guidelines, standards and best practices developed in line with the NCPF" [65] (pp. 18–19). Along with other defined roles, this role interconnects the water and wastewater sector as an actor with other stakeholders or actors/elements inside and outside the national cybersecurity system to achieve the nation's function or purpose of securing against cyberattacks. Additionally, cybercrimes and concomitant penalties from such cyberattacks are defined in the Cybercrimes Bill and ECT Act as supported by other mentioned key legislation and policies. These are the interconnections of the national cybersecurity and water and wastewater systems. Therefore, the water and wastewater system's cybersecurity purpose, stakeholders, and legislation and policies are only defined when the sector is an actor—public sector CSIRT—within the national cybersecurity system. The ramifications of these findings as they pertain to the aim of the study are therefore discussed in detail.

5. Discussion

The aim of this study was to contextualise the water and wastewater sector's cybersecurity responsibilities within the national cybersecurity legislative and policy environment. To achieve the aim, systems thinking was adopted to analyse the purpose or function of both the national cybersecurity and water and wastewater systems, stakeholders involved to achieve the functions, and stakeholder interrelation. The ramifications of the study findings are discussed under two headings: (i) National cybersecurity legislative and policy environment; and (ii) Water and wastewater legislative and policy environment.

National cybersecurity legislative and policy environment. The study findings indicate that the function of the national cybersecurity system is clearly defined in the NCPF. The purpose of the national cybersecurity strategy is therefore very clear. According to Meadows [84], altering the function of a system has the greatest impact on the entire system and may render it unrecognisable. This means that changing the purpose of the national cybersecurity strategy has the greatest impact on the entire national cybersecurity programme. The findings also indicated that the JCPS CRC was established to oversee the implementation of the national cybersecurity strategy by ensuring consistency with guidelines, standards and best practices developed in the NCPF. The JCPS CRC is the key stakeholder or element/actor in the national cybersecurity system. Although it is acknowledged that some key stakeholders can indeed be more important than others [84], systems thinking indicates that changing individual stakeholders should have the least impact on the national cybersecurity programme provided that the purpose and legislation and policies remain unaltered. This means that stakeholders implementing the national cybersecurity strategy, including individual members of the JCPS CRC, can be changed without having a noticeable impact on the overall purpose of the programme.

Furthermore, the findings indicated that the flow of information among and between the national cybersecurity stakeholders is governed by legislation and policies such as the Cybercrimes Bill, CIPA, ECT Act, NCPF, POPI Act, RICA, and PAIA. In terms of international cybersecurity cooperation, South Africa is yet to ratify the Budapest Convention of 2001 as of 10 November 2020 [35]. That leaves Interpol and extradition treaties between South Africa and other countries as the only available international cooperation mechanisms to fight cybercrimes perpetrated outside its jurisdiction. Systems thinking indicates that each legislation and/or policy interconnects stakeholders in such a way that it could generate its own characteristic or emergent behaviour, which may start to differ from the espoused or defined purpose of the national cybersecurity strategy. This means that amending or repealing cybersecurity-related legislation and government policy could have significant impact on the overall purpose and performance of the national cybersecurity programme. This is why it was important to dig deeper to understand the interconnected relationships among the stakeholders involved and the impact these relationships have on the overall purpose and performance of the national cybersecurity programme. What the findings show is that a seamless coordinated effort is required to implement the national cybersecurity strategy. The argument that government has a below par performance record when it comes to the implementation of policies involving several government stakeholders and requiring public-private partnerships [91] is not encouraging. It was also found that the no less that 37 different pieces of legislation and policies led to further implementation gaps and challenges. The ramifications of these gaps and challenges, which also impact on the water and wastewater sector's cybersecurity responsibilities, are fourfold.

Firstly, since the enactment of the ECT Act in 2002, the DCDT has failed to establish the Cyber Inspectorate unit and appoint cyber inspectors, failed to report any activities by the National Cybersecurity Advisory Council, if any, and progresses slowly to ensure the establishment of industry and sector CSIRTs as stipulated in the NCPF since it was gazetted in 2015. All these shortcomings point to a lack either of capacity or capability by the DCDT, or a combination of both.

Secondly, tasked to be the national structure dedicated to cybersecurity activities, including cybersecurity technical skills and user awareness campaigns and engagement with the private sector and civil society, the DCDT's Cybersecurity Hub is visibly absent in the coordination of these activities. As already alluded to by Detecon [37] and corroborated by Gcaza [92], cybersecurity awareness and education have proven to be effective in significantly reducing the risk of a security breach. This is because awareness and education prepare technical experts to put proactive safeguards in place, and ordinary end-users to be consciously alert. The case in point on the importance of cybersecurity awareness and education is the data breach at Experian South Africa, a credit records organisation, where a database containing personal details of approximately 24 million consumers and nearly 800,000 businesses was willingly handed over to a fraudster [93] as a result of a social engineering attack. Thus, the national government, and in particular the water and wastewater sector, should develop a strategy to embark on a coordinated effort to achieving the required sector cybersecurity skillset. This investment is fully supported and encouraged in Section 2.7 of the NCPF. This lack of visible and strategic coordination by the Cybersecurity Hub also points to a lack either of capacity or capability within the DCDT.

Thirdly, the regulations to promulgate the CIPA had not yet been gazetted by the SAPS at the time of writing. In terms of the transitional arrangements in the Act, Parliament must first approve the SAPS draft regulations. Until that happens, the Act is held in abeyance [94]. In this regard, it is not yet clear which national assets per sector, including the water and wastewater sector, will be identified and classified as national critical infrastructure. Perhaps when the CIPA regulations are gazetted, the roles, responsibilities, and accountability of different parties will be defined to also include cyber resilience. As argued by Mutemwa [66], a good cybersecurity strategy should also include cyber resilience in addition to cyber defence policies and capabilities. A cyber resilience strategy helps shift from a retroactive to a more proactive approach [95]. As matters currently stand, the CIPA merely promises to enable the protection and safeguarding of critical infrastructure to achieve resiliency. How that critical infrastructure resilience is going to be achieved with cooperation between government and the private sector remains unclear.

Lastly, the findings suggest a clear lack of capacity and capability by law enforcement agencies in fighting cybercrimes in the country. This might require a coordinated cybercrimes skills development collaboration programme with international stakeholders such as Interpol and similar others to help bridge the gaps in the short term. In addition to all the matters considered above relating to the national cybersecurity legislation and policy environment, there is another concern: It would appear that the national cybersecurity strategy is primarily more defensive [8], and thus retroactive, than offensive which requires proactiveness [96]. It is more passive and static than proactive. Under international laws, any sovereign state has the right to defend itself against adversarial actors [96]. As the national cybersecurity policy overarching both the DoD's Defence Review and Cyber Warfare Strategy, the NCPF does not explicitly state whether South Africa would execute cyber offence strategies in response to a cyberattack. Even in its delineation of the role and responsibilities of the DoD, the NCPF refers to the development of a "Cyber Defence Strategy, that is informed by the National Security Strategy of South Africa" [65] (p. 24). Defence (retroactive approach) seems to be our cybersecurity strategy as opposed to adopting an offensive (proactive approach) or a combination of both strategies.

In spite of these national cybersecurity challenges, the Cybercrimes Bill, CIPA, ECT Act, NCPF, POPI Act, RICA, and PAIA, together with other cybersecurity-relevant legislation and policies, are drafted in such a way as to address the cybersecurity requirements of the water and wastewater sector without the need to propose any new legislation and/or policies or amend existing ones. All the sector needs to do is to encourage member organisations to align their ICT policies and cybersecurity practices with the NCPF to address cyber risks and water-related cybersecurity implementation challenges such as those highlighted in Table 1.

Water and wastewater legislative and policy environment. The study findings indicate that the water and wastewater sector has two functions fulfilled through two different stakeholder responsibilities. The first function is that the water and wastewater sector is mandated to supply quality water and wastewater services to the nation. This function or purpose is achieved through the water and wastewater sector as an independent system comprised of its own stakeholders (system elements/actors)—such as DWS, water boards, and Trans-Caledon Tunnel Authority)—and legislation and policies (interconnections) —such as the National Water Act, Water Services Act, and National Water and Wastewater Master Plan. The second function is that the water and wastewater sector has national cybersecurity responsibilities. This function is achieved by the water and wastewater sector as a stakeholder—public sector CSIRT—in the bigger national cybersecurity system. The public sector CSIRT cybersecurity responsibilities of the water and wastewater sector are defined in Section 6.3.6 of the NCPF [65].

The findings also indicated that the public sector CSIRT will report to the national CSIRT or ECS-CSIRT in the SSA. It is not clear whether the ECS-CSIRT caters for both corporate IT and ICS cybersecurity services nor how, specifically, it helps the public sector CSIRTs as it claims on its website. The roles and responsibilities defined in the NCPF [65] (pp. 18–19) further require that the Cybersecurity Centre located in the SSA be consulted by public sector CSIRTs when establishing national security standards and best practices for their sectors. The question is, what is the relationship between the Cybersecurity Centre and ECS-CSIRT, both located in the SSA? Is COMSEC (Pty) Ltd. now the Cybersecurity Centre? Are they different? To reiterate Sutherland's [38] point, perhaps this is what contributes to the complex manner in which the national cybersecurity strategy of South Africa is being implemented. Nonetheless, it has already been proven that the existing national cybersecurity legislative and policy environment provides for the establishment of the water and wastewater sector-specific CSIRT without the need to propose any new laws or amend existing ones. However, this is based on the assumption that the DWS will host the CSIRT on behalf of the entire sector. Whether this is the best way to do it is a separate discussion. Alignment of the sector's ICT policies and cybersecurity practices with the NCPF is enough to establish a CSIRT that will be hosted at the DWS.

By understanding the dynamic nature of its interconnected relationships [23,85,97] among various stakeholders, the water and wastewater sector is therefore immediately able to develop its own cybersecurity governance framework and resilience strategy as illustrated in Figure 6.

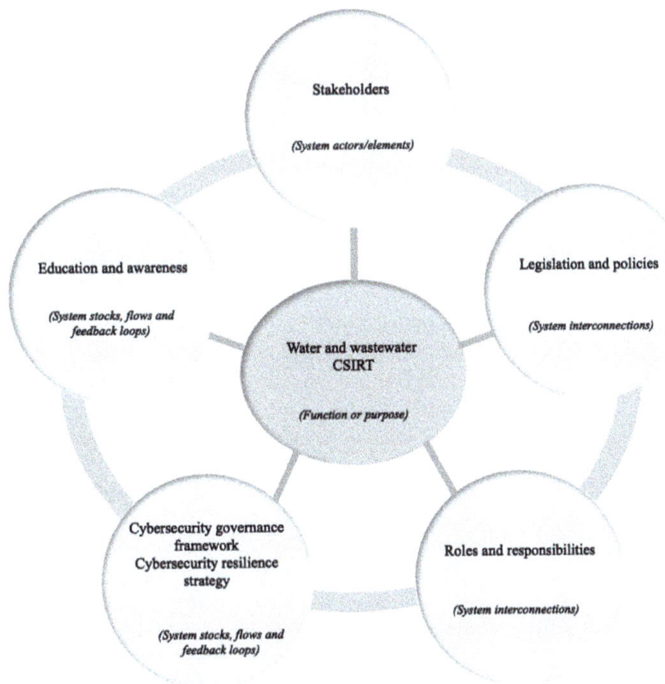

Figure 6. Water and wastewater cybersecurity system.

De Jong et al. [98] assert that outsiders usually offer creative and innovative policy inputs that can lead to a better understanding of societal challenges. This approach yields better policy decisions with more realistic judgements of the advantages and disadvantages of potential policy measures [98,99]. The water and wastewater sector should therefore be as collaborative with "outsiders" such as the JCPS CRC, Cybersecurity Hub in the DCDT, and Cybersecurity Centre in the SSA and as representative (among its member organisations) as possible in order to attain, through better policy decisions, the desired level of sector cybersecurity resiliency against cyber threats and attacks. In this regard, policy recommendations are proposed as outlined in the next section.

6. Recommendations

The study has a few recommendations regarding the national cybersecurity legislation and policy environment and the water and wastewater sector's cybersecurity responsibilities within this legal context. Firstly, regarding the national cybersecurity legislation and policy environment, the following are recommended:

- The National Cybersecurity Advisory Council, and/or Cybersecurity Hub, and/or Cyber Inspectorate unit should either be moved from the DCDT, or their operating models and mandates be reviewed, or a combination of both.
- The Critical Infrastructure Protection Act should be amended to explicitly include "cyber" and/or "digital or information" infrastructure in its definitions of "infrastructure" and "critical infrastructure" terms.
- To boost capacity and capability in fighting cybercrimes in the sort-term, South African law enforcement agencies may need to partner with international stakeholders such as Interpol and similar others to develop cybercrimes and digital forensics skills. For medium to long term solutions, the law enforcement agencies should recruit the best and brightest students with passion and a keen interest in cybercrimes and digital forensics from local universities.

Lastly, regarding the water and wastewater sector's cybersecurity responsibilities within the national cybersecurity legislation and policy environment, the following are recommended:

- *Establish a sector computer security incidents response team.* Establish the national water CSIRT that will have specialist teams serving both the IT and ICS cybersecurity requirements to help formulate and implement the cybersecurity governance framework, resilience strategy, and education and awareness campaigns. Although the establishment of the CSIRT to be hosted at the DWS requires no development of new legislation and/or policies or amendments of existing ones, the authors recommend that a sector-specific agency be established. This would indeed require either the development of a new piece of legislation or amendment of the CIPA and probably the National Water Act. The rationale behind this recommendation is based on international best practices where it would appear that sector-specific agencies for each classified critical infrastructure sector are the best way to look after the cybersecurity requirements of a sector.
- *Develop a sector cybersecurity governance framework.* Probably most of the sector stakeholders have a cybersecurity governance framework at organisational level based largely, if not solely, on corporate IT security requirements. Such stakeholders merely need to align these with the NCPF as stipulated in Section 16.7 of the policy and incorporate ICS cybersecurity requirements where applicable. At sector level, a governance framework would help with facilitating the exchange of cybersecurity information, sharing of knowledge and collaboration, skills development, and rapid responses to incidents.
- *Develop a sector cybersecurity resilience strategy.* Cybersecurity resilience refers to a critical infrastructure's capability to anticipate, withstand, adapt and/or rapidly recover from any cyber terrorism, cybercriminal activities, cyber vandalism, cyber sabotage, accidents, or naturally occurring threats or human error induced infrastructure failure. This refers more to the water and wastewater ICS as critical infrastructure. Likewise,

at sector level, a cybersecurity resilience strategy would help with ICS cybersecurity information exchange, knowledge sharing and collaboration, skills development, and rapid recovery from any deliberate cyberattacks, accidents, or naturally occurring threats or incidents.

- *Encourage sector members to have documented ICS cybersecurity policies and procedures.* The water and wastewater sector members who either own and/or operate a critical infrastructure (or water ICS) should be encouraged to have documented ICS cybersecurity policies and procedures separate from the corporate IT security policies and procedures in their security operations centres.
- *Develop a sector cybersecurity education and skills development strategy.* A coordinated skills development programme in collaboration with the Cybersecurity Hub in the DCDT, Cybersecurity Centre in the SSA, and other external stakeholders as stipulated in the NCPF should be initiated through the water CSIRT. The sector can partner with academic institutions such as the University of Johannesburg and ICS vendors to develop a formal but customised ICS cybersecurity training and certification programme. This could bolster the specialist domain of ICS cybersecurity in the country tremendously as IT security already has an established body of knowledge and certification programmes. Ultimately, though, the desired picture is to have a cross-functional team of cybersecurity experts in the CSIRT sector to share their varied domain knowledge and experiences to evaluate and mitigate risk in the sector. Thus, cybersecurity operation centres in member organisations should comprise both IT security and specialist ICS cybersecurity experts where applicable.
- *Develop a sector cybersecurity awareness campaign strategy.* Coordinated sector-wide cybersecurity education and awareness campaigns should become regular occurrences.

7. Conclusions

The national cybersecurity strategy is a system mainly comprising stakeholders from the justice, crime prevention, and security cluster of South Africa. However, industry, civil society, and other government entities such as the water and wastewater sector are recognised as important stakeholders in the national cybersecurity system. A systems thinking approach was employed to analyse the national cybersecurity and water and wastewater systems. Through the stated stakeholders (system elements/actors) and legislation and policies (system interconnections), the ultimate purpose (system function) of the national cybersecurity system was found to be the establishment of a conducive environment and the provision of guidelines, standards, and best-practices for key cybersecurity stakeholders in South Africa. The interconnected relationships among these key stakeholders were found to be determined largely by the Cybercrimes Bill, CIPA, ECT Act, NCPF, POPI Act, RICA and PAIA in particular, and other cybersecurity-relevant pieces of legislation and policies.

It is concluded that the water and wastewater sector can immediately address its cybersecurity requirements without the need to propose any new legislation and/or government policies or amend existing ones. The aim of the study has therefore been achieved. But the water and wastewater sector will need to identify where changes and concomitant actions in the underlying structure of the national cybersecurity system can result in significant and lasting improvements for the sector. This can only be achieved by establishing a sector CSIRT that should continuously monitor the changes in the underlying structure of the national cybersecurity programme. This is especially important as changing cybersecurity-relevant legislation and policies greatly impact the entire national cybersecurity system, including the water and wastewater sector's cybersecurity responsibilities.

Future research work could use systems thinking or system dynamics to analyse the impact of the national cybersecurity legislation and policies in South Africa since 2015. Other research projects could explore the recommendations discussed above. Moreover, a review of how other countries deal with cybersecurity in the water and wastewater sector in contrast to South Africa should form part of future research works. After all, the exchange of international experiences is crucial in the advancement of cybersecurity practices. As the country embarks on a digital transformation strategy future research could look at related challenges in the water and wastewater sector. For example, noting that some municipalities have already embarked upon installing smart meters, legislation and policies governing security and privacy of smart water meters and other Internet of Things (smart) devices could be explored.

Author Contributions: Conceptualization, All authors; methodology, M.M.; writing—original draft preparation, M.M.; review, A.L.M. and S.v.S.; visualization, M.M.; supervision, A.L.M.; project administration, A.L.M.; funding acquisition, All authors. All authors have read and agreed to the published version of the manuscript.

Funding: This research was funded by the Water Research Commission (WRC) of South Africa, grant number 2021/2023-00354, and the article processing charge was also funded by WRC.

Institutional Review Board Statement: Not applicable for studies not involving humans or animals.

Informed Consent Statement: Not applicable for studies not involving humans or animals.

Conflicts of Interest: The authors declare no conflict of interest.

Appendix A Analysis of the National Cybersecurity Policy Framework System

A literature review of the previous analysis work on the National Cybersecurity Policy Framework (NCPF) was conducted in this appendix. This looked at mainly the stakeholders involved, legislation and policies underpinning the national cybersecurity strategy, and challenges in the implementation of the NCPF.

Researcher	Stakeholders (Elements/Actors)	Legislation and Policies (Interconnections)	Gaps or Identified Challenges
[100]	Domestic • Cybersecurity centre (SSA) • Cyber Crime Centre (SAPS) • Cybersecurity Hub (Department of Telecommunications and Postal Services) • Cyber Command (DoD)		
[101]	Foreign • International Telecommunication Union (ITU)	• NCPF	
[102]	Foreign • African Union (AU)	• African Union Convention on Cybersecurity and Personal Data Protection	

Researcher	Stakeholders (Elements/Actors)	Legislation and Policies (Interconnections)	Gaps or Identified Challenges
[36]	Domestic • Justice, crime prevention and security (JCPS) cluster (SSA and others) • Cybersecurity Response Committee (CRC) • Department of Telecommunications and Postal Services (DTPS) • SITA • Department of Science and Technology • Department of International Relations and Cooperation (DIRCO) • South African Revenue Service (SARS) Foreign • International Criminal Police Organisation (Interpol)	• Constitution of the Republic of South Africa • Computer Evidence Act 57 of 1983 • Copyright Act 98 of 1978 • Critical Infrastructure Bill of 2017 • Cybercrimes and Cybersecurity Bill of 2017 • ECT Act 25 of 2002 • Electronic Communications Act 36 of 2005 • Films and Publications Act 65 of 1996 • Financial Intelligence Centre Act (FICA) 38 of 2001 • National Prosecutions Act 32 of 1998 • Monitoring and Prohibition Act 127 of 1992 • Prevention of Organised Crime Act 38 of 1999 • Promotion of Access to Information Act (PAIA) 25 of 2002 • Protection of Constitutional Democracy against Terrorism and Related Activities Act 33 of 2004 • Protection of Personal Information (POPI) Act 4 of 2013 • RICA 70 of 2002	• New laws and institutions are required in South Africa to effectively address cybersecurity requirements. • The military, intelligence agencies, and critical infrastructure experience the most cyber incidents in South Africa. It should, however, be noted that national critical infrastructure is mostly operated and managed by provincial and local governments as well as the private sector. • New cybersecurity capabilities have to be developed and acquired by South Africa.
[66]	Domestic • South African National Defence Force (SANDF) • JCPS cluster	• NCPF • Defence Review • Cybercrimes and Cybersecurity Bill	

Researcher	Stakeholders (Elements/Actors)	Legislation and Policies (Interconnections)	Gaps or Identified Challenges
[38]	Domestic • Department of State Security • SSA • SSA Cybersecurity Centre • Electronic Communications Security—Cyber Security Incidents Response Team (ECS-CSIRT) • Department of Justice and Constitutional Development • NPA • SAPS • DoD • Cyberwarfare Command • Centre Headquarters (HQ) • COMSEC Ltd. • Department of Telecommunications and Postal Services • National Cybersecurity Advisory Council • National Cybersecurity Hub • Cyber Inspectorate • Department of Trade and Industry • Public Service and Administration • SITA • Foreign • Forum for Incident Response and Security Teams (FIRST)	• Section 198 of the 1996 Constitution • NCPF • RICA 70 of 2002 • Protection of State Information Bill • POPI Act 4 of 2013 • Cybercrimes and Cybersecurity Bill • Cyber Warfare Strategy • ECT Act 25 of 2002 • Cryptography Regulations • E-government strategy and roadmap • Companies Act 71 of 2008 • PAIA 2 of 2000 • Corporate Governance of ICT Framework • E-government strategy for each province	• Establishment of a Cyber Inspectorate is provided for in Chapter 12 of the ECT Act. Its mandate includes the powers to inspect, search and seize electronic content in pursuit of illegal activities. However, no regulations were ever promulgated to establish this unit. • Coordination in government is generally an issue. Add to that the inadequacy of existing cybercrime and cybersecurity legal framework, and there is an even bigger issue. The National Cybersecurity Advisory Council was tasked with reducing these deficiencies but there is very little evidence of its activities. • The proposed coordination mechanisms in the NCPF are complex, thus making their management difficult. This is exacerbated by a poor track record of inter-ministerial coordination of programmes. Additionally, there are only limited review and oversight mechanisms, and many activities are shrouded in secrecy. • One of the major challenges for the South African government is the promotion of cybersecurity measures to the (i) national, provincial, and local governments; (ii) general public; (iii) private sector; (iv) civil society; and (v) special interest groups.

Researcher	Stakeholders (Elements/Actors)	Legislation and Policies (Interconnections)	Gaps or Identified Challenges
[103]	Domestic • SSA	• NCPF • ECT Act 25 of 2002 • RICA 70 of 2002 • POPI 4 of 2013 • Cybercrimes and Cybersecurity Bill	
[57]		• NCPF • National Key Points Act 102 of 1980 • ECT Act 25 of 2002 • King III Report on Corporate Governance	
[63]	Domestic • Department of Communications • National Cybersecurity Advisory Council (NCAC) *Foreign* • Council of Europe (CoE)	• NCPF • CoE's Cybercrime Convention	• South Africa was ranked in the top 10 countries most affected by internet crimes. The statistics were drawn from the Internet Crime Complaint Center that is managed by the USA's Federal Bureau of Investigation. The challenge is not a lack of cybercrime laws but enforcing them. There is a huge gap between enacted laws and practical enforcement capability on the ground in most emerging and developing countries such as SA.

Researcher	Stakeholders (Elements/Actors)	Legislation and Policies (Interconnections)	Gaps or Identified Challenges
[37]	Domestic • State Security Agency (SSA) • South African Policy Service (SAPS) • Department of Justice and Constitutional Development (DOJ & CD) • National Prosecuting Authority (NPA) • Department of Communications (DOC) • Department of Defence and Military Veterans (DoD & MV) • Department of Science and Technology (DST) • Foreign • African Union • Southern African Development Community (SADC) • Commonwealth	• Films and Publication Act 65 of 1996 • Protection from Harassment Act 17 of 2011 • Regulation of Interception of Communications and Provision of Communication-related Information Act (RICA) 70 of 2002 • Promotion of Equality and Prevention of Unfair Discrimination Act 4 of 2000 • Copyright Act 98 of 1978 • Consumer Protection Act 68 of 2008 • National Archives and Record Service of South Africa Act 43 of 1996 • Trade Marks Act 194 of 1993 • Designs Act 195 of 1993 • Electronic Communications Act 36 of 2005 • Electronic Communications and Transactions Act 25 of 2002 (ECT Act) • Independent Communications Authority of South Africa (ICASA) Act 13 of 2000 • Inter-Governmental Relations Framework of 2005 • Competition Act 89 of 1998 • Broadband Infraco Act 33 of 2007 • State Information Technology Agency (SITA) Act 88 of 1998 • Public Service Act: Regulation	• South Africa follows several global methods. However, a clear commitment towards existing conventions such as the Budapest, AU, SADC and Commonwealth conventions is still outstanding. • Advanced cybersecurity strategies include protection of critical infrastructure (CI) as a key element. The ECT Act also alludes to the protection of CI. However, the implementation of CI protection is still in abeyance. The country had planned for CI protection of the following priority sectors: (i) energy; (ii) information and communications technology; and (iii) transport. • Sector CSIRTs have not yet been established. These would be effective for incident responses and information exchange between sectors. • In the current configuration, the cybersecurity and cybercrime legal framework is spread among very different pieces of legislation. Aligning these would improve predictability and transparency of the policies. • There is a lack of technical cybersecurity skills in government to enable the Cybersecurity Hub to assume the role of a national CERT. Skills development must be prioritised by government in this regard. • A lack of user cybersecurity education and awareness in the general public exacerbates spoofing and phishing related cybercrimes as these are not generally associated with inadequate technical safeguards. • Implementation of a national cybersecurity programme requires sound expertise in several disciplines, and this is lacking in government. This includes commitment and guidance from the top echelons of government, availability and development of the required cybersecurity expert level, and continuous cybersecurity awareness campaigns for the general public.

Researcher	Stakeholders (Elements/Actors)	Legislation and Policies (Interconnections)	Gaps or Identified Challenges
[104]		• NCPF	• In South Africa, cybersecurity awareness initiatives are rolled out through a variety of independent and uncoordinated mechanisms. An integrated and coordinated approach would be effective.

References

1. UN. Make the SDGs a Reality. Available online: https://web.archive.org/web/20201110154701/https://sdgs.un.org/#goal_section (accessed on 10 November 2020).
2. White, R. Risk analysis for critical infrastructure protection. In *Critical Infrastructure Security and Resilience. Advanced Sciences and Technologies for Security Applications*; Gritzalis, D., Theocharidou, M., Stergiopoulos, G., Eds.; Springer: Cham, Switzerland, 2019; pp. 35–54.
3. Birkett, D.; Mala-Jetmarova, H. Plan, prepare and safeguard: Water critical infrastructure protection in Australia. In *Securing Water and Wastewater Systems: Protecting Critical Infrastructure*; Clark, R.M., Hakim, S., Eds.; Springer International Publishing: Cham, Switzerland, 2014; pp. 287–313.
4. Zabašta, A.; Juhna, T.; Tihomirova, K.; Rubulis, J.; Ribickis, L. Latvian practices for protecting water and wastewater infrastructure. In *Securing Water and Wastewater Systems: Protecting Critical Infrastructure*; Clark, R.M., Hakim, S., Eds.; Springer International Publishing: Cham, Switzerland, 2014; pp. 315–342.
5. Möderl, M.; Rauch, W.; Achleitner, S.; Lukas, A.; Mayr, E.; Neunteufel, R.; Perfler, R.; Neuhold, C.; Godina, R.; Wiesenegger, H.; et al. Austrian activities in protecting critical water infrastructure. In *Securing Water and Wastewater Systems: Protecting Critical Infrastructure*; Clark, R.M., Hakim, S., Eds.; Springer International Publishing: Cham, Switzerland, 2014; pp. 343–373.
6. Tiirmaa-Klaar, H. Building national cyber resilience and protecting critical information infrastructure. *J. Cyber Policy* **2016**, *1*, 94–106. [CrossRef]
7. Alexander, A.; Graham, P.; Jackson, E.; Johnson, B.; Williams, T.; Park, J. An analysis of cybersecurity legislation and policy creation on the state level. In *National Cyber Summit (NCS) Research Track. NCS 2019. Advances in Intelligent Systems and Computing*; Choo, K.K., Morris, T., Peterson, G., Eds.; Springer: Cham, Switzerland, 2020; Volume 1055, pp. 30–43.
8. Burmeister, O.; Phahlamohlaka, J.; Al-Saggaf, Y. Good governance and virtue in South Africa's cyber security policy implementation. *Int. J. Cyber Warf. Terror.* **2015**, *5*, 19–29. [CrossRef]
9. Jansen van Vuuren, J.; Leenen, J.; Zaaiman, J.J. Using an ontology as a model for the implementation of the national cybersecurity policy framework for South Africa. In Proceedings of the 9th International Conference on Cyber Warfare and Security 2014 (ICCWS 2014), West Lafayette, IN, USA, 24–25 March 2014; pp. 107–115.
10. Ismail, S.; Sitnikova, E.; Slay, J. Using integrated system theory approach to assess security for SCADA systems cyber security for critical infrastructures: A pilot study. In Proceedings of the 11th International Conference on Fuzzy Systems and Knowledge Discovery (FSKD 2014), Xiamen, China, 19–21 August 2014; pp. 1000–1006.
11. NIST. Framework for Improving Critical Infrastructure Cybersecurity, Version 1.1. Available online: https://web.archive.org/web/20201109030328/https://nvlpubs.nist.gov/nistpubs/CSWP/NIST.CSWP.04162018.pdf (accessed on 10 November 2020).
12. Rasekh, A.; Hassanzadeh, A.; Mulchandani, S.; Modi, S.; Banks, M.K. Smart water networks and cyber security. *J. Water Resour. Plan. Manag.* **2016**, *142*, 1–3. [CrossRef]
13. Hassanzadeh, A.; Rasekh, A.; Galelli, S.; Aghashahi, M.; Taormina, R.; Ostfeld, A.; Banks, M.K. A review of cybersecurity incidents in the water sector. *J. Environ. Eng.* **2020**, *146*, 03120003. [CrossRef]
14. Hahn, A. Operational technology and information technology in industrial control systems. In *Cyber-Security of SCADA and Other Industrial Control Systems: Advances in Information Security*; Colbert, E.J.M., Kott, A., Eds.; Springer International Publishing: Cham, Switzerland, 2016; Volume 66, pp. 51–68.
15. Sullivan, D.; Luiijf, E.; Colbert, E.J.M. Components of industrial control systems. In *Cyber-Security of SCADA and Other Industrial Control Systems: Advances in Information Security*; Colbert, E.J.M., Kott, A., Eds.; Springer International Publishing: Cham, Switzerland, 2016; Volume 66, pp. 15–28.
16. Weiss, J. Industrial Control System (ICS) cyber security for water and wastewater systems. In *Securing Water and Wastewater Systems: Protecting Critical Infrastructure*; Clark, R.M., Hakim, S., Eds.; Springer International Publishing: Cham, Switzerland, 2014; pp. 87–105.
17. White, R.; George, R.; Boult, T.; Chow, C.E. Apples to apples: RAMCAP and emerging threats to lifeline infrastructure. *Homel. Secur. Aff.* **2016**, *12*, 1–20.
18. Ramotsoela, D.; Abu-Mahfouz, A.; Hancke, G. A survey of anomaly detection in industrial wireless sensor networks with critical water system infrastructure as a case study. *Sensors* **2018**, *18*, 2491. [CrossRef] [PubMed]

19. Purvis, B.; Mao, Y.; Robinson, D. Three pillars of sustainability: In search of conceptual origins. *Sustain. Sci.* **2019**, *14*, 681–695. [CrossRef]
20. Clark, R.M.; Panguluri, S.; Nelson, T.D.; Wyman, R.P. Protecting drinking water utilities from cyberthreats. *J. Am. Water Works Assoc.* **2017**, *109*, 50–58. [CrossRef]
21. Barbier, E.B.; Burgess, J.C. The sustainable development goals and the systems approach to sustainability. *Econ. E-J.* **2017**, *11*, 1–22. [CrossRef]
22. Chowdhury, R. *Systems Thinking for Management Consultants: Flexible Systems Management*; Springer International Publishing: Singapore, 2019.
23. Fiksel, J. Systems Thinking. In *Resilient by Design: Creating Businesses that Adapt and Flourish in a Changing World*; Fiksel, J., Ed.; Island Press: Washington, DC, USA, 2015; pp. 35–50.
24. Chung, A.; Dawda, S.; Hussain, A.; Shaikh, S.A.; Carr, M. Cybersecurity: Policy. In *Encyclopedia of Security and Emergency Management*; Shapiro, L., Maras, M.H., Eds.; Springer: Cham, Switzerland, 2019.
25. Srinivas, J.; Das, A.K.; Kumar, N. Government regulations in cyber security: Framework, standards and recommendations. *Future Gener. Comput. Syst.* **2019**, *92*, 178–188. [CrossRef]
26. Brechbühl, H.; Bruce, R.; Dynes, S.; Johnson, M.E.E. Protecting critical information infrastructure: Developing cybersecurity policy. *Inf. Technol. Dev.* **2010**, *16*, 83–91. [CrossRef]
27. OECD. Cybersecurity Policy Making at a Turning Point: Analysing a New Generation of National Cybersecurity Strategies for the Internet Economy. Available online: https://web.archive.org/web/20201017142718/http://www.oecd.org/sti/ieconomy/cybersecurity%20policy%20making.pdf (accessed on 10 November 2020).
28. UNECE. Working Party on Regulatory Cooperation and Standardization Policies (WP.6)—Report on the Sectoral Initiative on Cyber Security. 2019. Available online: https://web.archive.org/web/20201217093616/https://undocs.org/pdf?symbol=en%2FECE%2FCTCS%2FWP.6%2F2019%2F9 (accessed on 17 December 2020).
29. Sabillon, R.; Cavaller, V.; Cano, J. National Cyber Security Strategies: Global Trends in Cyberspace. *Int. J. Comput. Sci. Softw. Eng.* **2016**, *5*, 67–81.
30. WMO. Water and Cyber Security—Protection of Critical Water-Related Infrastructure. Available online: https://web.archive.org/web/20201118080434/https://public.wmo.int/en/events/meetings/water-and-cyber-security-protection-of-critical-water-related-infrastructure-online (accessed on 17 December 2020).
31. Dlamini, I.Z.; Taute, B.; Radebe, J. Framework for an African policy towards creating cyber security awareness. In Proceedings of the Southern African Cyber Security Awareness Workshop (SACSAW 2011), Gaborone, Botswana, 12 May 2011; pp. 15–31.
32. Budapest Convention. Convention on Cybercrime, European Treaty Series—No. 185. Available online: https://web.archive.org/web/20200724075610/https://www.europarl.europa.eu/meetdocs/2014_2019/documents/libe/dv/7_conv_budapest_/7_conv_budapest_en.pdf (accessed on 10 November 2020).
33. Clough, J. A world of difference: The Budapest convention on cybercrime and the challenges of harmonisation. *Monash Univ. Law Rev.* **2014**, *40*, 698–736.
34. Wicki-Birchler, D. The Budapest Convention and the General Data Protection Regulation: Acting in concert to curb cybercrime? *Int. Cybersecur. Law Rev.* **2020**, *1*, 63–72. [CrossRef]
35. Budapest Convention. Chart of Signatures and Ratifications of Treaty 185: Convention on Cybercrime, Status as of 05/11/2020. Available online: https://web.archive.org/web/20201110182746/https://www.coe.int/en/web/conventions/full-list/-/conventions/treaty/185/signatures?p_auth=9UQ3WQaH (accessed on 10 November 2020).
36. Ntsaluba, N. Cybersecurity Policy and Legislation in South Africa. Master's Thesis, University of Pretoria, Pretoria, South Africa, 2017.
37. Detecon. E-Commerce, Cybercrime and Cybersecurity—Status, Gaps and the Road Ahead. Available online: https://web.archive.org/web/20201110182612/https://www.dtps.gov.za/index.php?option=com_phocadownload&view=category&download=121%3A20131126_policy-review_e-commerce-cybercrime-and-cybersecurity_final&id=39%3Ae-commerce-cyber-security&Itemid=143 (accessed on 10 November 2020).
38. Sutherland, E. Governance of cybersecurity-the case of South Africa. *Afr. J. Inf. Commun.* **2017**, *20*, 83–112. [CrossRef]
39. Coleman, D. Digital colonialism: The 21st century scramble for Africa through the extraction and control of user data and the limitations of data protection laws. *Mich. J. Race Law* **2019**, *24*, 417–439.
40. AU Convention. List of Countries Which Have Signed, Ratified/Acceded to the African Union Convention on Cyber Security and Personal Data Protection. Available online: https://au.int/sites/default/files/treaties/29560-sl-AFRICAN%20UNION%20CONVENTION%20ON%20CYBER%20SECURITY%20AND%20PERSONAL%20DATA%20PROTECTION.pdf (accessed on 5 November 2020).
41. UN. Resolution Adopted by the General Assembly on 23 December 2015. Available online: https://web.archive.org/web/20200913145346/https://undocs.org/en/A/RES/70/237 (accessed on 10 November 2020).
42. WSIS. Basic Information: About WSIS. Available online: https://web.archive.org/web/20201110172646/https://www.itu.int/net/wsis/basic/about.html (accessed on 10 November 2020).
43. UN. Open-Ended Working Group. Available online: https://web.archive.org/web/20201110172949/https://www.un.org/disarmament/open-ended-working-group/ (accessed on 10 November 2020).

44. UN. Group of Governmental Experts. Available online: https://web.archive.org/web/20201110173326/https://www.un.org/disarmament/group-of-governmental-experts/ (accessed on 10 November 2020).
45. Janke, R.; Tryby, M.; Clark, R.M. Protecting water supply critical infrastructure: An overview. In *Securing Water and Wastewater Systems: Protecting Critical Infrastructure*; Clark, R.M., Hakim, S.S., Eds.; Springer International Publishing: Cham, Switzerland, 2014; pp. 29–85.
46. Clark, R.M.; Hakim, S. *Securing Water and Wastewater Systems: Global Experiences*; Springer International Publishing: Cham, Switzerland, 2014.
47. Spathoulas, G.; Katsikas, S. Towards a secure Industrial Internet of Things. In *Security and Privacy Trends in the Industrial Internet of Things: Advanced Sciences and Technologies for Security Applications*; Alcaraz, C., Ed.; Springer International Publishing: Cham, Switzerland, 2019; pp. 29–45.
48. CISA. Alert (AA20-352A): Advanced Persistent Threat Compromise of Government Agencies, Critical Infrastructure, and Private Sector Organizations. Available online: https://web.archive.org/web/20201218004033/https://us-cert.cisa.gov/ncas/alerts/aa20-352a (accessed on 18 December 2020).
49. Galinec, D.; Možnik, D.; Guberina, B. Cybersecurity and cyber defence: National level strategic approach. *Automatika* **2017**, *58*, 273–286. [CrossRef]
50. Birkett, D.M. Water critical infrastructure security and its dependencies. *J. Terror. Res.* **2017**, *8*, 1–21. [CrossRef]
51. Chung, J.J. Critical infrastructure, cybersecurity, and market failure. *Or. Law Rev.* **2018**, *96*, 441–476.
52. Bernieri, G.; Pascucci, F. Improving Security in Industrial Internet of Things: A Distributed Intrusion Detection Methodology. In *Security and Privacy Trends in the Industrial Internet of Things: Advanced Sciences and Technologies for Security Applications*; Alcaraz, C., Ed.; Springer International Publishing: Cham, Switzerland, 2019; pp. 161–179.
53. Krotofil, M.; Kursawe, K.; Gollmann, D. Securing industrial control systems. In *Security and Privacy Trends in the Industrial Internet of Things: Advanced Sciences and Technologies for Security Applications*; Alcaraz, C., Ed.; Springer International Publishing: Cham, Switzerland, 2019; pp. 3–26.
54. Clark, R.M.; Hakim, S.; Panguluri, S. Protecting water and wastewater utilities from cyber-physical threats. *Water Environ. J.* **2018**, *32*, 384–391. [CrossRef]
55. Ranathunga, D.; Roughan, M.; Nguyen, H.; Kernick, P.; Falkner, N. Case Studies of SCADA firewall configurations and the implications for best practices. *IEEE Trans. Netw. Serv. Manag.* **2016**, *13*, 871–884. [CrossRef]
56. Panguluri, S.; Phillips, W.; Cusimano, J. Protecting water and wastewater infrastructure from cyber attacks. *Front. Earth Sci.* **2011**, *5*, 406–413. [CrossRef]
57. Pretorius, B.; Van Niekerk, B. Cyber-Security for ICS/SCADA: A South African perspective. *Int. J. Cyber Warf. Terror.* **2016**, *6*, 1–16. [CrossRef]
58. Stellios, I.; Kotzanikolaou, P.; Psarakis, M. Advanced persistent threats and zero-day exploits in industrial Internet of Things. In *Security and Privacy Trends in the Industrial Internet of Things: Advanced Sciences and Technologies for Security Applications*; Alcaraz, C., Ed.; Springer International Publishing: Cham, Switzerland, 2019; pp. 47–68.
59. Noble, T.; Manalo, C.; Miller, K.; Ferro, C. Cybersecurity assessments of 30 drinking water utilities. *J. N. Engl. Water Works Assoc.* **2017**, *131*, 219–227.
60. McNabb, J.K. Securing public drinking water utilities. *J. N. Engl. Water Works Assoc.* **2012**, *27*, 37–59.
61. Gourisetti, S.N.G.; Mylrea, M.; Patangia, H. Cybersecurity vulnerability mitigation framework through empirical paradigm: Enhanced prioritized gap analysis. *Future Gener. Comput. Syst.* **2020**, *105*, 410–431. [CrossRef]
62. Jansen van Vuuren, J.C.; Leenen, L.; Phahlamohlaka, J.; Zaaiman, J. An approach to governance of cybersecurity in South Africa. *Int. J. Cyber Warf. Terror.* **2014**, *2*, 13–27. [CrossRef]
63. Kshetri, N. Cybercrime and cybersecurity issues in the BRICS economies. *J. Glob. Inf. Technol. Manag.* **2015**, *18*, 245–249. [CrossRef]
64. Wolfpack. The South African Cyber Threat Barometer: A Strategic Public-Private Partnership (PPP) Initiative to Combat Cybercrime in SA. Available online: https://web.archive.org/web/20201110174518/http://docplayer.net/17043391-2012-3-the-south-african-cyber-threat-barometer-a-strategic-public-private-partnership-ppp-initiative-to-combat-cybercrime-in-sa.html (accessed on 10 November 2020).
65. South Africa. National Cybersecurity Policy Framework (NCPF). Available online: https://web.archive.org/web/20200701182327/https://www.gov.za/sites/default/files/gcis_document/201512/39475gon609.pdf (accessed on 10 November 2020).
66. Mutemwa, M.; Mtsweni, J.; Mkhonto, N. Developing a cyber threat intelligence sharing platform for South African organisations. In Proceedings of the Conference on Information Communication Technology and Society (ICTAS 2017), Durban, South Africa, 9–10 March 2017.
67. Government of South Africa. What Are the Government Clusters and Which Are They? Available online: https://web.archive.org/web/20201110180329/https://www.gov.za/faq/guide-government/what-are-government-clusters-and-which-are-they (accessed on 10 November 2020).
68. South Africa. South African Constitution (as amended). Available online: https://web.archive.org/web/20201101003156/https://justice.gov.za/legislation/constitution/SAConstitution-web-eng.pdf (accessed on 10 November 2020).
69. South Africa. Cybercrimes Bill (B6-2017). Available online: https://web.archive.org/web/20200805080827/https://pmg.org.za/bill/684/?via=homepage-card (accessed on 10 November 2020).

70. South Africa. Critical Infrastructure Protection Act 8 of 2019. Available online: https://web.archive.org/web/20201110181412/https://www.gov.za/sites/default/files/gcis_document/201911/4286628-11act8of2019criticalinfraprotectact.pdf (accessed on 10 November 2020).
71. South Africa. Protection of Personal Information Act. 2013. Available online: https://web.archive.org/web/20201014062057/https://www.justice.gov.za/inforeg/docs/InfoRegSA-POPIA-act2013-004.pdf (accessed on 10 November 2020).
72. South Africa. The Regulation of Interception of Communications and Provision of Communication-Related Information Act. Available online: https://web.archive.org/web/20200918041001/https://www.gov.za/documents/regulation-interception-communications-and-provision-communication-related-information--13 (accessed on 10 November 2020).
73. South Africa. Electronic Communications and Transactions Act. Available online: https://web.archive.org/web/20201106144645/https://www.gov.za/sites/default/files/gcis_document/201409/a25-02.pdf (accessed on 10 November 2020).
74. South Africa. Promotion of Access to Information Act No. 2 of 2000. Available online: https://web.archive.org/web/20200517051020/https://www.justice.gov.za/legislation/acts/2000-002.pdf (accessed on 10 November 2020).
75. Karabacak, B.; Yildirim, S.O.; Baykal, N. Regulatory approaches for cyber security of critical infrastructures: The case of Turkey. *Comput. Law Secur. Rev.* **2016**, *32*, 526–539. [CrossRef]
76. De Bruijn, H.; Janssen, M. Building cybersecurity awareness: The need for evidence-based framing strategies. *Gov. Inf. Q.* **2017**, *34*, 1–7. [CrossRef]
77. DWS. Department of Water and Sanitation 2018/19 Annual Report, Vote 36: Water Is Life—Sanitation Is Dignity. Available online: https://web.archive.org/web/20201110183936/http://www.dwa.gov.za/documents/AnnualReports/19213_Annual%20Report%20201819inhouse.pdf (accessed on 10 November 2020).
78. Government SA. National Water and Sanitation Master Plan, Volume 1: Call to Action, Version 10.1: Ready for the Future and Ahead of the Curve. Available online: https://web.archive.org/web/20200408080543/https://www.gov.za/sites/default/files/gcis_document/201911/national-water-and-sanitation-master-plandf.pdf (accessed on 10 November 2020).
79. Makaya, E.; Rohse, M.; Day, R.; Vogel, C.; Mehta, L.; McEwen, L.; Rangecroft, S.; Van Loon, A.F.E. Water governance challenges in rural South Africa: Exploring institutional coordination in drought management. *Water policy* **2020**, *22*, 519–540. [CrossRef]
80. Government SA. Water and Sanitation. Available online: https://www.gov.za/about-sa/water-affairs (accessed on 10 November 2020).
81. SERI. Water and Sanitation Legislation and Regulations. Available online: https://web.archive.org/web/20170301051254/https://www.seri-sa.org/index.php/links/policy-and-legislation/15-links/policy-and-legislation/87-water-and-sanitation (accessed on 10 November 2020).
82. Stuart-Hill, S.I.; Schulze, R.E. Does South Africa's water law and policy allow for climate change adaptation? *Clim. Dev.* **2010**, *2*, 128–144. [CrossRef]
83. Pedrosa, V.A. The necessity of IWRM: The case of San Francisco river water conflicts. In *Integrated Water Resource Management*; Vieira, E.O., Sandoval-Solis, S., Pedrosa, V.A., Ortiz-Partida, J.P., Eds.; Springer International Publishing: Cham, Switzerland, 2020; pp. 27–34.
84. Meadows, D. *Thinking in Systems: A Primer*; Earthscan: London, UK, 2008.
85. Senge, P.M. *The Fifth Discipline: The Art and Practice of the Learning Organization*; Doubleday, Random House: New York, NY, USA, 2006.
86. Ramos, H. Creativity and Systems Thinking. In *Encyclopedia of Creativity, Invention, Innovation and Entrepreneurship*; Carayannis, E.G., Ed.; Springer International Publishing: New York, NY, USA, 2013; pp. 12–58.
87. Schuster, S. *The art of Thinking in Systems: Improve Your Logic, Think More Critically, and Use Proven Systems to Solve Your Problems*; CreateSpace Independent Publishing Platform: Scotts Valley, CA, USA, 2018; ISBN 9781983847547.
88. SEBoK Editorial Board. *The Guide to the Systems Engineering Body of Knowledge (SEBoK)*; v. 2.3; BKCASE: Hoboken, NJ, USA, 2020.
89. Stroh, D.P. *Systems Thinking for Social Change: A practical Guide for Solving Complex Problems, Avoiding Unintended Consequences, and Achieving Lasting Results*; Chelsea Green Publishing: White River Junction, VT, USA, 2015.
90. Sterman, J.D. *Business Dynamics: Systems Thinking and Modeling for a Complex World*; Irwin/McGraw-Hill: New York, NY, USA, 2000; ISBN 13 9780072311358.
91. Sutherland, E. The Fourth Industrial Revolution–The Case of South Africa. *Politikon* **2020**, *47*, 233–252. [CrossRef]
92. Gcaza, N. Cybersecurity awareness and education: A necessary parameter for smart communities. In Proceedings of the Twelfth International Symposium on Human Aspects of Information Security and Assurance (HAISA 2018), Dundee, Scotland, 29–31 August 2018; pp. 80–90.
93. Mahlaka, R. Experian Offers Mea Culpa after Massive Data Breach Blunder. Available online: https://web.archive.org/web/20201110190306/https://www.dailymaverick.co.za/article/2020-08-23-experian-offers-mea-culpa-after-massive-data-breach-blunder/ (accessed on 10 November 2020).
94. Merten, M. SAPS Regulations: It's Crucial to Watch Critical Infrastructure Rules to Prevent a Power Grab. Available online: https://web.archive.org/web/20200808183846/https://www.dailymaverick.co.za/article/2020-04-03-saps-regulations-its-crucial-to-watch-critical-infrastructure-rules-to-prevent-a-power-grab/ (accessed on 10 November 2020).
95. Timmers, P. The European Union's cybersecurity industrial policy. *J. Cyber Policy* **2018**, *3*, 363–384. [CrossRef]
96. Flowers, A.; Zeadally, S. US policy on active cyber defense. *J. Homel. Secur. Emerg. Manag.* **2014**, *11*, 289–308. [CrossRef]

97. Van Woensel, L. Systems thinking and assessing cross-policy impacts. In *A Bias Radar for responsible POLICY-MAKING: Foresight-Based Scientific Advice*; Van Woensel, L., Ed.; Palgrave Macmillan: Cham, Switzerland, 2020; pp. 69–84.
98. De Jong, M.D.T.; Neulen, S.; Jansma, S.R. Citizens' intentions to participate in governmental co-creation initiatives: Comparing three co-creation configurations. *Gov. Inf. Q.* **2019**, *36*, 490–500. [CrossRef]
99. Karlsson, F.; Holgersson, J.; Söderström, E.; Hedström, K. Exploring user participation approaches in public e-service development. *Gov. Inf. Q.* **2012**, *29*, 158–168. [CrossRef]
100. Phahlamohlaka, J.; Hefer, J. The Impact of cybercrimes and cybersecurity Bill on South African national cybersecurity: An institutional theory analytic perspective. In Proceedings of the THREAT 2019 Cybersecurity Summit, Johannesburg, South Africa, June 2019; pp. 1–7.
101. De Barros, M.J.Z.; Lazarek, H.; Jennifer, M. Comparative study of cybersecurity policy among South Africa and Mozambique. In Proceedings of the 13th International Conference on Cyber Warfare and Security (ICCWS 2018), Reading, UK, 8–9 March 2018; pp. 521–529.
102. Dalton, W.; Jansen Van Vuuren, J.; Westcott, J. Building cybersecurity resilience in Africa. In Proceedings of the 12th International Conference on Cyber Warfare and Security (ICCWS 2017), Dayton, OH, USA, 2–3 March 2017; pp. 112–120.
103. Van Niekerk, B. An analysis of cyber-incidents in South Africa. *Afr. J. Inf. Commun.* **2017**, *20*, 113–132. [CrossRef]
104. Dlamini, I.Z.; Modise, M. Cyber security awareness initiatives in South Africa: A synergy approach. In *Case Studies in Information Warfare and Security for Researchers, Teachers and Students*; Warren, M., Ed.; ACPIL: London, UK, 2013; pp. 1–22.

MDPI
St. Alban-Anlage 66
4052 Basel
Switzerland
Tel. +41 61 683 77 34
Fax +41 61 302 89 18
www.mdpi.com

Sustainability Editorial Office
E-mail: sustainability@mdpi.com
www.mdpi.com/journal/sustainability

www.ingramcontent.com/pod-product-compliance
Lightning Source LLC
LaVergne TN
LVHW080503180726
843515LV00016B/1259

* 9 7 8 3 0 3 6 5 1 9 5 0 0 *